Series Editor
John M. Walker
School of Life and Medical Sciences,
University of Hertfordshire,
Hatfield, Hertfordshire, AL10 9AB, UK

For further volumes:
http://www.springer.com/series/7651

Super-Resolution Microscopy

Methods and Protocols

Edited by

Holger Erfle

High-Content Analysis of the Cell (HiCell) and Advanced Biological Screening Facility, BioQuant, Heidelberg University, Heidelberg, Germany

Humana Press

Editor
Holger Erfle
High-Content Analysis of the Cell (HiCell)
and Advanced Biological Screening Facility
BioQuant, Heidelberg University
Heidelberg, Germany

ISSN 1064-3745 ISSN 1940-6029 (electronic)
Methods in Molecular Biology
ISBN 978-1-4939-7264-7 ISBN 978-1-4939-7265-4 (eBook)
DOI 10.1007/978-1-4939-7265-4

Library of Congress Control Number: 2017950493

Printed on acid-free paper

This Humana Press imprint is published by Springer Nature
The registered company is Springer Science+Business Media LLC
The registered company address is: 233 Spring Street, New York, NY 10013, U.S.A.

Preface

Super-resolution microscopy also known as nanoscopy opens a new avenue in the success story of fluorescence microscopy in modern biology by surpassing Abbe's diffraction limit. It bridges the resolution gap to electron microscopy by resolving structures at the low nanometer scale. In addition, it maintains the advantages of optical fluorescence microscopy like simple sample preparation allowing breakthrough discoveries in diverse directions of life sciences.

This book aims to provide a current overview of the advanced technologies and methods in super-resolution microscopy addressing novices as well as experienced users in this field. We start with a mini-review, followed by a manuscript on how to introduce super-resolution microscopy into a core facility, and present an up-to-date set of protocols.

As there is no single gold standard for fluorescence-based super-resolution microscopy available, we try with the collection of chapters at hand to cover a wide range of technologies and methods. From the technological side we include the two major streams currently pursued in super-resolution imaging: ensemble techniques and single-molecule localization techniques. Further, we include chapters describing methods to establish correlations between fluorescence-based super-resolution imaging and other imaging techniques like confocal microscopy, atomic force microscopy (AFM) and electron microscopy.

The chapters offer state-of-the-art, reproducible and comprehensive protocols on a wide range of topics in super-resolution microscopy.

Heidelberg, Germany *Holger Erfle*

Contents

Preface . . . . *v*
Contributors . . . . *ix*

1 Super-Resolution Microscopy Techniques and Their Potential for Applications in Radiation Biophysics 1
Jan Philipp Eberle, Alexander Rapp, Matthias Krufczik, Marion Eryilmaz, Manuel Gunkel, Holger Erfle, and Michael Hausmann

2 Managing the Introduction of Super-Resolution Microscopy into a Core Facility 15
Jeffrey A. Kamykowski and Brian Storrie

3 Live-Cell STED Imaging with the HyPer2 Biosensor 21
Natalia M. Mishina and Vsevolod V. Belousov

4 Diffraction-Unlimited Fluorescence Imaging with an EasySTED Retrofitted Confocal Microscope 29
André Klauss and Carsten Hille

5 Two-Photon STED Microscopy for Nanoscale Imaging of Neural Morphology In Vivo 45
Mirelle J.T. ter Veer, Thomas Pfeiffer, and U. Valentin Nägerl

6 STED Imaging of Golgi Dynamics with Cer-SiR: A Two-Component, Photostable, High-Density Lipid Probe for Live Cells 65
Roman S. Erdmann, Derek Toomre, and Alanna Schepartz

7 Four-Channel Super-Resolution Imaging by 3-D Structured Illumination 79
Ulrike Engel

8 Correlative SIM-STORM Microscopy 95
O. Burri, T. Laroche, R. Guiet, and A. Seitz

9 Correlative Super-Resolution Fluorescence Imaging and Atomic Force Microscopy for the Characterization of Biological Samples 105
Patricia Bondia, Santiago Casado, and Cristina Flors

10 Quantitative Single-Molecule Localization Microscopy (qSMLM) of Membrane Proteins Based on Kinetic Analysis of Fluorophore Blinking Cycles 115
Franziska Fricke, Joel Beaudouin, Sebastian Malkusch, Roland Eils, and Mike Heilemann

11 Two-Color Single-Molecule Tracking in Live Cells 127
Siegfried Hänselmann and Dirk-Peter Herten

12 Fully Automated Targeted Confocal and Single-Molecule Localization Microscopy 139
Jan Philipp Eberle, Walter Muranyi, Holger Erfle, and Manuel Gunkel

13 Brain Slice Staining and Preparation for Three-Dimensional Super-Resolution Microscopy ... 153
Christopher L. German, Manasa V. Gudheti, Annette E. Fleckenstein, and Erik M. Jorgensen

14 Correlative In-Resin Super-Resolution Fluorescence and Electron Microscopy of Cultured Cells ... 163
Errin Johnson and Rainer Kaufmann

15 Synthesis of Janelia Fluor HaloTag and SNAP-Tag Ligands and Their Use in Cellular Imaging Experiments ... 179
Jonathan B. Grimm, Timothy A. Brown, Brian P. English, Timothée Lionnet, and Luke D. Lavis

16 Measuring Nanometer Distances Between Fluorescent Labels Step-by-Step ... 189
Susanna Maria Früh and Ingmar Schoen

17 Correlative Single-Molecule Localization Microscopy and Confocal Microscopy ... 205
Christian Soeller, Yufeng Hou, Isuru D. Jayasinghe, David Baddeley, and David Crossman

18 Correlative Fluorescence Super-Resolution Localization Microscopy and Platinum Replica EM on Unroofed Cells ... 219
Kem A. Sochacki and Justin W. Taraska

19 In Situ Super-Resolution Imaging of Genomic DNA with OligoSTORM and OligoDNA-PAINT ... 231
Brian J. Beliveau, Alistair N. Boettiger, Guy Nir, Bogdan Bintu, Peng Yin, Xiaowei Zhuang, and C.-ting Wu

20 Super-Resolution High Content Screening and Analysis ... 253
K. Soliman

Index ... *261*

Contributors

DAVID BADDELEY • *Department of Cell Biology, Yale University, New Haven, CT, USA; Auckland Bioengineering Institute, University of Auckland, Auckland, New Zealand*

JOEL BEAUDOUIN • *Department for Bioinformatics and Functional Genomics, Bioquant and Institute of Pharmacy and Molecular Biotechnology, Heidelberg University, Heidelberg, Germany; Institut de Biologie Structurale, Université Grenoble Alpes, Centre National de la Recherche Scientifique, Commissariat à l'Energie Atomique et aux Energies Alternatives, F-38000 Grenoble, France; Division of Theoretical Bioinformatics, German Cancer Research Center (DKFZ), Heidelberg, Germany*

BRIAN J. BELIVEAU • *Department of Genetics, Harvard Medical School, Boston, MA, USA; Wyss Institute for Biologically Inspired Engineering, Harvard University, Boston, MA, USA; Department of Systems Biology, Harvard Medical School, Boston, MA, USA*

VSEVOLOD V. BELOUSOV • *Shemyakin-Ovchinnikov Institute of Bioorganic Chemistry, Moscow, Russia*

BOGDAN BINTU • *Howard Hughes Medical Institute, Cambridge, MA, USA; Department of Physics, Harvard University, Cambridge, MA, USA*

ALISTAIR N. BOETTIGER • *Department of Developmental Biology, Stanford University, Stanford, CA, USA; Howard Hughes Medical Institute, Cambridge, MA, USA; Department of Chemistry and Chemical Biology, Harvard University, Cambridge, MA, USA*

PATRICIA BONDIA • *Madrid Institute for Advanced Studies in Nanosciences (IMDEA Nanoscience), Madrid, Spain; Nanobiotechnology Unit Associated to the National Center for Biotechnology (CSIC), Madrid, Spain*

TIMOTHY A. BROWN • *Janelia Research Campus, Howard Hughes Medical Institute, Ashburn, VA, USA*

O. BURRI • *Faculty of Life Sciences (SV), BioImaging and Optics Platform (BIOP), Ecole Polytechnique Fédérale de Lausanne (EPFL), Lausanne, Switzerland*

SANTIAGO CASADO • *Madrid Institute for Advanced Studies in Nanoscience (IMDEA Nanoscience), Madrid, Spain*

DAVID CROSSMAN • *Department of Physiology, University of Auckland, Auckland, New Zealand*

JAN PHILIPP EBERLE • *High-Content Analysis of the Cell (HiCell) and Advanced Biological Screening Facility, BioQuant, Heidelberg University, Heidelberg, Germany*

ROLAND EILS • *Department for Bioinformatics and Functional Genomics, Bioquant and Institute of Pharmacy and Molecular Biotechnology, Heidelberg University, Heidelberg, Germany; Division of Theoretical Bioinformatics, German Cancer Research Center (DKFZ), Heidelberg, Germany*

ULRIKE ENGEL • *Nikon Imaging Center at Heidelberg University and Centre for Organismal Studies (COS), Heidelberg University, Heidelberg, Germany*

BRIAN P. ENGLISH • *Janelia Research Campus, Howard Hughes Medical Institute, Ashburn, VA, USA*

ROMAN S. ERDMANN • *Department of Cell Biology, Yale University School of Medicine, New Haven, CT, USA; Department of Chemistry, Yale University, New Haven, CT, USA*

HOLGER ERFLE • *High-Content Analysis of the Cell (HiCell) and Advanced Biological Screening Facility, BioQuant, Heidelberg University, Heidelberg, Germany*
MARION ERYILMAZ • *Kirchhoff-Institute for Physics, Heidelberg University, Heidelberg, Germany*
ANNETTE E. FLECKENSTEIN • *School of Dentistry, University of Utah, Salt Lake City, UT, USA*
CRISTINA FLORS • *Madrid Institute for Advanced Studies in Nanoscience (IMDEA Nanoscience), Madrid, Spain; Nanobiotechnology Unit Associated to the National Center for Biotechnology (CSIC), Madrid, Spain*
FRANZISKA FRICKE • *Institute of Physical and Theoretical Chemistry, Goethe-University Frankfurt, Frankfurt am Main, Germany*
SUSANNA MARIA FRÜH • *Laboratory of Applied Mechanobiology, ETH Zurich, Zurich, Switzerland*
CHRISTOPHER L. GERMAN • *School of Dentistry, University of Utah, Salt Lake City, UT, USA*
JONATHAN B. GRIMM • *Janelia Research Campus, Howard Hughes Medical Institute, Ashburn, VA, USA*
MANASA V. GUDHETI • *Department of Biology, University of Utah, Salt Lake City, UT, USA; Vutara, Salt Lake City, UT, USA*
R. GUIET • *Faculty of Life Sciences (SV), BioImaging and Optics Platform (BIOP), Ecole Polytechnique Fédérale de Lausanne (EPFL), Lausanne, Switzerland*
MANUEL GUNKEL • *High-Content Analysis of the Cell (HiCell) and Advanced Biological Screening Facility, BioQuant, Heidelberg University, Heidelberg, Germany*
SIEGFRIED HÄNSELMANN • *Physikalisch-Chemisches Institut and CellNetworks Cluster, Heidelberg University, Heidelberg, Germany*
MICHAEL HAUSMANN • *Kirchhoff-Institute for Physics, Heidelberg University, Heidelberg, Germany*
MIKE HEILEMANN • *Institute of Physical and Theoretical Chemistry, Goethe-University Frankfurt, Frankfurt am Main, Germany; Single Molecule Biology, Bioquant, Heidelberg University, Heidelberg, Germany; Institute for Anatomy and Cell Biology, Heidelberg University, Heidelberg, Germany*
DIRK-PETER HERTEN • *Physikalisch-Chemisch Institut and CellNetworks Cluster, Heidelberg University, Heidelberg, Germany*
CARSTEN HILLE • *Department of Physical Chemistry/Applied Laser Sensing in Complex Biosystems (ALS ComBi), Institute of Chemistry, University of Potsdam, Potsdam, Germany*
YUFENG HOU • *Institute of Experimental Medical Research, Oslo University Hospital Ullevål, University of Oslo, Oslo, Norway; Department of Physiology, University of Auckland, Auckland, New Zealand*
ISURU D. JAYASINGHE • *School of Biomedical Sciences, University of Leeds, Leeds, UK*
ERRIN JOHNSON • *Sir William Dunn School of Pathology, University of Oxford, Oxford, UK*
ERIK M. JORGENSEN • *Department of Biology, University of Utah, Salt Lake City, UT, USA; Howard Hughes Medical Institute, Silver Springs, MD, USA*
JEFFREY A. KAMYKOWSKI • *Digital Microscopy Laboratory, Department of Physiology and Biophysics, University of Arkansas for Medical Sciences, Little Rock, AR, USA*
RAINER KAUFMANN • *Division of Structural Biology, Wellcome Trust Centre for Human Genetics, University of Oxford, Oxford, UK; Department of Biochemistry, University of Oxford, Oxford, UK*
ANDRÉ KLAUSS • *Department of Physical Chemistry/Applied Laser Sensing in Complex Biosystems (ALS ComBi), Institute of Chemistry, University of Potsdam, Potsdam, Germany*

MATTHIAS KRUFCZIK • *Kirchhoff-Institute for Physics, Heidelberg University, Heidelberg, Germany*
T. LAROCHE • *Faculty of Life Sciences (SV), BioImaging and Optics Platform (BIOP), Ecole Polytechnique Fédérale de Lausanne (EPFL), Lausanne, Switzerland*
LUKE D. LAVIS • *Janelia Research Campus, Howard Hughes Medical Institute, Ashburn, VA, USA*
TIMOTHÉE LIONNET • *Janelia Research Campus, Howard Hughes Medical Institute, Ashburn, VA, USA*
SEBASTIAN MALKUSCH • *Institute of Physical and Theoretical Chemistry, Goethe-University Frankfurt, Frankfurt am Main, Germany*
NATALIA M. MISHINA • *Shemyakin-Ovchinnikov Institute of Bioorganic Chemistry, Moscow, Russia*
WALTER MURANYI • *Department of Infectious Diseases and Virology, Heidelberg University Hospital, Heidelberg, Germany*
U. VALENTIN NÄGERL • *Interdisciplinary Institute for Neuroscience, UMR 5297 CNRS, Bordeaux, France; Université de Bordeoux, Bordeaux, France*
GUY NIR • *Department of Genetics, Harvard Medical School, Boston, MA, USA*
THOMAS PFEIFFER • *Interdisciplinary Institute for Neuroscience, UMR 5297 CNRS, Bordeaux, France; Université de Bordeoux, Bordeaux, France*
ALEXANDER RAPP • *Department of Biology, Technical Universität Darmstadt, Darmstadt, Germany*
ALANNA SCHEPARTZ • *Department of Chemistry, Yale University, New Haven, CT, USA; Department of Molecular, Cellular, and Developmental Biology, Yale University, New Haven, CT, USA*
INGMAR SCHOEN • *Laboratory of Applied Mechanobiology, ETH Zurich, Zurich, Switzerland*
A. SEITZ • *Faculty of Life Sciences (SV), BioImaging and Optics Platform (BIOP), Ecole Polytechnique Fédérale de Lausanne (EPFL), Lausanne, Switzerland*
KEM A. SOCHACKI • *National Heart Lung and Blood Institute, National Institutes of Health (NIH), Bethesda, MD, USA*
CHRISTIAN SOELLER • *Biomedical Physics, University of Exeter, Exeter, UK; Department of Physiology, University of Auckland, Auckland, New Zealand*
K. SOLIMAN • *College of Pharmacy, Arab Academy of Science and technology and Maritime Transport, Alexandria, Egypt; Göttingen Medical Center Neuropediatric Division, Göttingen University, Göttingen, Germany*
BRIAN STORRIE • *Digital Microscopy Laboratory, Department of Physiology and Biophysics, University of Arkansas for Medical Sciences, Little Rock, AR, USA*
JUSTIN W. TARASKA • *National Heart Lung and Blood Institute, National Institutes of Health (NIH), Bethesda, MD, USA*
MIRELLE J. T. TERVEER • *Interdisciplinary Institute for Neuroscience, UMR 5297 CNRS, Bordeaux, France; Université de Bordeaux, Bordeaux, France*
DEREK TOOMRE • *Department of Cell Biology, Yale University School of Medicine, New Haven, CT, USA*
C. -TING WU • *Department of Genetics, Harvard Medical School, Boston, MA, USA*
PENG YIN • *Wyss Institute for Biologically Inspired Engineering, Harvard University, Boston, MA, USA; Department of Systems Biology, Harvard Medical School, Boston, MA, USA*
XIAOWEI ZHUANG • *Howard Hughes Medical Institute, Cambridge, MA, USA; Department of Chemistry and Chemical Biology, Harvard University, Cambridge, MA, USA; Department of Physics, Harvard University, Cambridge, MA, USA*

Chapter 1

Super-Resolution Microscopy Techniques and Their Potential for Applications in Radiation Biophysics

Jan Philipp Eberle, Alexander Rapp, Matthias Krufczik, Marion Eryilmaz, Manuel Gunkel, Holger Erfle, and Michael Hausmann

Abstract

Fluorescence microscopy is an essential tool for imaging tagged biological structures. Due to the wave nature of light, the resolution of a conventional fluorescence microscope is limited laterally to about 200 nm and axially to about 600 nm, which is often referred to as the Abbe limit. This hampers the observation of important biological structures and dynamics in the nano-scaled range ~10 nm to ~100 nm. Consequentially, various methods have been developed circumventing this limit of resolution. Super-resolution microscopy comprises several of those methods employing physical and/or chemical properties, such as optical/instrumental modifications and specific labeling of samples. In this article, we will give a brief insight into a variety of selected optical microscopy methods reaching super-resolution beyond the Abbe limit. We will survey three different concepts in connection to biological applications in radiation research without making a claim to be complete.

Key words Super-resolution microscopy, Spatially modulated illumination (SMI) microscopy, Stimulated emission depletion (STED), Single-molecule localization microscopy (SMLM)

1 Introduction

Resolution criteria such as the Abbe limit [1] state that the minimum resolvable distance is proportional to the detection wavelength. A popular method utilizing this relationship is electron microscopy [2] which achieves atomic resolution due to the small de Broglie wavelength of electrons transmitting or reflecting the sample. Recent applications demonstrate the usefulness also for radiation research (*see*, e.g., [3]). However, sample preparation is quite complex and does not allow live cell imaging, since measurements are performed in a vacuum. Near-field scanning optical microscopy (NSOM) [4–6] overcomes diffraction by the use of evanescent waves under total reflection conditions, achieving a resolution in the tens of nanometers. NSOM offers the possibility of observing nano-structures on surfaces such as chromosome

Holger Erfle (ed.), *Super-Resolution Microscopy: Methods and Protocols*, Methods in Molecular Biology, vol. 1663,
DOI 10.1007/978-1-4939-7265-4_1,

morphologies in fluorescently labeled metaphase spreads or optical cell imaging of plasma membranes under vital conditions (*see*, e.g., [7, 8]). Recent developments in far-field fluorescence microscopy enjoy growing popularity for addressing biological questions without being restricted to imaging of surfaces. This is due to advantages of standard preparation methods taken from fluorescence microscopy, such as maintenance of 3D morphology and simultaneous multi-color labeling of different structures. By using spatially modulating illumination, beam shaping optics, and/or employing optical characteristics of fluorophores, these methods were able to circumvent the Abbe diffraction limit toward improved resolution in the sub-wavelength range. In the following, we will describe three selected far-field optical concepts spatially modulated illumination (SMI) microscopy, stimulated emission depletion (STED) microscopy, and single-molecule localization microscopy (SMLM) in more detail and demonstrate their potential for applications in radiation research.

2 Spatially Modulated Illumination

An important far-field scanning microscopy method, suppressing out of focus signals with a pinhole, is confocal microscopy [9, 10], which facilitates axial sectioning of the sample. Built on that, 4Pi-confocal microscopy [11, 12] is able to enhance axial resolution down to 75 nm. In this method, two opposing objective lenses are utilized to increase the aperture angle while illumination beam interference with or without detection beam interference is applied. 4Pi-confocal microscopy is commonly used in combination with two-photon excitation (2PE). A similar approach to this 2PE 4Pi-imaging mode is the interference microscopy method I^5 microscopy (I^5M) [13–15]. Like 4Pi microscopy, I^5M takes advantage of two opposing objectives and the interference of illumination and detection beam for a higher resolution. I^5M differs in the usage of incoherent light sources and homogeneous Köhler-illumination instead of a focussed laser beam. A comparison of both methods [16] showed better image quality of 2PE 4Pi-images, whereas I^5M produced brighter images at faster frame rates. This is due to stronger side lope artifacts in I^5M, which are reduced by 2PE and the need of 2PE 4Pi microscopy to scan the sample. Anyhow, these approaches offer no significant improvement in lateral resolution.

There are several other methods using interference patterns for illumination to obtain resolution beyond the Abbe limit. One of them is standing-wave fluorescence microscopy (SWFM) [17, 18]. SWFM achieves higher axial resolution for distance measurements [19] by exciting the sample with an interference pattern of two coherent plane-wave beams. In addition, the quantitative analysis of the resulting modulation pattern obtained from a specifically labeled region of a cell allows volume and shape measurements also in 3D conserved cell nuclei [19–23]. Super-resolution structured illumination microscopy (SR-SIM) [24, 25], on the other side, superposes lateral line patterns for illumination, resulting in the expansion of the "observable region" in reciprocal space and thus enhanced lateral resolution. For a simultaneous resolution improvement along the optical axis either an additional pattern in the axial direction or a lateral pattern with adapted frequency is used. Latter approach utilizes the possibility of enhancing axial resolution by applying a lateral pattern, like in optical sectioning SIM (OS-SIM) [26, 27]. Also, combinations of SR-SIM with other microscopy methods such as I^5M, named I^5S [28], are possible reaching near-isotropic resolution of 100 nm. Lateral resolution can further be improved down to about 50 nm using saturated SIM (SSIM) [29, 30]. SR-SIM is restricted in resolution by diffraction limited spatial frequencies of the illumination pattern. If the emission rate depends nonlinearly on the illumination intensity, the illumination pattern will contain higher order harmonics. By these means multiples of the previously achievable spatial frequencies occur increasing resolution.

2.1 Applications with Spatially Modulated Illumination

Numerous examples of the application of SIM and 3D-SIM have been published demonstrating the benefit of enhanced resolution in live sciences [31–34]. For example, the enhanced resolution enables identifying replication sites [31, 35] in unprecedented clarity as well as the nuclear structure and chromatin composition [36, 32]. 3D-SIM is widely applied in combination with indirect immunofluorescence and a broad range of standard dyes can be used. In addition, FISH-labeling [37] is suitable for 3D-SIM applications as well as the use of fluorescent proteins. Even live cell imaging is possible, although for a limited number of time points due to fluorescence bleaching and phototoxicity effects on the specimen [31, 38, 39].

Figure 1 shows an example of resolution improvement in the detection of replication foci [31] obtained by 3D-SIM microscopy. The example shows the pseudo-wide-field images reconstructed from the individually modulated images obtained during 3D-SIM acquisition as well as the resulting image of the same after software deconvolution. This step alone improves the resolution and allows the segmentation of additional focal structures. 3D-SIM imaging further enhances the resolution and allows even more substructures to be visualized and segmented.

A similar experimental approach was described by Lopez-Perez et al. [40] and Natale et al. [41] where the substructure of γH2AX foci induced by carbon ion and X-ray radiation, respectively, was analyzed. The authors demonstrated that classical γH2AX foci as analyzed so far by wide-field or confocal microscopy can be further

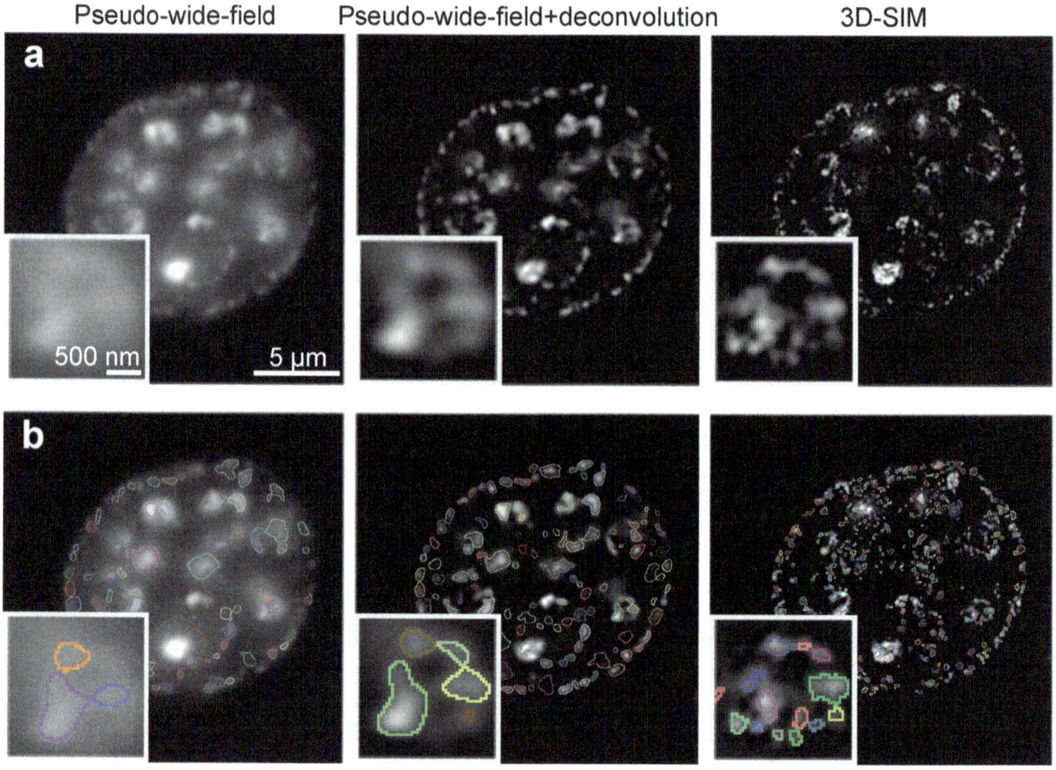

Fig. 1 Resolution improvements obtained by 3D-SIM imaging. (**a**) Mid nuclear z-section of a mid-S-phase HeLa cell, labeled for 30 min with EdU and subsequent EdU-Click staining using Alexa594. Pseudo-wide-field images of replication foci reconstructed from individual phase shifted images obtained during 3D-SIM imaging. The same image after deconvolution reveals additional structures. 3D-SIM acquisition further enhances the number of resolvable elements. (**b**) Foci segmentation from the images shown in (**a**). With the increase of resolving power more and finer substructures can be segmented. Replication foci as seen in wide-field microscopy can be further resolved to substructures. For details *see* [31]

resolved into substructures if imaged by 3D-SIM microscopy. Also the clustering of the subdomains can be seen and the authors question whether the traditional view of one γH2AX focus representing a single DNA double-strand break holds true. In addition, this work also studied the accordance between structures identified by 3D-SIM and SMLM by using a microscopy setup that allowed 3D-SIM and SMLM performed on the same sample and microscope [42]. With this approach they found the γH2AX foci to be represented by both the methods in the same described substructured appearance. Furthermore, the authors speculate that the discontinuous signal from γH2AX represents the looping of the chromatin and the non-discontinuous presence of H2AX in chromatin. Similar substructures have been reported also by 4Pi microscopy imaging, where the authors analyzed the substructure of γH2AX foci induced by 3 Gy of γ-irradiation [43]. This work reported also clustered appearance of γH2AX nano-foci in response to DNA double-strand break induction. Even in electron microscopy studies on heavy ion induced radiation tracks [44], loose clusters of γH2AX were reported. In summary, novel high resolution optical microscopy allows a new view on the DNA damage response in the context of chromatin.

3 STED Microscopy

Using the concept of stimulated emission to switch fluorophores between an electronically excited singlet state and the ground state, STED microscopy [45, 46] was able to further improve lateral resolution. Here, a focussed laser beam excites fluorophores in a diffraction limited focal spot. The excited fluorophores can then be stimulated with a second laser wavelength for reversal into the ground state, resulting in the depletion of the excited state before fluorescence occurs. This is achieved with a doughnut-shaped STED beam to reduce the effective spot size with fluorescing molecules. Moreover, STED uses the nonlinearity between STED-beam intensity and depletion efficiency to sharpen the depletion edge. By scanning both beams simultaneously over the sample, images on biological samples with a lateral resolution down to about 20 nm also in living cells [47] can be acquired. Enhanced 3D resolution of about 50 nm lateral and 110 nm axial is possible by interfering the STED beam with a beam modulated in the z-direction [48]. Instead of depleting the excited state, also the ground state can be depleted by shelving fluorophores in a metastable dark state like a triplet state. This principle is realized in the method ground state depletion (GSD) microscopy [49, 50].

3.1 Applications of STED

As mentioned above, many applications have been published that demonstrate how improved resolution enables novel insights in nearly all fields of cell biology. This also applies to STED microscopy. Ultrafine structure analysis of the cytoskeleton (reviewed in [51]), vesicles [52], protein complexes [53], or sites of DNA replication [54] have shown the benefits. Besides structural information, also size and volumetric (in 3D STED) information can be obtained. So far, the multi-color capabilities of STED are limited, but novel dyes and depletion lasers are overcoming this limitation.

Recently, the ultra-structure of the DNA damage response marked by γH2AX has also been analyzed using STED microscopy [55]. In this work, the authors compared the expansion of γH2AX spreading after exposure to both low LET photons and high LET carbon ions. The use of STED microscopy allowed the authors to identify substructures in macroscopic radiation-induced foci. They showed that the γH2AX signal spreading is more disperse compared to 53BP1. In a two-colour STED approach they showed that both proteins are found colocalized in photon induced foci as well as in carbon ion-induced tracks. The authors reported that γH2AX-defined foci are less well defined but show an intensive central structure that is also highest colocalized with 53BP1. Simultaneous analysis of Rad51 and 53BP1 showed a special exclusion of these two factors when the radiation-induced foci are analyzed by optical super-resolution microscopy. This suggests an underlying structure of the DNA damage response tightly associated with the chromatin conformation and the process of damage processing.

4 SMLM

SMLM is a method based on optical isolation of single fluorophores by exploiting their optical characteristics. To this end, some methods of the still ongoing development process of SMLM are described (*see* also [56]).

First single-molecule detection (SMD) experiments on fluorophores recorded their absorption spectra [57], followed by observing their fluorescence [58]. Based on that, single-molecule localization was found to be applicable which opens new possibilities apart from ensemble-measuring concepts, which were described before. In this connection, a method was proposed that is based on isolating single molecules by unique optical characteristics, locating them separately and accumulating all localized positions to a final image [59]. Moreover, optical characteristics were found such as photo-activation for fluorescence increase [60] and optical switching between fluorescent and dark states based on dye pairs [61]. Hereupon, several methods have been published using different optical characteristics but employing the same principle resulting in a lateral resolution of about 20 nm. Photo-activated localization microscopy (PALM) [62] and fluorescence-PALM (fPALM) [63] used photo-activatable fluorescent proteins (PA-FPs) and stochastic optical reconstruction microscopy (STORM) [64] initially utilized the cyanine dye-pair Cy5-Cy3 as optical switch. In this process, sparse random subsets of fluorophores were imaged iteratively to localize individual fluorophores by centering their point-spread functions (PSFs). For activating fluorophore subsets in a fluorescing state and imaging them, two laser wavelengths were needed for their initial implementations. These methods and their further developments are referred to here as SMLM. Progress of SMLM has been made in the choice of optical characteristics of dyes by using iterative cycles of fluorescence, called "blinking," for image acquisition. Initially, this property has been observed for the green fluorescent protein (GFP) [65]. A benefit of inducing fluorophore blinking is that only one laser wavelength is needed for image acquisition which results in a measurement acceleration [66]. Several conventional fluorophores were found to blink as well and were applied for SMLM [67–69]. For instance, for the cyanine dyes Cy5, Cy5.5, C7 and derivatives, a blinking mechanism was found that depends on the pH value and the concentration of primary thiols like β-mercaptoethanol (βME), which form the dark state in an adduct with the dye molecule [70]. Furthermore, a dark state recovery via illumination with UV-light was observed for these cyanine dyes, like it has been reported earlier for GFP.

There exist several methods for 3D SMLM imaging with enhanced resolution along the optical axis. One of them employs astigmatism, introduced by a cylindrical lens [71], gaining a resolution in the lateral direction of 20–30 nm and in the axial direction of 50–60 nm. For example, other approaches use two detection planes [72], double-helix (DH) PSFs [73], focussed front and side views [74], interferometric fluorescence imaging [75], self-bending (SB) PSFs [76], or supercritical angle detection [77] to achieve super-resolution along the *z*-axis. The combination of spatially modulated illumination and SMLM has also been successfully applied to many biological tasks [78].

4.1 Applications of SMLM

There exists a large spectrum of systematic studies in biological and biomedical research that is based on SMLM. Receptor clustering in membranes (*see*, e.g., [79, 78]) has been studied as well as protein trafficking after virus infection [80]. In the cell nucleus, chromatin organisation and re-arrangements after irradiation treatment [81, 82] or chromatin looping formation have given new insights [83] into the nano-cosmos of the genome to obtain a better understanding of mechanisms behind spatial genome organisation. Thus, SMLM has opened novel perspectives on many parts of cellular organization and mechanisms and is closing the gap between conventional light microscopy and electron microscopy.

Here, we will again show some applications that are typical for radiation research [84]. In order to better understand the mechanisms behind DNA and chromatin repair processes, conformational changes of chromatin were studied and the arrangements of proteins and DNA in and around repair foci were investigated.

For this, breast cancer cells (SkBr3) were irradiated with X-rays for inducing double-strand breaks (DSBs). In Fig. 2a, a fluorescence wide-field image is shown where H3K9me3 (green) marks constitutive heterochromatin (HC) and γH2AX (red) labels DNA-DSB repair sites. Figure 2b shows a zoomed region and Fig. 2c the corresponding SMLM image revealing complex structures of HC and γH2AX. The coordinates from SMLM data analysis allowed determining γH2AX clusters (for detailed procedure *see* [79]) and the HC density in relation to the cluster centers. The principle of how γH2AX clusters were determined is shown in Fig. 2d–f. In this process, the points surrounded by a minimum number of neighbors within a maximum distance (Fig. 2d) were

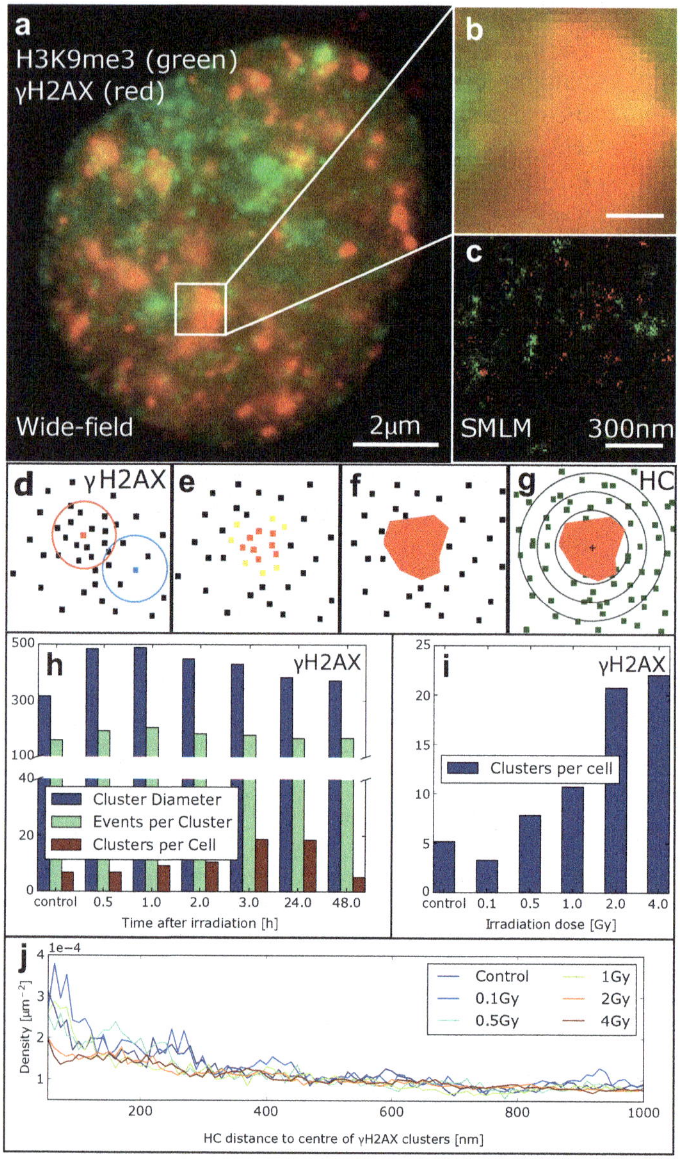

Fig. 2 SMLM imaging and data evaluation. (**a**) Fluorescence wide-field image of a breast cancer cell nucleus (SKBR3) irradiated with 1 Gy where H3K9me3 (*green*) marks constitutive HC and γH2AX (*red*) DNA-DSB repair foci, (**b**) a zoom in the wide-field image, (**c**) the corresponding SMLM image section revealing a complex structure of γH2AX and HC, (**d–g**) schematic representation of how γH2AX clusters and their center distances to HC signals were analyzed using SMLM data points, (**h**, **i**) γH2AX cluster properties in dependency of time after irradiation and irradiation dose gained via cluster analysis, (**j**) HC density in spatial dependency to γH2AX clusters from distance analysis

identified as cluster points (Fig. 2e, red and yellow). From the point distribution the cluster area and diameter were deduced (Fig. 2f, h). Furthermore, the number of cluster points (events per cluster) and the number of clusters per cell in dependency of time after irradiation and irradiation dose (Fig. 2h, i) were derived. Also, the HC signal distribution (green points) around γH2AX clusters (red area) were observed via counting points in concentric circles around cluster centers (Fig. 2g, j).

The analyzed data showed a typical size of γH2AX clusters of about 400 nm after irradiation and 300 nm for non-irradiated controls (Fig. 2h). The number of events per cluster did not change significantly during repair, whereas the number of clusters per cell was time and dose dependent (Fig. 2h, i). The density measurements showed a higher increase of HC signals for short distances to γH2AX cluster centers at low irradiation doses than for cells irradiated with 2 or 4 Gy (Fig. 2j). This may indicate that damages for higher doses are repaired through different processes than for lower doses (*see* also [83, 84]).

The examples illustrated here show some possibilities SMLM imaging offers. Further, algorithms of quantification are available, leading to new findings in research. In conclusion, calculations from pointillist measurements open new insights in mechanisms behind radiation damaging and repair processes.

References

1. Abbe E (1873) Beiträge zur Theorie des Mikroskops und der mikroskopischen Wahrnehmung. Arch Mikrosk Anat 9(1):413–418
2. Knoll M, Ruska E (1932) Das Elektronenmikroskop. Z Phys 78(5):318–339
3. Lorat Y, Brunner CU, Schanz S, Jakob B, Taucher-Scholz G, Rübe CE (2015) Nanoscale analysis of clustered DNA damage after high-LET irradiation by quantitative electron microscopy – the heavy burden to repair. DNA Repair 28:93–106
4. Betzig E, Trautman JK (1992) Near-field optics: microscopy, spectroscopy, and surface modification beyond the diffraction limit. Science 257(5067):189–195
5. Lewis A, Isaacson M, Harootunian A, Muray A (1984) Development of a 500 Å spatial resolution light microscope. Ultramicroscopy 13 (3):227–231
6. Pohl DW, Denk W, Lanz M (1984) Optical stethoscopy: image recording with resolution λ/20. Appl Phys Lett 44(7):651–653
7. Perner B, Rapp A, Dressler C, Wollweber L, Beuthan J, Greulich KO, Hausmann M (2002) Variations in cell surfaces of estrogen treated breast cancer cells detected by a combined instrument for far-field and near-field microscopy. Anal Cell Pathol 24:89–100
8. Winkler R, Perner B, Rapp A, Durm M, Cremer C, Greulich KO, Hausmann M (2003) Labelling quality and chromosome morphology after low temperature FISH analysed by scanning far-field and near-field optical microscopy. J Microsc 209(1):23–33
9. Brakenhoff GJ, Blom P, Barends P (1979) Confocal scanning light microscopy with high aperture immersion lenses. J Microsc 117 (2):219–232
10. Sheppard C, Choudhury A (1977) Image formation in the scanning microscope. Opt Acta 24(10):1051–1073
11. Hell SW, Lindek S, Cremer C, Stelzer EHK (1994) Measurement of the 4Pi-confocal point spread function proves 75 nm axial resolution. Appl Phys Lett 64(11):1335–1337
12. Hell SW, Stelzer EHK (1992) Fundamental improvement of resolution with a 4Pi-confocal fluorescence microscope using two-photon excitation. Optic Commun 93(5):277–282
13. Gustafsson MGL, Agard DA, Sedat JW (1995) Sevenfold improvement of axial resolution in 3D wide-field microscopy using two objective lenses. Proc SPIE 2412:147–156
14. Gustafsson MGL, Agard DA, Sedat JW (1999) I5M: 3D widefield light microscopy with better than 100 nm axial resolution. J Microsc 195 (1):10–16
15. Gustafsson MGL, Sedat JW, Agard DA (1997) Method and apparatus for three-dimensional microscopy with enhanced depth resolution. US 5671085 A

16. Bewersdorf J, Schmidt R, Hell SW (2006) Comparison of I5M and 4Pi-microscopy. J Microsc 222(2):105–117
17. Bailey B, Farkas DL, Taylor DL, Lanni F (1993) Enhancement of axial resolution in fluorescence microscopy by standing-wave excitation. Nature 366(6450):44–48
18. Schneider B, Upmann I, Kirsten I, Bradl J, Hausmann M, Cremer C (1999) A dual-laser, spatially modulated illumination fluorescence microscope. Eur. Microsc Anal 57:5–7
19. Albrecht B, Failla AV, Heintzmann R, Cremer C (2001) Spatially modulated illumination microscopy: online visualization of intensity distribution and prediction of nanometer precision of axial distance measurements by computer simulations. J Biomed Opt 6(3):292–299
20. Failla AV, Spoeri U, Albrecht B, Kroll A, Cremer C (2002) Nanosizing of fluorescent objects by spatially modulated illumination microscopy. Appl Optics 41(34):7275–7283
21. Grossmann C, Schwarz-Finsterle J, Schmitt E, Birk U, Hildenbrand G, Cremer C, Trakhtenbrot L, Hausmann M (2010) Variations of the spatial fluorescence distribution in ABL gene chromatin domains measured in blood cell nuclei by SMI microscopy after COMBO – FISH labelling. In: Microscopy: science, technology, applications and education, vol 1. Formatex, Badajoz, pp 688–695
22. Hildenbrand G, Rapp A, Spöri U, Wagner C, Cremer C, Hausmann M (2005) Nano-sizing of specific gene domains in intact human cell nuclei by spatially modulated illumination light microscopy. Biophys J 88(6):4312–4318
23. Wagner C, Hildenbrand G, Spöri U, Cremer C (2006) Beyond nanosizing: an approach to shape analysis of fluorescent nanostructures by SMI-microscopy. Optik 117(1):26–32
24. Gustafsson MGL (2000) Surpassing the lateral resolution limit by a factor of two using structured illumination microscopy. J Microsc 198(2):82–87
25. Heintzmann R, Cremer CG (1999) Laterally modulated excitation microscopy: improvement of resolution by using a diffraction grating. Proc SPIE 3568:185–196
26. Neil MAA, Juškaitis R, Wilson T (1997) Method of obtaining optical sectioning by using structured light in a conventional microscope. Opt Lett 22(24):1905–1907
27. Neil MAA, Juškaitis R, Wilson T (1998) Real time 3D fluorescence microscopy by two beam interference illumination. Optic Commun 153 (1-3):1–4
28. Shao L, Isaac B, Uzawa S, Agard DA, Sedat JW, Gustafsson MGL (2008) I5S: wide-field light microscopy with 100-nm-scale resolution in three dimensions. Biophys J 94 (12):4971–4983
29. Gustafsson MGL (2005) Nonlinear structured-illumination microscopy: wide-field fluorescence imaging with theoretically unlimited resolution. Proc Natl Acad Sci U S A 102 (37):13081–13086
30. Heintzmann R, Jovin TM, Cremer C (2002) Saturated patterned excitation microscopy—a concept for optical resolution improvement. J Opt Soc Am A 19(8):1599–1609
31. Chagin VO, Casas-Delucchi CS, Reinhart M, Schermelleh L, Markaki Y, Maiser A, Bolius JJ, Bensimon A, Fillies M, Domaing P, Rozanov YM, Leonhardt H, Cardoso MC (2016) 4D Visualization of replication foci in mammalian cells corresponding to individual replicons. Nat Commun 7:11231. doi:10.1038/ncomms11231
32. Schermelleh L, Carlton PM, Haase S, Shao L, Winoto L, Kner P, Burke B, Cardoso MC, Agard DA, Gustafsson MG, Leonhardt H, Sedat JW (2008) Subdiffraction multicolor imaging of the nuclear periphery with 3D structured illumination microscopy. Science 320(5881):1332–1336. doi:10.1126/science.1156947
33. Schermelleh L, Heintzmann R, Leonhardt H (2010) A guide to super-resolution fluorescence microscopy. J Cell Biol 190 (2):165–175. doi:10.1083/jcb.201002018
34. Smeets D, Markaki Y, Schmid VJ, Kraus F, Tattermusch A, Cerase A, Sterr M, Fiedler S, Demmerle J, Popken J, Leonhardt H, Brockdorff N, Cremer T, Schermelleh L, Cremer M (2014) Three-dimensional super-resolution microscopy of the inactive X chromosome territory reveals a collapse of its active nuclear compartment harboring distinct Xist RNA foci. Epigenetics Chromatin 7:8. doi:10.1186/1756-8935-7-8
35. Baddeley D, Chagin VO, Schermelleh L, Martin S, Pombo A, Carlton PM, Gahl A, Domaing P, Birk U, Leonhardt H, Cremer C, Cardoso MC (2010) Measurement of replication structures at the nanometer scale using super-resolution light microscopy. Nucleic Acids Res 38(2):e8. doi:10.1093/nar/gkp901
36. Mitchell-Jordan S, Chen H, Franklin S, Stefani E, Bentolila LA, Vondriska TM (2012) Features of endogenous cardiomyocyte chromatin revealed by super-resolution STED microscopy. J Mol Cell Cardiol 53(4):552–558. doi:10.1016/j.yjmcc.2012.07.009
37. Markaki Y, Smeets D, Fiedler S, Schmid VJ, Schermelleh L, Cremer T, Cremer M (2012) The potential of 3D-FISH and super-

resolution structured illumination microscopy for studies of 3D nuclear architecture: 3D structured illumination microscopy of defined chromosomal structures visualized by 3D (immuno)-FISH opens new perspectives for studies of nuclear architecture. Bioessays 34 (5):412–426. doi:10.1002/bies.201100176

38. Fiolka R, Shao L, Rego EH, Davidson MW, Gustafsson MG (2012) Time-lapse two-color 3D imaging of live cells with doubled resolution using structured illumination. Proc Natl Acad Sci U S A 109(14):5311–5315. doi:10.1073/pnas.1119262109
39. Shao L, Kner P, Rego EH, Gustafsson MG (2011) Super-resolution 3D microscopy of live whole cells using structured illumination. Nat Methods 8(12):1044–1046. doi:10.1038/nmeth.1734
40. Lopez Perez R, Best G, Nicolay NH, Greubel C, Rossberger S, Reindl J, Dollinger G, Weber KJ, Cremer C, Huber PE (2016) Superresolution light microscopy shows nanostructure of carbon ion radiation-induced DNA double-strand break repair foci. FASEB J 30 (8):2767–2776. doi:10.1096/fj.201500106R
41. Natale F, Rapp A, Yu W, Maiser A, Harz H, Scholl A, Grulich S, Anton T, Hörl D, Chen W, Durante M, Taucher-Scholz G, Leonhardt H, Cristina Cardoso M (2017) Identification of the elementary structural units of the DNA damage response. Nat Commun 8:15760
42. Rossberger S, Ach T, Best G, Cremer C, Heintzmann R, Dithmar S (2013) High-resolution imaging of autofluorescent particles within drusen using structured illumination microscopy. Br J Ophthalmol 97(4):518–523. doi:10.1136/bjophthalmol-2012-302350
43. Bewersdorf J, Bennett BT, Knight KL (2006) H2AX chromatin structures and their response to DNA damage revealed by 4Pi microscopy. Proc Natl Acad Sci U S A 103 (48):18137–18142. doi:10.1073/pnas.0608709103
44. Lorat Y, Schanz S, Schuler N, Wennemuth G, Rube C, Rube CE (2012) Beyond repair foci: DNA double-strand break repair in euchromatic and heterochromatic compartments analyzed by transmission electron microscopy. PLoS One 7(5):e38165. doi:10.1371/journal.pone.0038165
45. Hell SW, Wichmann J (1994) Breaking the diffraction resolution limit by stimulated emission: stimulated-emission-depletion fluorescence microscopy. Opt Lett 19 (11):780–782
46. Klar TA, Hell SW (1999) Subdiffraction resolution in far-field fluorescence microscopy. Opt Lett 24(14):954–956
47. Göttfert F, Wurm CA, Mueller V, Berning S, Cordes VC, Honigmann A, Hell SW (2013) Coaligned dual-channel STED nanoscopy and molecular diffusion analysis at 20 nm resolution. Biophys J 105(1):L01–L03
48. Wildanger D, Medda R, Kastrup L, Hell SW (2009) A compact STED microscope providing 3D nanoscale resolution. J Microsc 236(1):35–43
49. Bretschneider S, Eggeling C, Hell SW (2007) Breaking the diffraction barrier in fluorescence microscopy by optical shelving. Phys Rev Lett 98:218103
50. Hell SW, Kroug M (1995) Ground-state-depletion fluorscence microscopy: a concept for breaking the diffraction resolution limit. Appl Phys B 60(5):495–497
51. Finkenstaedt-Quinn SA, Qiu TA, Shin K, Haynes CL (2016) Super-resolution imaging for monitoring cytoskeleton dynamics. Analyst 141 (20):5674–5688. doi:10.1039/c6an00731g
52. Miller SE, Mathiasen S, Bright NA, Pierre F, Kelly BT, Kladt N, Schauss A, Merrifield CJ, Stamou D, Honing S, Owen DJ (2015) CALM regulates clathrin-coated vesicle size and maturation by directly sensing and driving membrane curvature. Dev Cell 33(2):163–175. doi:10.1016/j.devcel.2015.03.002
53. Lau L, Lee YL, Sahl SJ, Stearns T, Moerner WE (2012) STED microscopy with optimized labeling density reveals 9-fold arrangement of a centriole protein. Biophys J 102(12): 2926–2935. doi:10.1016/j.bpj.2012.05.015
54. Cseresnyes Z, Schwarz U, Green CM (2009) Analysis of replication factories in human cells by super-resolution light microscopy. BMC Cell Biol 10:88. doi:10.1186/1471-2121-10-88
55. Reindl J, Drexler GA, Girst S, Greubel C, Siebenwirth C, Drexler SE, Dollinger G, Friedl AA (2015) Nanoscopic exclusion between Rad51 and 53BP1 after ion irradiation in human HeLa cells. Phys Biol 12(6):066005. doi:10.1088/1478-3975/12/6/066005
56. Cremer C, Masters BR (2013) Resolution enhancement techniques in microscopy. Eur Phys J H 38(3):281–344. doi:10.1140/epjh/e2012-20060-1
57. Moerner WE, Kador L (1989) Optical detection and spectroscopy of single molecules in a solid. Phys Rev Lett 62:2535–2538
58. Orrit M, Bernard J (1990) Single pentacene molecules detected by fluorescence excitation in a p-terphenyl crystal. Phys Rev Lett 65:2716–2719
59. Betzig E (1995) Proposed method for molecular optical imaging. Opt Lett 20(3):237–239

60. Patterson GH, Lippincott-Schwartz J (2002) A photoactivatable GFP for selective photolabeling of proteins and cells. Science 297 (5588):1873–1877
61. Bates M, Blosser TR, Zhuang X (2005) Short-range spectroscopic ruler based on a single-molecule optical switch. Phys Rev Lett 94:108101–108101
62. Betzig E, Patterson GH, Sougrat R, Lindwasser OW, Olenych S, Bonifacino JS, Davidson MW, Lippincott-Schwartz J, Hess HF (2006) Imaging intracellular fluorescent proteins at nanometer resolution. Science 313(5793):1642–1645
63. Hess ST, Girirajan TPK, Mason MD (2006) Ultra-high resolution imaging by fluorescence photoactivation localization microscopy. Biophys J 91(11):4258–4272. doi:10.1529/biophysj.106.091116
64. Rust MJ, Bates M, Zhuang X (2006) Sub-diffraction-limit imaging by stochastic optical reconstruction microscopy (STORM). Nat Methods 3(10):793–796
65. Dickson RM, Cubitt AB, Tsien RY, Moerner WE (1997) On/off blinking and switching behaviour of single molecules of green fluorescent protein. Nature 388(6640):355–358
66. Geisler C, Schönle A, von Middendorff C, Bock H, Eggeling C, Egner A, Hell SW (2007) Resolution of λ/10 in fluorescence microscopy using fast single molecule photo-switching. Appl Phys A 88(2):223–226
67. Fölling J, Bossi M, Bock H, Medda R, Wurm CA, Hein B, Jakobs S, Eggeling C, Hell SW (2008) Fluorescence nanoscopy by ground-state depletion and single-molecule return. Nat Methods 5(11):943–945
68. Heilemann M, van de Linde S, Schüttpelz M, Kasper R, Seefeldt B, Mukherjee A, Tinnefeld P, Sauer M (2008) Subdiffraction-resolution fluorescence imaging with conventional fluorescent probes. Angew Chem Int Ed 47 (33):6172–6176
69. Lemmer P, Gunkel M, Baddeley D, Kaufmann R, Urich A, Weiland Y, Reymann J, Müller P, Hausmann M, Cremer C (2008) SPDM: light microscopy with single-molecule resolution at the nanoscale. Appl Phys B 93(1):1–1
70. Dempsey GT, Bates M, Kowtoniuk WE, Liu DR, Tsien RY, Zhuang X (2009) Photoswitching mechanism of cyanine dyes. J Am Chem Soc 131(51):18192–18193. doi:10.1021/ja904588g
71. Huang B, Wang W, Bates M, Zhuang X (2008) Three-dimensional super-resolution imaging by stochastic optical reconstruction microscopy. Science 319(5864):810–813
72. Juette MF, Gould TJ, Lessard MD, Mlodzianoski MJ, Nagpure BS, Bennett BT, Hess ST, Bewersdorf J (2008) Three-dimensional sub-100 nm resolution fluorescence microscopy of thick samples. Nat Methods 5(6):527–529
73. Pavani SRP, Thompson MA, Biteen JS, Lord SJ, Liu N, Twieg RJ, Piestun R, Moerner WE (2009) Three-dimensional, single-molecule fluorescence imaging beyond the diffraction limit by using a double-helix point spread function. Proc Natl Acad Sci 106(9):2995–2999
74. Tang J, Akerboom J, Vaziri A, Looger LL, Shank CV (2010) Near-isotropic 3D optical nanoscopy with photon-limited chromophores. Proc Natl Acad Sci 107 (22):10068–10073
75. Shtengel G, Galbraith JA, Galbraith CG, Lippincott-Schwartz J, Gillette JM, Manley S, Sougrat R, Waterman CM, Kanchanawong P, Davidson MW, Fetter RD, Hess HF (2009) Interferometric fluorescent super-resolution microscopy resolves 3D cellular ultrastructure. Proc Natl Acad Sci 106(9):3125–3130
76. Jia S, Vaughan JC, Zhuang X (2014) Isotropic three-dimensional super-resolution imaging with a self-bending point spread function. Nat Photon 8(4):302–306
77. Deschamps J, Mund M, Ries J (2014) 3D superresolution microscopy by supercritical angle detection. Opt Express 22 (23):29081–29091
78. Kaufmann R, Müller P, Hildenbrand G, Hausmann M, Cremer C (2011) Analysis of Her2/neu membrane protein clusters in different types of breast cancer cells using localization microscopy. J Microsc 242(1):46–54
79. Boyd PS, Struve N, Bach M, Eberle JP, Gote M, Schock F, Cremer C, Kriegs M, Hausmann M (2016) Clustered localization of EGFRvIII in glioblastoma cells as detected by high precision localization microscopy. Nanoscale 8:20037–20047
80. Müller P, Lemmermann NA, Kaufmann R, Gunkel M, Paech D, Hildenbrand G, Holtappels R, Cremer C, Hausmann M (2014) Spatial distribution and structural arrangement of a murine cytomegalovirus glycoprotein detected by SPDM localization microscopy. Histochem Cell Biol 142(1):61–67
81. Bohn M, Diesinger P, Kaufmann R, Weiland Y, Müller P, Gunkel M, von Ketteler A, Lemmer P, Hausmann M, Heermann DW, Cremer C (2010) Localization microscopy reveals expression-dependent parameters of chromatin nanostructure. Biophys J 99(5):1358–1367
82. Zhang Y, Máté G, Müller P, Hillebrandt S, Krufczik M, Bach M, Kaufmann R, Hausmann

M, Heermann DW (2015) Radiation induced chromatin conformation changes analysed by fluorescent localization microscopy, statistical physics, and graph theory. PLoS One 10 (6):1–23

83. Stuhlmüller M, Schwarz-Finsterle J, Fey E, Lux J, Bach M, Cremer C, Hinderhofer K, Hausmann M, Hildenbrand G (2015) In situ optical sequencing and nano-structure analysis of a trinucleotide expansion region by localization microscopy after specific COMBO-FISH labelling. Nanoscale 7:17938–17946

84. Falk M, Hausmann M, Lukasova E, Biswas A, Hildenbrand G, Davidkova M, Krasavin E, Kleibl Z, Falkova I, Jezkova L, Stefancikova L, Sevcik J, Hofer M, Bacikova A, Matula P, Boreyko A, Vachelova J, Michaelidesova A, Kozubek S (2014) Determining omics spatiotemporal dimensions using exciting new nanoscopy techniques to assess complex cell responses to DNA damage: Part B structuromics. Crit Rev Eukaryot Gene Expr 24 (3):225–247

85. Ceccaldi R, Rondinelli B, D'Andrea AD (2016) Repair pathway choices and consequences at the double-strand break. Trends Cell Biol 26(1):52–64

86. Mladenov E, Magin S, Soni A, Iliakis G (2016) DNA double-strand-break repair in higher eukaryotes and its role in genomic instability and cancer: cell cycle and proliferation-dependent regulation. Semin Cancer Biol 37-38:51–64

Chapter 2

Managing the Introduction of Super-Resolution Microscopy into a Core Facility

Jeffrey A. Kamykowski and Brian Storrie

Abstract

Super resolution techniques place the resolution of fluorescence microscopy closer to the size of the underlying cell structure or molecular machine being studied. Structured illumination techniques will give users a set of tools that are close to their past experience and relatively simple and quick to learn. The present dyes can be used. Resolution approaching 100 nm XY can be achieved. In contrast, stochastic methods such as PALM/STORM typically require the choice of new dyes and a much greater learning curve to master the technology and calculations. However, a further fivefold resolution improvement is possible. Stimulated depletion techniques such as STED offer a third set of approaches that will again require the use of new dyes. All these approaches require substantial investment in new equipment and in user training. There is no free lunch in the search for better resolution.

Key words Core facility, Super resolution microscopy, 3D-SIM, Structured illumination, PALM, STORM, Management

1 Introduction

Super-resolution fluorescence microscopy represents a change in resolution standards that requires core facilities to re-fit themselves. The equipment brings a large improvement in resolution and hence yields images that now much better approximate the size of the subcellular organelle or molecular complex. New biology is possible. This transition to higher resolution now possible may well have the same level of impact as commercialization of confocal microscopy in the early 1990s. Confocal microscopy's ability to "haze bust" was important. Core facilities must evolve to provide and maintain state-of-the-art equipment that individual laboratories would have a difficult time acquiring alone. The benefit of a core facility is lost if it does not update. Spreading acquisition and operating costs over multiple laboratories can create a virtuous cycle in which the core facility is able to keep up with the latest equipment and techniques and hence enhance research capabilities.

Holger Erfle (ed.), *Super-Resolution Microscopy: Methods and Protocols*, Methods in Molecular Biology, vol. 1663, DOI 10.1007/978-1-4939-7265-4_2,

For a microscopy core facility, the commercial introduction of super-resolution microscopy systems now presents a new opportunity and standard in both equipment and expertise in supporting quality research results. When expanding, or re-fitting core microscopy facilities to include super-resolution capabilities, several key aspects have to be examined and addressed.

2 Considerations

The first aspect of super-resolution microscopy that has to be considered is the choice of what type of system will be purchased. There are several factors that are highly important when choosing a super-resolution system (*see* Table 1). The primary consideration is the determination of what type of super-resolution system will fulfill the needs of researchers. A group of super-resolution candidate users will likely not have the same needs. Super-resolution systems with faster image acquisition allow for live cell imaging, while others have much longer imaging times that can give better image resolution. Limits on fluorescent dye choices may well be important. Hardware and software options may need to be added to the purchase to increase the research capabilities. Familiarity with one or more microscope manufacturers will help with getting the needed information to weigh the base models and optional additions that each system offers. Demonstrations onsite or at the manufacturer's locations with multiple sample types should be

Table 1
Comparison of super-resolution microscopes

	Structured Illumination Microscopy (SIM)	Instant SIM (iSIM)	Stimulated Depletion (STED)	PALM/STORM (stochastic sampling)
Lateral resolution	110–150 nm	120 nm	50–70 nm or better	10–20 nm possible
Axial resolution	220–250 nm	300 nm	100 nm or better	50 nm
Dyes and fluorescent proteins	Conventional	Conventional	STED dyes	Photoactivable Photoswitchable
Live cell imaging	Yes	Yes	Yes	No
Photobleaching Phototoxicity	Yes	Low	Yes, can be minimized by RESOLFT technique	Yes
Image processing	Yes	Not required but can be done	Not required but can be done	Yes

conducted to gauge the overall capabilities and ease of use. Most software is designed for user friendliness but prior familiarity may benefit in easing the transition to a new system. In our experience, a hybrid system that will support both structured illumination microscopy (SIM) and stochastic sampling (PALM/STORM) is often a good choice. Most core facility directors comment that the hybrid system is more used, in practice, for SIM than PALM/STORM, techniques that demand even more user expertise. An important point for any core facility to remember is that the goal is to meet a general need. A core itself is not doing "discovery" science in super-resolution microscopy; rather a core is a resource to others.

After the choice of the super-resolution system is made, consideration for the system location must be made. Super-resolution systems can require a fair amount of floor space and hence can put limitations on location. Even though systems are typically enclosed to shield from air drafts and mounted on high quality vibration tables to minimize vibration, actual potential interferences in the chosen room should be considered. Since moving super-resolutions microscopes can cause disruptions in the alignment of components, the system should be spaced adequately from walls to allow access to the rear of the system without the need to move. Space for multiple operators should also be considered. Any room modifications, i.e., electrical, air and vacuum, networking, should be done prior to installation. Optical cabling, if required, should be pre-installed to speed the transfer of large image files, 4–6 Gb or larger.

Delivery and installation can pose problems. As with most microscopy systems, super-resolutions systems are delivered in large boxes and wooden crates that most fire safety concerns prevent from being kept in hallways until installation. Temporary space separate from the actual installation space is beneficial but may not be a must. Unpacking boxes and crates will leave a substantial amount of packing material to remove, so having a large trash bin on hand is useful. Once the system installation begins, delays and problems should be expected. Issues with provided computer hardware and software may arise that cause disruptions in proper system function. Some components may not work optimally and replacements will need to be ordered. Occasionally, these components are manufactured to order, which could then result in a delay of several days to weeks depending on manufacturing schedules and location. All these unforeseen problems can prolong installation and routine operation. The purchaser should insist that all be fully rectified by the manufacturer before any warranty period begins.

Once installation is complete, initial training must be conducted. A manufacturer's trainer will usually spend a day checking the system to make sure everything is functioning properly. Training core personnel and the primary users should occur first, followed by the next group of users that will use the system less frequently. The minor user population will often require training and re-training to keep skills current. Having preset hardware configurations will make the process flow smoother and forgo the need to cover a lot of setup during user training. Providing "cheat" sheet guides on basic operations can be helpful. Both these steps will save time for core personnel to train future users. Core facility personnel must monitor for proper microscope use. For additional training and hands on assistance, there are courses in super-resolution microscopy provided by manufacturers and government sponsored agencies (i.e., FAES BioTech 53 at NIH, EMBO courses in Europe).

Routine operations have several key technical points that must be addressed during continued training of new users. Super-resolution microscopy is particularly sensitive to immersion oil refractive index; proper immersion oil must be used for microscope systems at room temperature (23 °C), in a closed enclosure (30 °C), or when using a heated stage insert (37 °C). These will initially be provided at installation. Any residue of incorrect oil must be thoroughly removed from a slide or it will disrupt proper resolution and require extensive cleaning of the objective(s). If users will be using more than one microscope, multiple slides may be necessary to prevent any crossover of oil between microscopes. Coverslips on slides or mounted to dishes must be high quality and very flat. Likewise, stages must be maintained at proper leveling for best focus and resolution. This is a particular issue regarding the used and exchange of regular versus live-cell incubation stages. Antifade mountant is recommended with super-resolution techniques because they are prone to photobleaching, particularly because of the need to take repeated images in SIM and PALM/STORM.

Most super-resolution methods require post-imaging processing. These processing algorithms are often very sensitive to bright objects and having images in which the detector range is not saturated is important. Because of the extended time for image acquisition and image processing, there will likely be the need for an external workstation with full software capabilities to allow for image processing, manipulation, and analysis independent of the microscope system computer. Depending on intensity of use and convenience, individual laboratories may consider purchasing their own software packages. Because of the large file size of super-resolution images, i.e., SIM or PALM/STORM images, connecting the microscope computer and a workstation through a fiber optic connection is a practical necessity for faster file transfer compared to using external USB drives. Attaching external drives to the microscope runs the hazard of introducing virus contamination. Most microscope operation computers do not have virus protection software due to possible interference with microscope usage.

The facility must be aware of the risks that it is taking. Monitoring hard drive space on the core facility computers is important to have adequate space available. It is best to purge drives to remove user image files on a regularly scheduled basis.

3 Concluding Remarks

In conclusion, the introduction of super-resolution microscopy introduces both a new world of resolution but also a set of techniques that are more demanding on core expertise and user expertise. Any core facility needs to be prepared not only physically but also mentally for the increased demands of this level of technology. The facility in making choices on equipment must also realize that the technology is presently fluid. Hence, the reality is to make a practical choice that serves the investigator community. There is no perfect choice.

In a facility that supports both 3D-SIM and PALM/STORM, our experience is that most usage will be 3D-SIM, the simpler technology. Finally, we note that many users find PALM/STORM approaches to be counterintuitive because the blinking incidence needs to be low to avoid spatially close events that occur in the same time window being binned together by the algorithm(s). Time-binned blinking can substantially alter resolution. In sum, please be forewarned that the introduction of these techniques will require considerably more user training than past approaches to light microscopy.

Acknowledgment

Supported by US NIH grants S10 OD018065 and R01 HL119393.

Chapter 3

Live-Cell STED Imaging with the HyPer2 Biosensor

Natalia M. Mishina and Vsevolod V. Belousov

Abstract

Stimulated emission depletion (STED) microscopy is a popular super resolution imaging technique. Not only synthetic dyes and fluorescent proteins can be utilized as STED fluorophores, but also genetically encoded biosensors. Fusing the biosensor with proteins of interest allows subdiffraction imaging of intracellular macromolecular architecture with simultaneous extraction of functional information about cellular activities. Here, we describe a protocol for live-cell STED microscopy of the HyPer2 biosensor fused to cytoskeletal filaments.

Key words Super-resolution microscopy, STED, Live-cell imaging, Biosensors, HyPer2, Hydrogen peroxide

1 Introduction

HyPer2 is a genetically encoded biosensor for the detection of H_2O_2 in living cells [1]. It is an improved variant of biosensor HyPer [2] with twice expanded dynamic range. The biosensor consists of fluorescent protein cpYFP and a bacterial H_2O_2-sensitive domain OxyR. Previously published protocols [3, 4] cover the HyPer usage in live-cell conventional fluorescent microscopy, all the recommendations on immobilizing biosensor within cell, choosing cell lines and stimuli. Most of these protocols can be straightforwardly applied to HyPer2. HyPer2 was neither specifically designed nor optimized for STED microscopy; nevertheless, biosensor had good performance in the STED super-resolution microscopy, allowing for simultaneous H_2O_2 detection and a resolution improvement [5].

Here, we present a detailed protocol for live-cell STED imaging of intracellular H_2O_2 production including detailed step-by-step sample preparation instructions, experimental setup, and image analysis guidelines. This protocol may be used as a starting point to adapt live-cell imaging of HyPer2 and similar biosensors for the requirements of STED microscopy.

Holger Erfle (ed.), *Super-Resolution Microscopy: Methods and Protocols*, Methods in Molecular Biology, vol. 1663,
DOI 10.1007/978-1-4939-7265-4_3,

2 Materials

2.1 Reagents

1. Vector pHyPer2-C (Addgene, cat. no. 42211) for HyPer2 fusion protein construction by means of molecular cloning.
2. DMEM.
3. Minimum Essential Medium Eagle (EMEM).
4. OPTI-MEM.
5. Hanks Balanced Salt Solution (HBSS).
6. Dulbecco-PBS.
7. Fetal calf serum, FCS.
8. Penicillin/streptomycin.
9. L-Glutamine.
10. Trypsin/EDTA solution.
11. HEPES.
12. Bovine Serum Albumin (BSA), cell culture grade.
13. Paraformaldehyde.
14. Mowiol.
15. FuGene6 transfection reagent or similar.
16. Mammalian cells. This protocol describes the use of HeLa-Kyoto and NIH-3T3 cell lines. However, the overall technique is applicable to any transfectable or transducible cell culture.
17. Sodium bicarbonate.
18. H_2O_2. Caution. It may cause severe skin burns and eye damage. Follow the safety guide.
19. PDGF.

2.2 Equipment

1. Glass-bottom dishes.
2. Cover slips #1.5 (optimal: 0.170 ± 0.01 mm thick).
3. Microscopy slides.
4. Standard equipment for eukaryotic cell culturing.
5. Custom-built STED microscope or a commercial STED microscope (e.g., Leica Microsystems GmbH, Mannheim, Germany) equipped with a thermostating box.
6. Image processing software: ImageJ or FIJI (National Institutes of Health).
7. Data quantification and analysis software: Excel (Microsoft) or Origin (OriginLab Corporation).

2.3 Reagent Setup

1. Complete growth medium: We use DMEM supplemented with penicillin/streptomycin, L-glutamine, and 10% FCS.
2. Cell starvation and imaging medium: EMEM supplemented with either 2.2 g/l sodium bicarbonate and 20 mM HEPES (if CO_2 supply for thermostating box available) or just 20 mM HEPES (use bicarbonate free media in case no CO_2 supply for thermostating box is available in order to prevent alkalinization of the imaging media), penicillin/streptomycin, L-glutamine, and 0.5% BSA.
3. Preparation of growth factors for experiment: Growth factor suppliers recommend the optimal protocol for storage. Just before the experiment, dilute a small aliquot of the growth factor in EMEM to prepare a 1000 stock solution (10 mg/ml PDGF). Keep on blue ice.

2.4 Equipment Setup

We used Leica TCS SP8 microscope with a time-gated stimulated emission depletion (STED) module (Leica Microsystems GmbH, Mannheim, Germany), equipped with a white light laser (WLL), STED 592-nm laser, HCX PL APO 100×/1.40 oil objective, and a hybrid detection system (Leica HyD).

3 Methods

3.1 Plasmid Construction

Clone the protein of interest in-frame with HyPer2 using the appropriate cloning vector (*see* **Note 1**).

3.2 Transfection

1. Plate the cells on glass-bottom dishes for live-cell imaging or grow them directly on cover slips for subsequent cell fixation.
2. Transfect the cells with the HyPer2 fusion encoding vector by any appropriate method. We used FuGene6 transfection reagent, after 4-h incubation with the liposome:DNA complex, we recommend rinsing the cells with HBSS and changing the medium to fresh prewarmed liposome-free complete culture medium.
3. Culture the cells for 24–48 h at 37 °C, 5% CO_2.

3.3 Preparation of Cell Samples for an Experiment

We suggest three types of samples: choose cell fixation (**step 1**) for image settings adjustment and verifying HyPer2-fusion construct localization. Choose live-cell H_2O_2 addition (**step 2**) for HyPer2-fusion sensor performance tests and finally prepare for live-cell experiment in physiological conditions (**step 3**).

1. Cell fixation. Rinse the cover slip in PBS. Fix the cells by incubating in 4% paraformaldehyde in PBS for 15 min at room temperature. Wash the cells twice with PBS. Mount cover slip with the sample on a microscope slide with a small

drop (~15 μl) of Mowiol [6]. Leave slide for several hours in the dark to allow the Mowiol to harden.

2. Live-cell experiment with H_2O_2 addition. Replace the complete medium with 1.5 ml preheated (37 °C) HBSS or PBS.
3. Cell stimulation with growth factor. Replace the complete medium with 1.5 ml preheated (37 °C) MEM. Incubate for 2 h (for HeLa-Kyoto) or 4–6 h (for NIH-3T3) at 37 °C. Keep the cells out of the CO_2 incubator if using bicarbonate-free MEM.

3.4 Imaging Settings (General Considerations for Leica TCS SP8 STED Users)

Use fixed cells while setting up the microscope. HyPer2 can be excited with the 488 nm or 496 nm laser lines, emission can be detected in the 500–550 nm range (*see* **Note 2** and Table 1). As the high power of the lasers could induce phototoxicity, minimize white light laser (WLL) and STED laser power and reuse the line average number in live-cell experiments with growth factors.

1. Activate WLL, set the laser power to 70%. Choose 488 or 496 nm line.
2. Activate STED 592 nm laser, set the laser power to 100%.
3. Align STED and imaging lasers at the beginning of imaging session, then once again after 20 min and every 3–4 h thereafter.
4. Set the scanning resolution to 1456 × 1456 pixels.
5. Set the number of line averages equal to 8.
6. Set pinhole size to 1 Airy units and the pixel size to 20 nm.
7. For confocal imaging set the intensity of the WLL to 5%.
8. For STED imaging set the intensity of the WLL to 25% and the intensity of STED 592 nm laser to 100%.
9. Choose HyD as active detectors.
10. Use gating, starting with 0.7–1 ns.
11. Balance WLL power and gating.

3.5 Time Series Acquisition

1. At least 2 h before imaging, heat up the thermal incubator box to 37 °C.
2. Place the dish with cells onto the microscope stage. Set up focus.
3. Switch the system to a scanning mode.
4. Perform a single scan in confocal mode. Zoom selected cell.
5. Perform several single scans in confocal and STED modes to set the final focus, adjust laser power and gating. Adjust the delay between frames (*see* **Note 3**).

Table 1
Example of image acquisition settings for short and long experiments

Parameter	Short experiment (H_2O_2 addition), STED	Short experiment (H_2O_2 addition) confocal	Long experiment (stimulation with PDGF), STED	Long experiment (stimulation with PDGF), confocal
Scan speed, Hz	1000	1000	400	400
Line average	8	8	5	5
Filter - notchFW2	NF 594	–	NF 594	NF 594
Filter - polarization FW	NF 488	NF 488	–	–
STED (592 nm), output power %	100	0	91.2	0
STED (592 nm), intensity %	100	off	36.2	off
Intensity (488 nm), %	15	5	0.01	–
Intensity (496 nm), %	–	–	23.8	9
Detector	HPD (500–550 nm)	HPD (500–550 nm)	HPD (562–616 nm)	HPD (562–616 nm)
Gain	380	380	421	421
Gate start, ns	0.75	0.75	0.61	0.61
Gate end, ns	12	12	4.39	4.39
Time delay between frames	12 s	–	2 min	–

6. Start time series acquisition in STED mode. Pay special attention for the microscope focus stability during the entire experiment. The use of an automatic focus adjustment accessories is highly advisable.
7. H_2O_2 addition: After 1 min of imaging, add H_2O_2 diluted in 100 μl of pre-warmed HBSS or PBS dropwise. Changes in the signal occur immediately after H_2O_2 addition. Continue data acquisition for 2–3 min.
8. Growth factor stimulation: After 1–2 min of imaging, add the growth factor dropwise (for NIH 3T3 we add 10 ng/ml PDGF) diluted in 100 μl of prewarmed MEM or 100 μl of serum-free MEM as negative control. Continue data acquisition for 15–30 min (*see* **Note 4**).

3.6 Image Analysis

One of the great advantages of STED over the most of super-resolution techniques is that image processing is not needed to obtain super-resolution images. Similarly, HyPer2 intensity change can be monitored during imaging. Nevertheless, we recommend performing image postprocessing described below for the representation improvement and more precise image analysis if needed.

Analyze acquired data offline in ImageJ software (http://imagej.nih.gov/ij/) (*see* **Note 5**).

1. To aid the visualization convolve the image with Gaussian kernel (**Process/Filters/Gaussian blur**). Use kernel size of 1 or 2 pixels.
2. If sample drift is observed, apply "**StackReg**" plug-in (http://bigwww.epfl.ch/thevenaz/stackreg/) or any other appropriate plug-in for image stack registration.
3. Depict the resulting stack in pseudocolors using a "**Ratio**" lookup table for better visualization of intensity changes.
4. Choose a ROI for the time profiles quantification. The resultant table can be saved and used in .xls format for Excel or Origin.

3.7 Results

Using the method described above, we recorded STED images of vimentin filaments with a resolution of 70–80 nm [5]. Typical results are presented in Figs. 1, 2, and 3. Figure 1 shows a live HeLa-Kyoto cell expressing vimentin-HyPer2 imaged in confocal and STED mode. A line profile plot (Fig. 1c) shows the resolution improvement. Figure 2 demonstrates that upon the addition of H_2O_2 HyPer2 increases brightness. Figure 3 demonstrates STED images of microtubules labeled with EB3–HyPer2 in NIH 3T3 fibroblast. Stimulation of the fibroblast with PDGF resulted in cellular H_2O_2 production reflected by the increase of EB3–HyPer2 brightness.

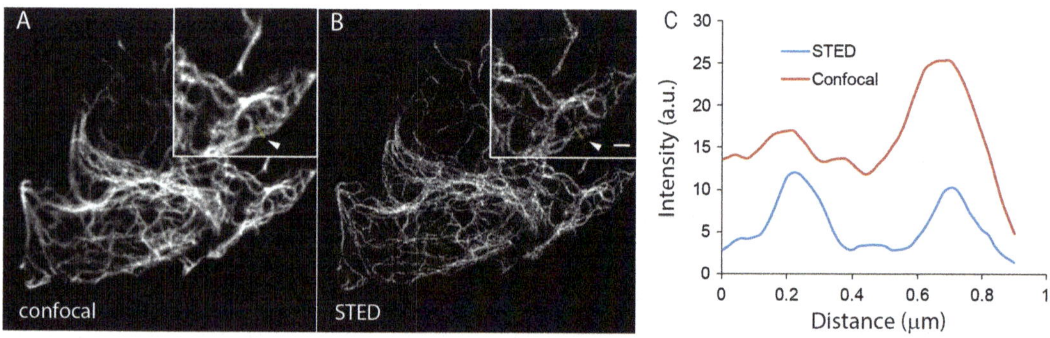

Fig. 1 A representative confocal (**a**) and STED (**b**) images and (**c**) intensity profiles along the linear ROI (see the *arrow*) of the HeLa-Kyoto cell expressing vimentin-HyPer2. Scale bar is 1 μm

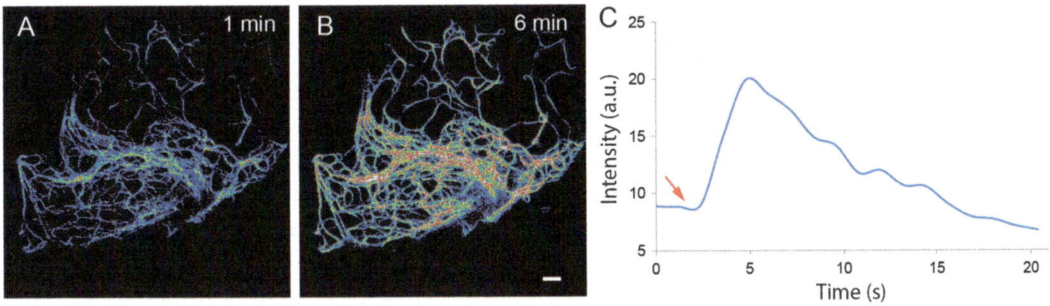

Fig. 2 Pseudocolored STED images of the HeLa-Kyoto cell expressing vimentin-HyPer2 in the course of H_2O_2 addition. Increase in fluorescence is clearly visible shortly after hydrogen peroxide addition (**a**, **b**). Scale bar is 5 μm. (**c**) Time course of HyPer2 intensity change. The *arrow* indicates H_2O_2 addition

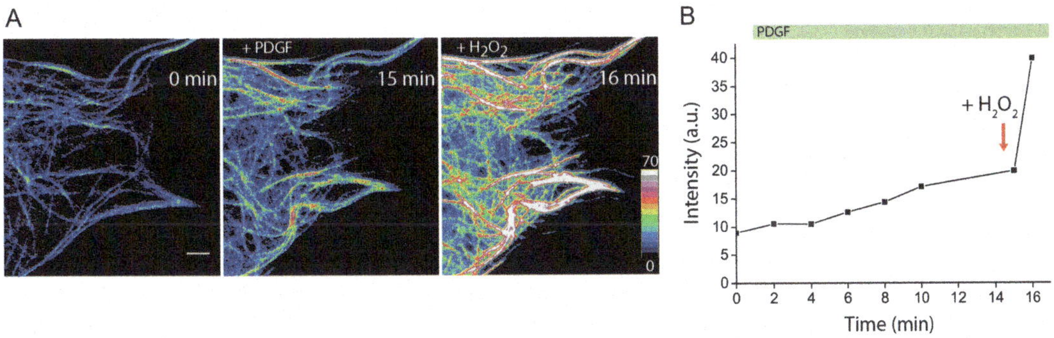

Fig. 3 Imaging of EB3-HyPer2 in live NIH 3T3 cells. (**a**) STED images of EB3-Hyper2 fluorescence intensity change in NIH 3T3 cell at indicated time points after stimulation with 10 ng/ml PDGF and subsequent addition of 200 μM H_2O_2. Scale bar: 2 μm. (**b**) Timing of fluorescence intensity change in cell shown on panel (**a**). Adapted from [5]. Copyright (2015) American Chemical Society

4 Notes

1. In our experiments fusion constructs of vimentin-HyPer2 (labeling of intermediate filaments) and EB3-HyPer2 (labeling of microtubules) worked well for STED microscopy. General considerations for creating HyPer2 fusion constructs can be found here ([3]).
2. The fluorescent core of HyPer2 biosensor is based on cpYFP, thus STED imaging settings for YFP (eYFP, citrine) should be applicable for HyPer2 too. It is important to note that HyPer2 is relatively dim probe; therefore, excitation laser intensity should be set to the level 3–5 times higher than for citrine. When setting the intensity of excitation laser, keep in mind that long exposure at high intensity is phototoxic for living cells and also causes rapid photobleaching of fluorescent probe. Thus,

the balance is needed between laser power, gating, and time delay between frames.

3. For short experiments (H_2O_2 addition) choose "no delay" between frames during time series acquisition. For long experiments (stimulation with growth factor) delay should be about 1–3 min.
4. At the end of the session you can add an H_2O_2 aliquot to the final concentration of 200 μM, in order to completely saturate HyPer2. This step can serve as a positive control for the biosensor performance, and will also help to evaluate the results of the experiment.
5. To compare STED and corresponding confocal images all the postprocessing steps of these images should be identical.

Acknowledgment

The work was supported by the EMBL-RFBR grant 15-54-74003.

References

1. Markvicheva KN, Bilan DS, Mishina NM, Gorokhovatsky AY, Vinokurov LM, Lukyanov S, Belousov VV (2011) A genetically encoded sensor for H_2O_2 with expanded dynamic range. Bioorg Med Chem 19:1079–1084
2. Belousov VV, Fradkov AF, Lukyanov KA, Staroverov DB, Shakhbazov KS, Terskikh AV, Lukyanov S (2006) Genetically encoded fluorescent indicator for intracellular hydrogen peroxide. Nat Methods 3:281–286
3. Mishina NM, Markvicheva KN, Fradkov AF, Zagaynova EV, Schultz C, Lukyanov S, Belousov VV (2013) Imaging H_2O_2 microdomains in receptor tyrosine kinases signaling. Methods Enzymol 526:175–187
4. Mishina NM, Markvicheva KN, Bilan DS, Matlashov ME, Shirmanova MV, Liebl D, Schultz C, Lukyanov S, Belousov VV (2013) Visualization of intracellular hydrogen peroxide with HyPer, a genetically encoded fluorescent probe. Methods Enzymol 526:45–59
5. Mishina NM, Mishin AS, Belyaev Y, Bogdanova EA, Lukyanov S, Schultz C, Belousov VV (2015) Live-Cell STED microscopy with genetically encoded biosensor. Nano Lett 15:2928–2932
6. Quick Guide to STED Sample Preparation by Wernher Fouquet, Ph.D. https://www.leica-microsystems.com/science-lab/quick-guide-to-sted-sample-preparation/

Chapter 4

Diffraction-Unlimited Fluorescence Imaging with an EasySTED Retrofitted Confocal Microscope

André Klauss and Carsten Hille

Abstract

The easySTED technology provides the means to retrofit a confocal microscope to a diffraction-unlimited stimulated emission depletion (STED) microscope.

Although commercial STED systems are available today, for many users of confocal laser scanning microscopes the option of retrofitting their confocal system to a STED system ready for diffraction-unlimited imaging may present an attractive option. The easySTED principle allowing for a joint beam path of excitation and depletion light promises some advantages concerning technical complexity and alignment effort for such an STED upgrade. In the one beam path design of easySTED the use of a common laser source, either a supercontinuum source or two separate lasers coupled into the same single-mode fiber, becomes feasible. The alignment of the focal light distribution of the STED beam relative to that of the excitation beam in all three spatial dimensions is therefore omitted respectively reduced to coupling the STED laser into the common single-mode fiber. Thus, only minor modifications need to be applied to the beam path in the confocal microscope to be upgraded. Those comprise adding polarization control elements and the easySTED waveplate, and adapting the beamsplitter to the excitation/STED wavelength combination.

Key words Super-resolution, Nanoscopy, Fluorescence microscopy, Confocal laser scanning microscopy, Time gating, Stimulated emission depletion

1 Introduction

Fluorescence microscopy, combining noninvasiveness, high sensitivity, and selectivity, is a very important and widespread method in cell biology. For decades, one serious restriction of optical far-field microscopy was seen in the diffraction limit, stating that two adjacent features may not be dissolved optically if they are closer than half a wavelength [1]. Even an idealized optical system will focus light into a finite sized spot called point spread function (PSF) due to the wave-like nature of light. In the case of confocal laser scanning microscopy, the resolution limit is the direct consequence of the finite PSF size of the focused light with a minimum full width at half maximum (FWHM) diameter of approximately half a

Holger Erfle (ed.), *Super-Resolution Microscopy: Methods and Protocols*, Methods in Molecular Biology, vol. 1663, DOI 10.1007/978-1-4939-7265-4_4,

wavelength, as the PSF illuminates adjacent molecules at the same time during the scanning process.

However, today, several techniques exist that allow us to bypass the diffraction limit in far-field fluorescence localization [2–5]. All those methods rely on the common principle of discerning adjacent fluorophores by having them transiently occupy different (discernable) states [6], a non-fluorescent dark OFF-state and a fluorescent bright ON-state, in the simplest case.

In stimulated emission depletion (STED) microscopy, for example, a high-power depletion laser, red-shifted in its wavelength by typically ~100 nm relative to the excitation laser, is used to force excited fluorophores (ON-state) back to their dark ground state (OFF-state) by stimulated emission in a predefined focal region. Typically, the depletion light is modulated to form a ring or doughnut-shaped focal intensity distribution featuring a central zero intensity that is brought to an exact spatial overlap with the intensity maximum of the focused excitation light. As both the diameters of the excitation and STED laser PSFs are diffraction-limited, subdiffraction resolution is only achieved as the applied STED intensity starts saturating the STED transition. Molecules that are excited by the focused excitation light then are efficiently de-excited by the STED laser in the crest of the ring and spontaneous emission, forming the signal to be detected, is only emitted from molecules close to the STED zero. With rising STED intensity, the central area where the fluorophores are not completely depleted diminishes, resulting in an effective PSF of fluorescence emission whose size is no longer ruled by diffraction. Its diameter and thus the spatial resolution of the STED microscope scales as a function of the applied STED laser intensity, I_{STED}, according to [7]:

$$\mathrm{FWHM}_{\mathrm{STED}} \approx \mathrm{FWHM}_{\mathrm{confocal}} / \sqrt{1 + I_{\mathrm{STED}}/I_{\mathrm{S}}} \tag{1}$$

where the term I_{S} may be interpreted as a threshold intensity that depends on the STED wavelength and the emission spectral properties of the fluorophore.

In the most common realization of the STED principle, the depletion laser beam is phase-modulated by a helical phase plate to form a doughnut-shaped light distribution in the focus of the objective lens (e.g., [8–14]). Following STED beam shaping, both the laser beams need to be combined with high accuracy to ensure precise overlap of the two PSFs in all three dimensions in the focus of the objective lens. This results in some alignment effort of such a two-beam setup.

Alternatively, the easySTED waveplate was developed to reduce the alignment effort by delivering a chromatic optical device that forms the doughnut intensity profile of the STED beam while leaving the excitation beam pass almost unaltered [15]. This allows for an STED setup with a common laser source for excitation and depletion realized by a supercontinuum source or the coupling of different lasers into a common single-mode optical fiber. The demand of relative alignment of two laser beams is reduced to the joint coupling into the common fiber resulting in a stable setup configuration.

For the realization of the chromatic PSF shaping, the easySTED waveplate relies on polarization shaping rather than phase shaping. In short, four segments of birefringent material modulate the polarization of the incident STED beam such that opposite segments of the beam profile at any time have opposite polarization direction (Fig. 1a), leading to destructive interference in the center of the focused beam (Fig. 1b). The relative strong

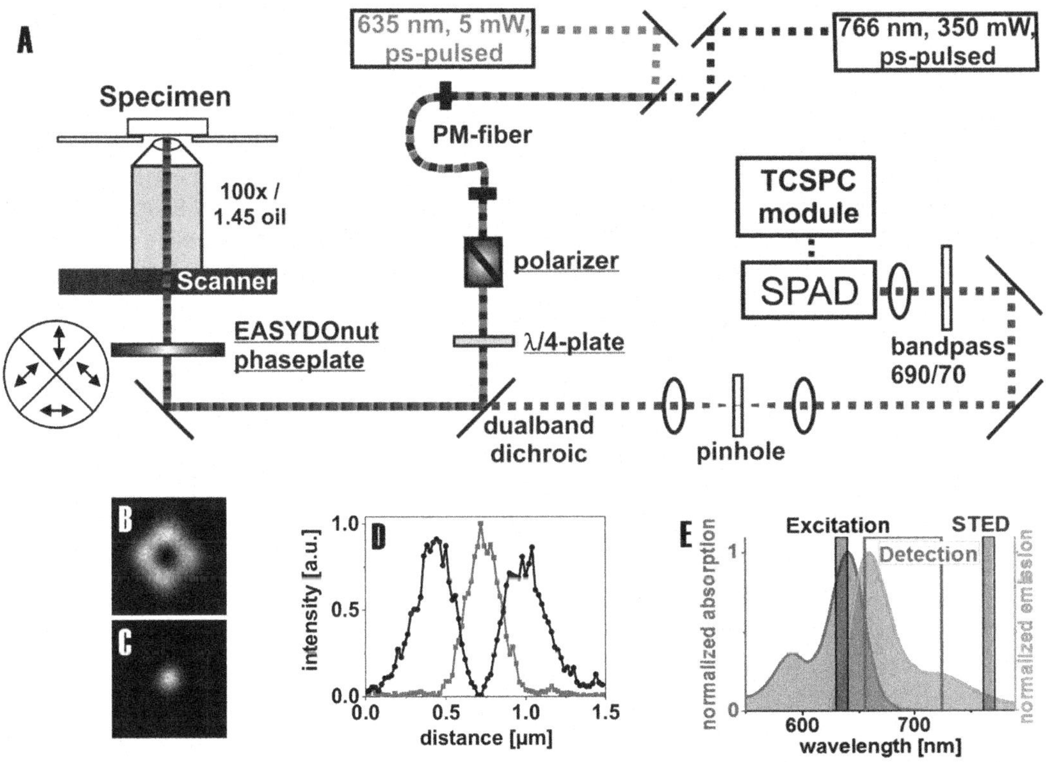

Fig. 1 An easySTED upgraded confocal laser scanning microscope. (**a**) Schematic representation of the basic elements realizing an easySTED microscope. The detection system consists in our setup of a single-photon avalanche diode (SPAD) and a time-correlated single-photon counting (TCSPC) module. The schematic drawing of the segmented easySTED waveplate (EASYDOnut) indicates the four segments of birefringent material. Arrows indicate the direction of the fast axis. (**b**) Focal intensity distribution of the STED laser modulated by the easySTED waveplate, visualized by imaging the backscattered light from a gold nanobead. (**c**) Excitation intensity distribution in the focal plane. (**d**) Profile plots through the center of **b** and **c** along a horizontal line. (**e**) Absorption and emission spectra of the fluorescent dye STAR 635P (Abberior) suitable for STED imaging with red excitation (~635 nm). Possible excitation and STED wavelength are indicated as well as the spectral detection window. Reproduced with modifications from [18] originally published under the CC BY license

wavelength dependence of the birefringent material is taken advantage of to realize effective polarization modulation for the STED wavelength only. The polarization of the excitation wavelength is only rotated in a way that avoids the appearance of interference effects when focused. The drawback of this chromaticity is that the waveplate is restricted to a single wavelength pair, where the bandwidth of STED wavelengths tolerated is only a few tens of nanometer, while restrictions for the excitation wavelength may (with reduced performance) be less stringent, in the range of ±40 nm [16]. The easySTED technology has been successfully applied for protein localization in dendritic spines [17], secretory epithelia [18], and fibroblasts [16]. Here, a hand-on protocol is presented for cost- and time-efficient upgrading of a confocal laser scanning microscope with easySTED technology for diffraction-unlimited imaging achieving lateral resolutions <50 nm with the wavelength pair 635 nm/766 nm.

2 Materials

2.1 Instrumentation: Microscope, Laser, and Optical Elements

Given a research grade scanning confocal microscope with the software for scanning and data acquisition as a basis (*see* **Notes 1** and **2**), the following items are needed for the upgrade.

1. ps-pulsed laser with pulse widths in the range of 100–1000 ps and repetition rates of 1–80 MHz. Average laser powers should be >100 mW (pulse energy >5 nJ). However, easySTED can be also implemented with other laser modalities (*see* **Notes 3** and **4**).
2. Laser beamsplitter for the combination of excitation and depletion laser beams and joint fiber coupling (H 643 LPXR beamsplitter, AHF Analysentechnik; or other filter manufacturers, e.g., Chroma) (*see* **Note 5**).
3. High power polarization maintaining (PM) single-mode fiber (P-FAnskFAnsk-4/125/5s-2 PM 630 fiber, Coastal Connections; or NKT Photonics, Thorlabs). In laser microscopy, single-mode fiber coupling is commonly applied for spatial filtering of the transmitted light with the convenient side effect of uncoupling laser sources and microscope and augmenting laser-safety (Fig. 1a).
4. High extinction rate ($<5 \times 10^{-6}$) polarizer (α-BBO Glan-Thompson polarizer no. 50.230.00010, Artifex Engineering; or other optics distributors: e.g., Newport, Qioptiq, Edmund Optics, Thorlabs) to clean the polarization state of the STED light at the exit of the PM fiber (*see* **Note 6**).
5. easySTED waveplate (EASYDOnut Phaseplate 640 nm/765 nm (e.g., Abberior)). We realized easySTED in a well-established

spectral range with red excitation (~640 nm) in combination with near infrared (NIR) depletion (~770 nm). Nevertheless, easySTED waveplates are also available for the wavelength combination of 500 nm/590 nm and can be customized to other wavelength pairs as well (e.g., Abberior).

6. Dual-band dichroic mirror with reflection bands at excitation and STED wavelength (zt 635/766 rpc beamsplitter, AHF Analysentechnik; or other filter manufacturers). Instead of standard beamsplitters with 1 mm thickness, thicker beamsplitters (3 mm or 5 mm) are recommended for STED.
7. Achromatic quarter-wave plate with broad wavelength range (500–800 nm, no. 50.260.00198, Artifex) in rotational mount (or other optics distributors) to change the polarization state of the laser light to circular (*see* **Note 7**).
8. High quality bandpass emission filter (e.g., OD6 blocking) to define the detection band (Fig. 1e) and further reject scattered excitation and STED light (690/70 H bandpass, AHF Analysentechnik; or other filter manufacturers).
9. Microscopic objective with high numerical aperture (NA) and sufficient correction of chromatic aberrations in the respective wavelength range (e.g., oil-immersion objectives Nikon CFI P-Apo 100× Lambda NA1.45 WD0.13; Olympus UPLSAPO 100× NA1.4 WD0.13). Standard objectives for confocal laser scanning microscopes may lack in chromatic correction in the NIR range as required for the STED laser (e.g., 765 nm) preventing optimal axial alignment of both the laser beams.

2.2 Materials for Bead and F-Actin Sample Preparation

1. Glass cover slips of thickness satisfying the demands of the objective lens, e.g., size no. 1 0.13–0.16 mm thick or high precision coverslips with well-defined thickness of 0.17 mm ± 0.005 mm (e.g., Carl Roth).
2. Gold nanobeads with diameter of 80 nm and approx. 1×10^{10} particles/mL, stored at 2 8 °C (BBI Solutions).
3. Crimson fluorescence beads with diameter of 20 nm, $\lambda_{ex} = 625$ nm, $\lambda_{em} = 645$ nm, solution of 2% solids, stored at 2–8 °C (Life Technologies).
4. Preformed F-actin filaments from rabbit skeletal muscle; protein is reconstituted from lyophilized powder in buffer solution to 0.4 mg/mL by adding bidest. water (=10 μM); stored as aliquots at −20 °C (no. AKF99, Cytoskeleton).
5. Poly-L-lysine, 0.1% (w/v) aqueous solution, stored at 2–8 °C.
6. Phosphate-buffered saline (PBS): 137 mM NaCl, 10 mM Na_2HPO_4, 2.7 mM KCl, 1.8 mM KH_2PO_4, pH 7.4.
7. F-actin buffer: 5 mM Tris THAM, 0.2 mM $CaCl_2$, pH 8.0.

8. Embedding medium Mowiol 4-88 prepared according to the instructions for hydrophilic embedding by the manufacturer and stored as aliquots at −20 °C: 2.4 g Mowiol 4-88, 6 g glycerol, 6 mL bidest. water, 12 mL Tris–HCl (0.2 M, pH 8.4).
9. Anti-bleaching agent DABCO (1,4-Diazabicyclo[2.2.2] octan): 25 mg per 1 mL embedding medium.
10. Blocking solution 2% (w/v) bovine serum albumin (BSA) in PBS.
11. Phalloidin-coupled fluorescent dyes for F-actin labeling, stored as aliquots at −20 °C (e.g., STAR 635P-phalloidin, Abberior; Atto 647N-phalloidin, Atto-Tec).

3 Methods

3.1 Bead Sample Preparation

1. Clean glass coverslip with absolute ethanol or acetone.
2. Treat with 0.1% poly-L-lysine for 15 min (apply a ~100 μL droplet or bath full coverslip). Rinse with bidest. water and allow it to dry on air.
3. Dilute gold nanobeads 1:1,000 in bidest. water and Crimson fluorescence beads 1:10,000 in bidest. water.
4. Sonicate bead suspensions in ultrasonic bath for 5–10 min.
5. Apply a small droplet (10–20 μL) of bead suspension on glass coverslip and allow it to incubate for 10 min.
6. Rinse carefully with water and let it dry on air.
7. Mount with the embedding medium (e.g., Mowiol 4-88 containing 25 mg/mL DABCO).
8. Bead samples are stable for several weeks if stored in the dark at 2–8 °C.

3.2 F-Actin Sample Preparation

1. Clean glass coverslip with absolute ethanol or acetone.
2. Treat with 0.1% poly-L-lysine for 15 min (apply a ~100 μL droplet or bath full coverslip). Rinse with bidest. water and allow it to dry on air.
3. Dilute 30 μL preformed F-actin (0.4 mg/mL) in 570 μL F-actin buffer to reach a protein concentration of 0.02 mg/mL (=0.5 μM). This volume is sufficient for the preparation of four coverslips.
4. Apply ~150 μL of diluted F-actin on coverslip and allow it to incubate for 30 min.
5. Rinse carefully with F-actin buffer. Avoid falling dry of the F-actin at any time of the following steps.

6. Incubate for at least 5 min in blocking solution (2% (w/v) BSA in PBS).
7. Wash three times with F-actin buffer.
8. Dilute phalloidin-coupled fluorescent dye in F-actin buffer to reach a final concentration of 0.1–0.5 μM (e.g., 10 μL dye with a concentration of 10 μM diluted in 590 μL F-actin buffer).
9. Apply ~150 μL of phalloidin-coupled dye on coverslip and incubate for 60 min at room temperature in the dark.
10. Wash three times for 5 min with PBS.
11. Mount with the embedding medium (e.g., Mowiol 4-88 containing 25 mg/mL DABCO) after the reduction of the droplet volume on the coverslip.
12. Fresh preparation of F-actin samples is highly recommended, since filaments start to depolymerize immediately (significant degradation can be already observed at the second day of storage).

3.3 Alignment and Optimization

1. After the implementation of the additional optical elements (quarter-wave plate, polarizer, dual-band dichroic mirror, and bandpass emission filter) the excitation path of the microscope is aligned and optimized as usual for conventional fluorescence imaging. That means the excitation beam path is aligned to the microscope's optical axis and is centered to the back aperture of the objective lens. The *xy*-position of the confocal pinhole is optimized to allow maximal transmission of fluorescence light. For the excitation wavelength of 635 nm we use a pinhole with 50 μm in diameter. For this, the excitation laser light is focused onto a droplet of an appropriate fluorescent dye (e.g., 1 μM Atto 655, Atto-Tec) placed on a glass cover slip and then the emission beam path can be aligned by detecting the fluorescence light.
2. The orientation of the quarter-wave plate is fine adjusted to optimize the circular polarization of the STED laser light (*see* **Notes 8–10**).
3. The easySTED waveplate is positioned in the beam path, preferably close to the back aperture of the objective. Depending on the available microscope body, a free filter slot (such as for Wollaston prisms for DIC microscopy) can be used to access the beam path. Alternatively, one could implement a *xy*-translation mount between objective and nosepiece (*see* optics distributers). This eases the lateral centering of the waveplate to the laser beam.
4. Fine lateral positioning of the easySTED waveplate may be controlled by imaging the backscattered STED laser light from gold nanobeads (Subheadings 3.1 and 3.4). An

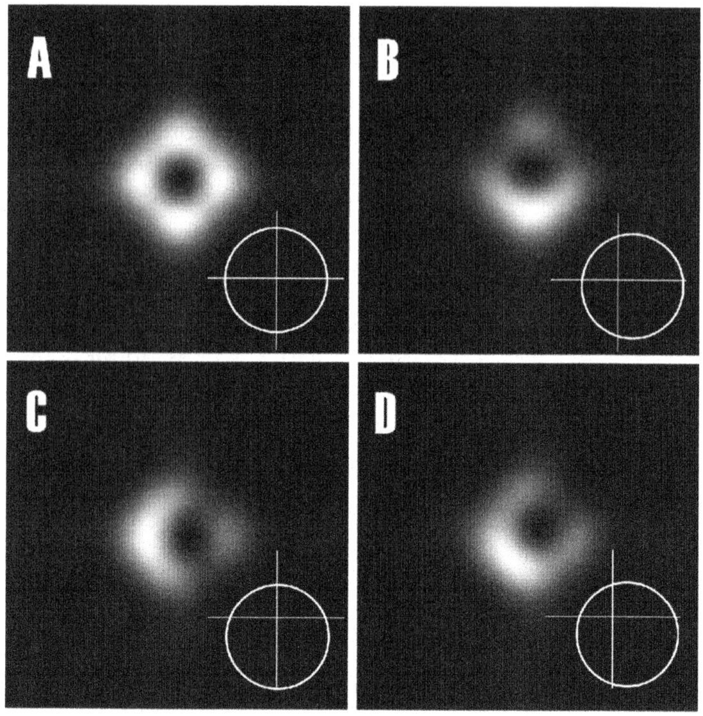

Fig. 2 Focal intensity distribution of the easySTED PSF for pupil centered (**a**) and laterally shifted segmented waveplate (**b–d**). Crosses and circles in the right lower corner indicate the direction of displacement relative to the pupil. Amount of displacement is one third of the pupil radius

asymmetric light distribution of the STED ring (Fig. 2) is an indicator for suboptimal lateral alignment of segmented waveplate and STED laser beam. In an iterative optimization process, the symmetry can be improved by moving the central cross of the easySTED waveplate relative to the STED laser beam to the particular lateral direction (*see* **Notes 11** and **12**).

5. Control the spatial superposition of excitation laser PSF and STED laser PSF by imaging the corresponding backscattered light from a gold nanobead (Subheading 3.4.); for example, start an overview scan (e.g., 50 μm × 50 μm) and then perform *xy*-, *xz*-, and *yz*-scans of a chosen gold nanobead (e.g., 2 μm × 2 μm). A certain displacement of the two laser PSFs is tolerable (*see* **Notes 12–14**).
6. Synchronize the excitation and STED pulse by rather electrical or optical delay units to guarantee efficient interaction of the STED photons with the excited molecules (*see* **Notes 15** and **16**).

3.4 Verifying the easySTED Alignment by Imaging Gold Nanobeads

1. The confocal pinhole should be removed for these measurements. If not possible, adjust to the largest available size.
2. The emission bandpass filter should be removed. The average laser power of both, the excitation and especially the STED laser needs to be (strongly) attenuated to avoid damage to the detector by overexposure (e.g., use-defined neutral density filters)!
3. A single isolated gold nanobead (Subheading 3.1.) at least 2 μm away from other beads is localized by applying only the excitation laser line. The *z*-position is adjusted precisely by maximizing the intensity count rate of an online point measurement at the center of the gold nanobead.
4. Imaging of the backscattered excitation light at a pixel size of ~20 nm allows us to control the diameter of the PSF in the focal plane. For an optimally aligned system, the FWHM of the Gaussian intensity distribution should be diffraction-limited and therefore not exceed half a wavelength (Fig. 1c, d).
5. Remember to reduce the STED laser power to an amount tolerable by the detector before imaging the backscattered light of the STED laser. If the easySTED waveplate is well centered the focal STED light distribution will appear as a symmetric ring with a zero intensity minimum surrounded by high intensities to all the directions in the focal plane (Fig. 1b, d).
6. If time-resolved detection is available, the PSFs at the excitation and the STED wavelength may be monitored in a single image acquisition run using the pulsed interleaved excitation (PIE) principle. By delaying the two laser pulses (excitation and STED) by some nanoseconds relative to each other and using time-gating to distinguish backscattered light from the one and the other laser source respectively, the different PSFs are measured quasi instantaneously. Thus, with PIE inaccuracy due to specimen shifting can be excluded.
7. An overlay of the gold nanobead images shows the relative displacement of the PSFs at both applied wavelengths (Fig. 1d).

3.5 Diffraction-Unlimited Imaging with easySTED

From the practical point of view, easySTED imaging is identical to imaging with a confocal laser scanning microscope at a reduced pixel size. To check the improved spatial resolution after easySTED alignment proceed as follows:

1. Remember to insert the pinhole and the emission bandpass filter after gold nanobead imaging (Subheading 3.4).
2. Start an overview scan (e.g., 50 μm × 50 μm) of a Crimson fluorescence bead sample in confocal mode using only the

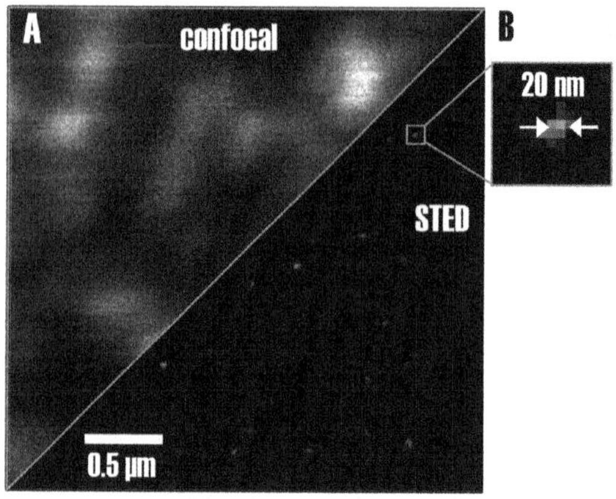

Fig. 3 Diffraction-unlimited easySTED imaging of Crimson fluorescence beads. (**a**) Confocal and STED image of the same preparation of 20 nm beads on a glass cover slip. Average STED laser power was 35 mW at a repetition rate of 2.5 MHz (measured at the position of the objective back aperture). (**b**) An example bead imaged under STED conditions with a diameter of approximately 20 nm. Pixel size was 10 nm for STED imaging, pixel dwell time 1 ms

excitation laser (e.g., 635 nm). If necessary, adjust the *z*-position by maximizing the intensity count rate of an online point measurement at the center of one fluorescence bead.

3. Choose a smaller region with homogenously distributed beads (e.g., 3 µm × 3 µm) and start a new scan, now with lower pixel size (e.g., ~20 nm). The confocal spatial resolution can be calculated from intensity profiles of such recorded beads (Fig. 3).
4. Repeat **step 3** at exactly the same region, but now also apply the STED laser. Thus, diffraction-unlimited resolution is achieved (Fig. 3).
5. For proving easySTED imaging at biological samples, repeat **steps 2–4** by using now an F-actin sample (Subheading 3.2). Anticipated results are shown in Fig. 4.

The number of parameters that can be optimized in STED image acquisition is still enlarged compared to the confocal case. In addition to the excitation laser energy, pixel dwell time, laser pulse repetition rate, and pixel size, in easySTED also the STED laser power and the pulse delay may be varied in order to optimize image quality, where quality criteria may comprise brightness, contrast, resolution, noise, bleaching, etc. Some general rules may be considered:

1. Pixel size to be chosen according to the Nyquist-Shannon theorem to not exceed half the anticipated resolution of the STED image. As the pixel size correlates with the signal to

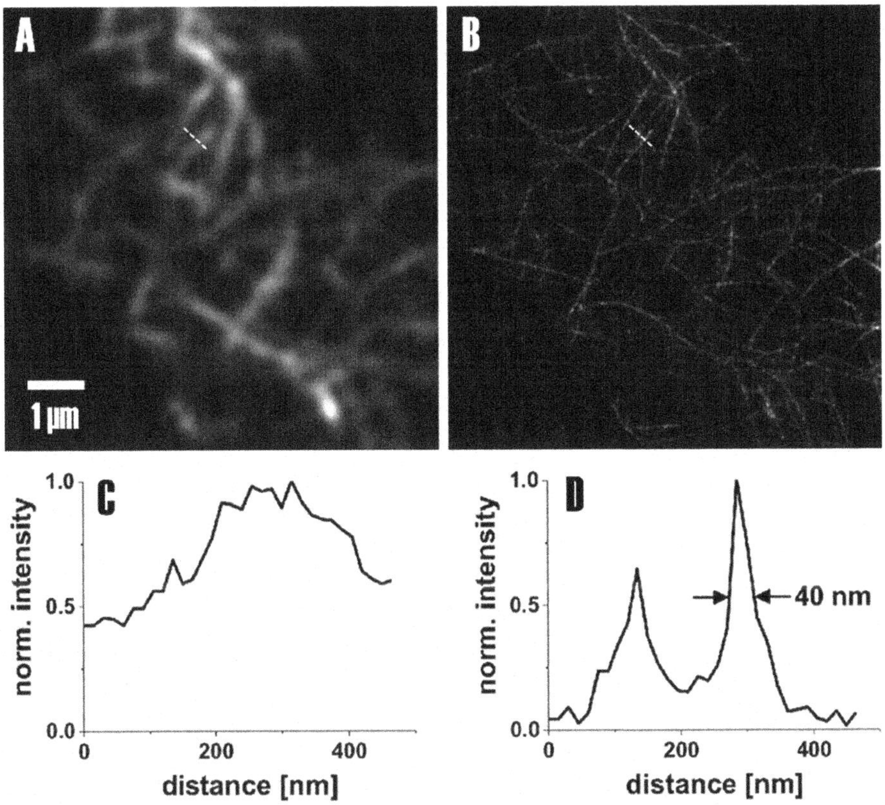

Fig. 4 Fluorescence images of F-actin labeled with STAR 635P-phalloidin. (**a**) Confocal and (**b**) easySTED image of the same filament structure. (**c**, **d**) Intensity profile plots along the dashed lines shown in **a** and **b**, respectively. Pixel size was 15 nm for STED imaging, pixel dwell time 1 ms, STED laser power 12 mW at a repetition rate of 2.5 MHz (measured at the position of the objective back aperture)

noise ratio excessive oversampling is not recommended, though.

2. Shorter pixel dwell times in combination with image or line averaging reduce the rate of photobleaching. For some fluorophores, bleaching may be reduced by choosing lower laser pulse repetition rates.
3. At increasing STED laser powers the remaining fluorescence signal will diminish due to an imperfect zero intensity of the STED ring and the achievable STED resolution will increasingly be signal-to-noise limited. Further increment of the excitation laser power may improve the resulting images (at the cost of increased photobleaching).
4. In STED microscopy diffraction-unlimited resolution is achieved without the need for any image postprocessing. However, STED images may benefit concerning signal to noise ratio from processing techniques as Wiener filtering or nonlinear deconvolution known from conventional microscopy (*see* **Note 17**).

4 Notes

1. Image acquisition of the easySTED-retrofitted microscope does not differ from a standard confocal laser scanning microscope, except for the smaller pixel/stepping size necessary. In general, any scanning image acquisition software where the pixel size can be chosen as low as 10 nm to 20 nm should therefore be suitable for STED imaging. Nanometer precise raster scanning for easySTED has been performed using galvanometric beam scanning [19] as well as objective scanning by a piezo nanopositioning stage (e.g., Physik Instrumente) as described in this protocol [18].
2. Time-resolved detection (time-correlated single-photon counting, TCSPC) is not mandatory for easySTED implementation. Allowing among other for offline time-gating, however, it provides a helpful tool for further improving contrast and resolution of STED images and enables also access to luminescence lifetime information [20, 21].
3. While in the early STED setups Titanium Sapphire (Ti:Sa) laser systems were used to generate high-power depletion light in the orange/red and near infrared region [22, 23], today specialized compact turnkey STED lasers are on the market, rendering STED setups more compact and less complex. Especially in the pulsed STED modality where the fs-pulses from the Ti:Sa laser needed stretching by 2–3 orders of magnitude to fulfill the demands of effective fluorescence depletion, the availability of high-power laser sources generating pulses of several hundred picoseconds length at repetition rates in the MHz regime, has facilitated STED imaging enormously. Examples for applied pulsed STED lasers are the VisIR-765 "STED" (PicoQuant) [24] and the Katana HP (onefive) [25], whereas the VFL592 (MPB Communications Inc.) is a continuous-wave (cw) laser applied for cwSTED [9]. Supercontinuum laser sources (e.g., NKT Photonics) have also been successfully applied in STED microscopy [8].
4. Longer STED pulses with reduced peak power in comparison to short (<100 ps) pulses are expected to reduce photostress to the specimen. Additionally, they reduce the demands regarding the temporal synchronization of excitation and depletion pulses and reduce problems with jitter. In cwSTED temporal synchronization becomes even redundant, but much higher average powers are needed to achieve the same depletion efficiency and thus the same spatial resolution as in the pulsed STED modality [22, 26].
5. Two moveable mirrors per laser (*see* Fig. 1a) deliver sufficient degrees of freedom for fiber coupling. The combining dichroic

mirror may serve as beam moving element, thereby, for one laser. Adding a beam displacer, a plane glass that may be tilted horizontally and vertically for parallel displacement of a passing laser beam, gives additional degrees of freedom to the user rendering efficient laser coupling to the fiber more straightforward. Achromatic collimation at the fiber exit guarantees stable aligned laser beams.

6. Cleaning the polarization state of the collimated STED laser light may be useful as bending of the PM fiber influences the degree of linear polarization maintenance in the fiber.
7. In contrast to STED laser beam shaping by a helical phase plate, the quality of the central zero intensity of the easySTED doughnut does not depend on the initial polarization state of the STED laser light. However, to avoid isotropic effects on the STED efficiency due to the dipole moment of fix molecules, easySTED (as normal STED) works best with circularly polarized light [15].
8. The polarization state of the STED laser light should be controlled as close to the microscopic objective (back aperture) as possible, as optical elements in the beam path may influence the polarization. We usually remove the objective for polarization analysis at the position of the objective back aperture.
9. Polarization analyzers (e.g., Thorlabs, Schäfter + Kirchhoff) are optimal to control the polarization state of the STED laser beam but they are rather expensive. A power meter or a photodiode in combination with an oscilloscope and a liner polarizer can also be used to optimize circular polarization. Rotating the linear polarizer in front of the power meter, the detected intensity will show sinusoidal behavior with an amplitude providing a measure of the ellipticity of the light. For linearly polarized light, the detected intensity will oscillate between zero and the maximum intensity of the laser. In contrast, for perfect circularly polarized light the detected intensity behind the polarizer becomes constant and is no longer modulated by the polarizer's rotation. We typically work with polarization states characterized by a relative deviation of <1% between minimum and maximum of the light intensity passing the rotating polarizer.
10. Not only rotation perpendicular to the beam but also tilt of the quarter-wave plate may influence the resulting polarization state of the passing laser light.
11. Considerable amount of optical aberration in the system may lead to distortion of the symmetry of the ring-shaped STED laser intensity distribution. Lateral alignment of the easySTED waveplate that maximizes the symmetry of the

STED ring may in this case be misleading and result in suboptimal alignment.

12. The quality of the central intensity zero of the easySTED doughnut is moderately robust to lateral displacement of the easySTED waveplate in the pupil (up to some tenths of mm). A certain modulation of the doughnut's rim intensity, resulting from lateral misalignment (and/or optical aberration), is also tolerable in practice, as the crest intensity of the doughnut is far above saturation anyway.
13. For a perfectly chromatically corrected system, there should not be a displacement of excitation maximum and central depletion minimum, at all. In practice, we experience a displacement of about 50 nm in the lateral direction and about 200 nm in the axial direction.
14. We experienced worse, mainly axial displacement of focal light distributions of 635 nm and 766 nm light for older high-NA objectives (not chromatically corrected up to the NIR region). If the axial displacement exceeds several hundred nanometers, the objective is obviously not suitable for easySTED, as effective depletion becomes impossible.
15. The preferable delay between excitation laser pulse and STED laser pulse depends on various factors such as the lengths and temporal shape of the pulses and the fluorescence lifetime of the used fluorophore [27]. The available STED laser pulse energy is optimally used for fluorescence depletion if the major part of the STED laser pulse interacts with the specimen after the major part of the excitation laser pulse has already passed. On the other hand, as the time delay becomes larger, an increasing part of the excited fluorophores emits spontaneous fluorescence with low spatial information already before the interaction with the STED laser pulse takes part. Such early low-resolution fluorescence may be removed by time-gated detection or offline time gating of time-resolved data, rendering early low-resolution photons less critical [18, 28]. If time gating cannot be applied, the user will want to work at a shorter delay for a reduced amount of early low spatial resolution photons.
16. A way to optimize the time delay between excitation laser pulse and STED laser pulse is to use the fluorescent dye required as label for the biological sample in a diluted aqueous solution of approx. 1–5 nM (or a dye with similar fluorescence lifetime). Apply a droplet on a glass coverslip to measure the remaining fluorescence signal in the solution as a function of the delay-time while applying excitation and STED pulses. The highest depletion efficiency is reached for the time delay that minimizes the remaining fluorescence signal [29].

17. A very basic processing step to reduce noise and pixelated appearance in STED images may be upscaling the resolution and applying bilinear interpolation. More effective noise suppression without loss of resolution can be achieved by means of Wiener filtering. Deconvolution methods even promise increasing the resolution of STED images. Special deconvolution software for STED images is available commercially (Huygens, Scientific Volume Imaging b.v.). One example for a free deconvolution software package is the ImageJ plugin DeconvolutionLab (Biomedical Imaging Group, EPFL, Switzerland; http://bigwww.epfl.ch/) [30]. A python implementation of the Richards-Lucy-deconvolution for time-gated STED may be found in reference [31].

Acknowledgments

This work was funded by the German Research Foundation (grant number 1850/30001355, www.dfg.de), and the Federal Ministry of Education and Research ("ALSComBi," grant number 03IPT517Y, www.bmbf.de).

References

1. Abbe E (1873) Beiträge zur Theorie des Mikroskops und der mikroskopischen Wahrnehmung. Arch für Mikrosk Anat 9:413–468. doi:10.1007/BF02956173
2. Hell SW (2007) Far-field optical nanoscopy. Science 316:1153–1158. doi:10.1126/science.1137395
3. Rust MJ, Bates M, Zhuang X (2006) Sub-diffraction-limit imaging by stochastic optical reconstruction microscopy (STORM). Nat Methods 3:793–795. doi:10.1038/nmeth929
4. Betzig E, Patterson GH, Sougrat R et al (2006) Imaging intracellular fluorescent proteins at nanometer resolution. Science 313:1642–1645. doi:10.1126/science.1127344
5. Hell SW, Sahl SJ, Bates M et al (2015) The 2015 super-resolution microscopy roadmap. J Phys D Appl Phys 48:443001. doi:10.1088/0022-3727/48/44/443001
6. Hell SW (2009) Microscopy and its focal switch. Nat Methods 6:24–32. doi:10.1038/nmeth.1291
7. Eggeling C, Willig KI, Sahl SJ, Hell SW (2015) Lens-based fluorescence nanoscopy. Q Rev Biophys 48:178–243. doi:10.1017/S0033583514000146
8. Wildanger D, Rittweger E, Kastrup L, Hell SW (2008) STED microscopy with a supercontinuum laser source. Opt Express 16:9614–9621. doi:10.1364/OE.16.009614
9. Moneron G, Medda R, Hein B et al (2010) Fast STED microscopy with continuous wave fiber lasers. Opt Express 18:1302–1309. doi:10.1364/OE.18.001302
10. Lukinavičius G, Reymond L, D'Este E et al (2014) Fluorogenic probes for live-cell imaging of the cytoskeleton. Nat Methods 11:731–733. doi:10.1038/nmeth.2972
11. Sidenstein SC, D'Este E, Böhm MJ et al (2016) Multicolour multilevel STED nanoscopy of actin/Spectrin Organization at Synapses. Sci rep 6:26725. doi:10.1038/srep26725
12. Kukat C, Wurm CA, Spahr H et al (2011) Super-resolution microscopy reveals that mammalian mitochondrial nucleoids have a uniform size and frequently contain a single copy of mtDNA. Proc Natl Acad Sci 108:13534–13539. doi:10.1073/pnas.1109263108
13. Li Q, Wang Y, Chen D et al (2014) 2PE-STED microscopy with a single Ti: sapphire laser for reduced illumination. PLoS One 2(9):e88464. doi:10.1371/journal.pone.0088464
14. Li D, Hérault K, Zylbersztejn K et al (2015) Astrocyte VAMP3 vesicles undergo Ca^{2+}-independent cycling and modulate glutamate

transporter trafficking. J Physiol 593:2807–2832. doi:10.1113/JP270362
15. Reuss M, Engelhardt J, Hell SW (2010) Birefringent device converts a standard scanning microscope into a STED microscope that also maps molecular orientation. Opt Express 18:1049–1058. doi:10.1364/OE.18.001049
16. Gorlitz F, Hoyer P, Falk H et al (2014) A STED microscope designed for routine biomedical applications. Prog Electromagn Res 147:57–68. doi:10.2528/PIER14042708
17. Westin L, Reuss M, Lindskog M et al (2014) Nanoscopic spine localization of Norbin, an mGluR5 accessory protein. BMC Neurosci 15:45. doi:10.1186/1471-2202-15-45
18. Klauss A, König M, Hille C (2015) Upgrade of a scanning confocal microscope to a single-beam path STED microscope. PLoS One 10(6): e0130717. doi:10.1371/journal.pone.0130717
19. Bingen P, Reuss M, Engelhardt J, Hell SW (2011) Parallelized STED fluorescence nanoscopy. Opt Express 19:23716–23726. doi:10.1364/OE.19.023716
20. Auksorius E, Boruah BR, Dunsby C et al (2008) Stimulated emission depletion microscopy with a supercontinuum source and fluorescence lifetime imaging. Opt Lett 33:113–115. doi:10.1364/OL.33.000113
21. Bückers J, Wildanger D, Vicidomini G et al (2011) Simultaneous multi-lifetime multicolor STED imaging for colocalization analyses. Opt Express 19:3130–3143. doi:10.1364/OE.19.003130
22. Willig KI, Harke B, Medda R, Hell SW (2007) STED microscopy with continuous wave beams. Nat Methods 4:915–918. doi:10.1038/nmeth1108
23. Donnert G, Keller J, Medda R et al (2006) Macromolecular-scale resolution in biological fluorescence microscopy. Proc Natl Acad Sci U S A 103:11440–11445. doi:10.1073/pnas.0604965103
24. Niehörster T, Löschberger A, Gregor I et al (2016) Multi-target spectrally resolved fluorescence lifetime imaging microscopy. Nat Methods 13:257–262. doi:10.1038/nmeth.3740
25. Bottanelli F, Kromann EB, Allgeyer ES et al (2016) Two-colour live-cell nanoscale imaging of intracellular targets. Nat Commun 7:1–5. doi:10.1038/ncomms10778
26. Vicidomini G, Moneron G, Han KY et al (2011) Sharper low-power STED nanoscopy by time gating. Nat Methods 8:571–573. doi:10.1038/nmeth.1624
27. Moffitt JR, Osseforth C, Michaelis J (2011) Time-gating improves the spatial resolution of STED microscopy. Opt Express 19:4242–4254. doi:10.1364/OE.19.004242
28. Vicidomini G, Schönle A, Ta H et al (2013) STED nanoscopy with time-gated detection: theoretical and experimental aspects. PLoS One 8(1):e54421. doi:10.1371/journal.pone.0054421
29. Galiani S, Harke B, Vicidomini G et al (2012) Strategies to maximize the performance of a STED microscope. Opt Express 20:7362–7374. doi:10.1364/OE.20.007362
30. Vonesch C, Unser M (2008) A fast Thresholded Landweber algorithm for wavelet-regularized multidimensional Deconvolution. IEEE Trans Image Process 17:539–549. doi:10.1109/TIP.2008.917103
31. Ingaramo M, York AG, Hoogendoorn E et al (2014) Richardson-Lucy deconvolution as a general tool for combining images with complementary strengths. ChemPhysChem 15:794–800. doi:10.1002/cphc.201300831

Chapter 5

Two-Photon STED Microscopy for Nanoscale Imaging of Neural Morphology In Vivo

Mirelle J.T. ter Veer, Thomas Pfeiffer, and U. Valentin Nägerl

Abstract

The advent of super-resolution microscopy offers to bridge the gap between electron and light microscopy. It has opened up the possibility of visualizing cellular structures and dynamic signaling events on the "mesoscale" well below the classic diffraction barrier of light microscopy (10–200 nm), while essentially retaining the advantages of fluorescence microscopy concerning multicolor labeling, detection sensitivity, signal contrast, live-cell imaging, and temporal resolution.

From among the new super-resolution techniques, STED microscopy stands out as a laser-scanning imaging modality, which enables nanoscale volume-metric imaging of cellular morphology. In combination with two-photon (2P) excitation, STED microscopy facilitates the visualization of the highly complex and dynamic morphology of neurons and glia cells deep inside living brain slices and in the intact brain in vivo.

Here, we present an overview of the principles and implementation of 2P-STED microscopy in vivo, providing the neurobiological context and motivation for this technique, and illustrating its capacity by showing images of dendritic spines and microglial processes obtained from living brain tissue.

Key words STED microscopy, Two-photon microscopy, Super-resolution imaging in vivo, Mouse cortex, Cranial window, Acute brain slices, Nanoscale neural morphology, Dendritic spines, Microglial processes, Microglia-synapse interactions

1 Introduction

Ever since Ramon y Cajal's foundational studies on the micro-anatomical organization of the brain, much of neuroscience research has been devoted to reconstructing brain cells and to understanding how their anatomical complexity and diversity may serve brain physiology and becomes compromised by disease.

Neurons receive input from thousands of neighboring cells via synapses, which are the physical junctions that mediate fast and flexible communication between neurons in the central and peripheral nervous system. Most excitatory synapses terminate on

Mirelle J. T. ter Veer and Thomas Pfeiffer contributed equally to this work.

Holger Erfle (ed.), *Super-Resolution Microscopy: Methods and Protocols*, Methods in Molecular Biology, vol. 1663, DOI 10.1007/978-1-4939-7265-4_5, © Springer Science+Business Media LLC 2017

dendritic spines that are formed by small protrusions in the postsynaptic membrane of the dendrite and that typically feature a bulbous head and an elongated neck, providing a veritable channel to the parent dendrite for the passage of electrical and biochemical signals.

Conspicuous and unique, spine morphology is thought to play a key role for synaptic physiology, shaping the ability of spines to compartmentalize electro-chemical signaling and enable synapse-specific regulatory mechanisms. Moreover, structural remodeling of spines is thought to be a mechanism for rewiring synaptic circuits and tuning synaptic potentials [1, 2]. Mounting evidence indicates that spine structural changes are an important anatomical substrate of brain plasticity that underlies adaptive processes such as learning and memory. Recent studies indicate that microglia, the immune-competent cells of the brain, contribute to structural remodeling of synapses, possibly using their hyperfine and motile processes to sculpt dendritic spines [3].

While electron microscopy (EM) revealed the ultrastructural organization of synapses [4], featuring a presynaptic active zone, synaptic cleft, and postsynaptic density, two-photon (2P) microscopy made it possible to study their morphological and functional properties in live brain tissue [5, 6]. Together with transgenic expression of fluorescent proteins for bright and sparse labeling of neurons [7] and "cranial window" surgery techniques for high-resolution optical access to superficial layers of mouse cortex [8], 2P microscopy has become the method of choice for imaging neurons and synapses in the intact brain in vivo.

However, the spatial resolution of fluorescence microscopy, including 2P microscopy, is limited by diffraction to around half the wavelength of the excitation light (~400 nm), which in many instances is insufficient for properly resolving crucial details of synaptic morphology and nearby glial cells such as astrocytes and microglia.

Fortunately, the advent of super-resolution microscopy, in particular STED microscopy [9, 10], which overcame the classic diffraction barrier of light microscopy, has made it possible to visualize synapses with nanoscale spatial resolution in living brain slices [11, 12] and even in vivo [13, 14].

1.1 Basic Principles of 2P-STED Microscopy

More recently, a combination of STED and 2P microscopy (2P-STED) was developed to reconcile diffraction-unlimited imaging with high optical sectioning and depth penetration in thick living brain tissue [15–17], building on proof-of-principle studies that demonstrated that 2P excitation can indeed be used together with stimulated emission to improve the spatial resolution of a microscope [18, 19].

As a laser-scanning fluorescence imaging technique, this new approach inherits most of its features from regular 2P microscopy, except the diffraction handicap. Therefore, the types of fluorescent dyes that are "STED-able" and the useful labeling strategies to visualize interesting biological structures are virtually identical to those of regular 2P microscopy.

2P excitation, which is based on femtosecond lasers, offers superior penetration into biological tissue because the wavelength of the excitation light is about twice as long as compared with one photon excitation, which greatly reduces the amount of light scattering in biological tissue [20]. Moreover, the nonlinearity of the 2P excitation process provides intrinsic optical sectioning and reduces photo toxicity and bleaching because the generation of the fluorescence stays restricted to the focal spot [21].

By simultaneously absorbing two near-infrared photons, the fluorophore is elevated from the ground state (S_0) to the excited state (S_1), emitting a fluorescence photon during its subsequent relaxation back to the ground state, which occurs spontaneously within a few nanoseconds after excitation in the case of regular 2P microscopy (*see* Fig. 1a). By contrast, in 2P-STED microscopy the excited fluorophore is expressly forced back to the ground state right after excitation, i.e., before spontaneous emission has a chance to occur. This transition is induced by stimulated emission by the STED laser.

Because the wavelength of the STED laser is set to the long wavelength end of the fluorophore's emission spectrum, the STED light produces very little direct excitation. The STED light, including the "copy photons" from stimulated emission that have the exact same wavelength as the STED light, can be discarded by standard optical filters and dichroic mirrors without losing too much of the fluorescence signal.

The phase of the STED beam is spatially modulated in a way that its intensity distribution resembles a doughnut in x-y cross sections. By bringing the STED doughnut into perfect registry with the focus of the Gaussian excitation laser beam (irrespective of whether one or two-photon excitation is used), the fluorescent spot can be cut down in size in a center-symmetric way because of the quenching effect by the STED light on the edge of the excitation spot. It is this spatial constriction that underlies the gain in spatial resolution achieved in a STED microscope.

To achieve a high quenching efficiency, and thus a high gain in resolution, the femtosecond 2P excitation pulses need to be optimally overlaid in time with the picosecond pulses of the STED lasers, which makes it necessary to be able to synchronize laser sources running at high repetition rates (80 MHz).

Whereas the classic "resolution formula" of light microscopy [22] is given by:

$$\Delta r = \frac{\lambda}{2 \times \mathrm{NA}},$$

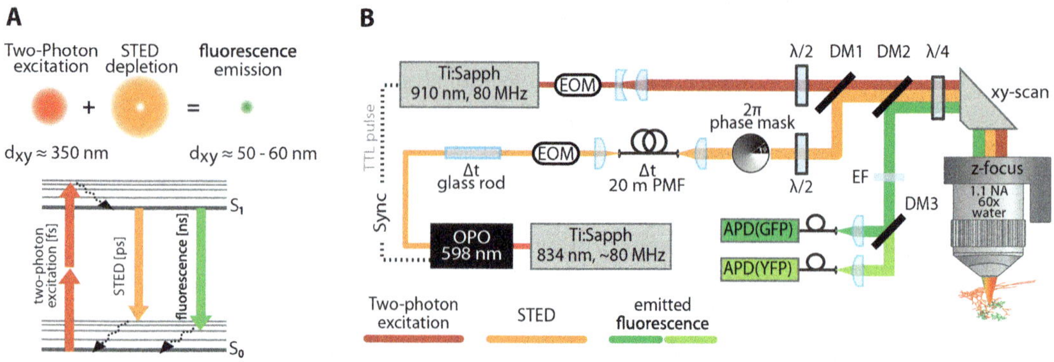

Fig. 1 2P-STED microscopy: Basic principle and implementation. (**a**) Simplified Jablonski diagram of the energy transitions during a fluorescence cycle. Simultaneous absorption of two infrared photons excites the fluorophore from the ground state (S_0) to the first excited singlet state (S_1). After vibrational relaxation to the lowest energy level of S_1, spontaneous emission of a photon usually occurs within a few nanoseconds after excitation, which returns the fluorophore back to S_0. Stimulated emission intercepts this transition by forcing the fluorophore back to S_0 and emitting a photon identical to the incident one. The doughnut-shaped STED beam quenches the fluorescence only in the outer rim of the excitation beam, which effectively enhances the spatial resolution. (**b**) Scheme of homebuilt 2P-STED microscope, designed for two-color super-resolution imaging of green and yellow fluorophores in living brain tissue. A Titanium:Sapphire (Ti:Sapph) laser tuned to 910 nm at ~80 MHz provides the 2P excitation light. The intensity and beam size are adjusted by an electrical optical modulator (EOM) and a telescope. The polarization of the 2P light is modulated by λ/2-wave plate. The 2P and STED beams are combined by a dichroic mirror (DM 1). The STED beam originates from a Ti:Sapph laser tuned to 834 nm at ~80 MHz pumping an optical parametric oscillator (OPO) to produce light at 598 nm. The light pulses are stretched by a glass rod and a 20 m long polarization-maintaining single-mode fiber to approximately 70 ps. The characteristic doughnut-shaped STED beam is generated by a helical vortex phase plate and its circular polarization can be adjusted by λ/2 and λ/4-wave plates. The combined beams are scanned across the sample using a beam scanner and a fast nano z-positioner for image acquisition through a water-immersion objective with a numerical aperture (NA) of 1.1 and 60× magnification. De-scanned detection and spectral separation of the emitted fluorescence is done with a dichroic mirror (DM 2 and 3) and avalanche photodiodes (APDs)

where λ is the wavelength of the excitation light and NA the numerical aperture of the objective, in the STED case, the denominator of the formula contains the term I_{STED}, which is the adjustable intensity of the STED laser:

$$\Delta r = \frac{\lambda}{2 \times \text{NA} \times \sqrt{1 + \frac{I_{\text{STED}}}{I_S}}},$$

where I_S is the intensity value at which half of the fluorescent molecules are effectively quenched [23].

By implication, as the intensity of the STED laser is increased, the STED "curtain fire" invades successively into the central region of the excitation spot, effectively shrinking it along the radial direction. As the transition from the excited to the ground state reaches saturation, or in other words, as the number of excited molecules gets depleted quantitatively, the relationship between excitation and fluorescence generation becomes highly nonlinear, which effectively decouples the achievable spatial resolution from diffraction.

1.2 Applications of STED Microscopy in Neuroscience

The STED concept was proposed in 1994 [10], confirmed experimentally 6 years later [24], and acknowledged by the Nobel Prize in 2014 [25]. Initially, the use of STED microscopy in neuroscience was limited to simple immuno-histochemical analyses of the distribution of proteins in fixed neuronal cell cultures and their synapses [26–28], delivering a "confocal plus" performance, which became available to a broad user community through imaging core facilities. Aided by technical improvements concerning optics and fluorescence labeling, the scope of STED microscopy soon expanded to include more sophisticated live-cell applications [29] and intact tissue preparations [12]. Since then, a series of papers have demonstrated the capacity of STED microscopy to image synapses deep inside living brain slices [30] and in the intact brain in vivo [13, 14]. Notably, the STED approach can be combined with other imaging and physiology techniques, such as 2P uncaging, fluorescent recovery after photobleaching (FRAP), and patch-clamp electrophysiology [31], making it possible to closely correlate super-resolution imaging data with physiological measurements obtained from the exact same cells or structures of interest.

Despite all the advance credit and excitement, STED microscopy has not yet been embraced by a large number of biology labs. However, ongoing technical improvements (concerning user-friendliness and stability of the microscope, signal-to-noise and sensitivity, temporal resolution and depth penetration) and commercialization of the technology are inspiring wider adoption of STED microscopy, turning STED from a fancy curiosity into an effective mainstream research tool.

In this protocol, we present the principle and operation of a homebuilt 2P-STED microscope, describing its main optical architecture and pieces of instrumentation, and providing a how-to guide for reliable performance and maintenance of the microscope. Finally, we demonstrate the capacity of 2P-STED microscopy by showing a couple of specific examples of dual-color imaging in brain tissue, including the intact mouse brain in vivo and thick living acute brain slices, focusing on nanoscale dendritic spines and microglial processes.

2 Materials

In the following, we list the main components and tools needed for constructing a 2P-STED microscope and performing imaging experiments in living brain tissue.

2.1 Microscope Equipment

1. Pulsed laser light sources
   (a) 2P excitation: 2P laser tuned to a wavelength of 910 nm at a repetition rate of 80 MHz (Millennia eV combined with a Tsunami, SpectraPhysics).
   (b) STED depletion: 2P laser (Chameleon, Coherent) tuned to 834 nm, which is converted by an optical parametric oscillator (OPO, APE) to generate light at 598 nm at 80 MHz.
2. Microscope base and objective
   (a) Modular research-grade upright microscope (e.g., Olympus BX-51WI).
   (b) Motorized sample stage (e.g., 380FM, Luigs & Neumann).
   (c) High NA water-dipping objective with correction collar (e.g., Olympus LUMFI 60x, NA 1.1, working distance 1.5 mm).
3. Optical elements
   (a) Pulse stretching: SF 6 glass rod, 12 cm, to stretch femtosecond STED pulses to ~1 ps. 20 m long polarization-maintaining single-mode fiber to further stretch the STED pulses to ~70 ps (Pure silica core, NA 0.12, Schäfter & Kirchhoff).
   (b) Fiber coupler (Schäfter & Kirchhoff).
   (c) Electro-optical modulator for light intensity modulation: Pockels cell (e.g., Conoptics).
   (d) 2π delay vortex phase mask (RPC Photonics).
   (e) $\lambda/2$-wave plate and $\lambda/4$-wave plate (e.g., Qioptiq).
   (f) Beam expansion and collimation: STED beam collimation after the fiber with a $f = 40$ mm convex lens; emitted fluorescence is focused with $f = 300$ mm convex lenses; 2P beam expansion = 1:2 ($f = 75$ mm concave and $f = 150$ mm convex), scan and tube lens combination with beam expansion = 1:5 ($f = 40$ mm convex and $f = 200$ mm convex).
   (g) "Telecentric" beam scanner for x,y plane with beam expansion = 2:1, due to the action of a focusing lens and spherical mirror inside the scanner (Yanus IV digital scan head, Till Photonics).
   (h) Piezo-driven objective positioner for z-scans (P-725, Physik Instrumente).
   (i) Mirrors with the proper coating adapted to the light in the respective beam paths: dielectric mirrors for the STED and 2P excitation light, and silver mirrors for the emitted

fluorescence to optimize performance (spectral reflectivity, polarization insensitivity) and price considerations.

(j) Pellicle beam splitter (55%/45%, Thorlabs).

(k) Dichroic mirrors and filters:

- Band-pass filter (593/40, F37-593, AHF Analysentechnik).
- Short-pass dichroic (700/10, T700dcspxruv, AHF Analysentechnik).
- Long-pass dichroic (580/10, 580dcxru, AHF Analysentechnik).
- Short-pass filter (680, FF01-680/SP-25, Semrock).
- Single Notch filter (594, NF02-594S-25, Semrock).
- Emission band-filter (520/70, FF01-520/70-25, Semrock).
- Long-pass dichroic (514, LP02-514RS-25, Semrock).

(l) 2× Ø 62.5 μm multi-mode fibers.

(m) Detectors: Photodiode (3932 BSM, SpectraPhysics) with a spectral responsiveness ranging from 400 to 1100 nm to detect the repetition rate of the STED laser and allow for synchronization with the 2P laser. Photomultiplier tube (PMT, Photosensor modules H10722 series, Hamamatsu Photonics K.K.) with a detection range suited for STED and 2P wavelengths to visualize the reflection of gold beads. Two avalanche photodiodes (APDs, SPCM-AQR-13, PerkinElmer) with a spectral responsiveness ranging from 400 to 1060 nm to detect the emitted fluorescence.

4. Opto-mechanical components

(a) Optical mounts, post assemblies, optical rails, and other useful standard optical equipment (e.g., from Thorlabs, Newport and Qioptiq).

(b) 2× motorized mirror mounts for high-precision, hands-off steering of the STED and 2P beams (Newport).

5. Additional hardware/tools

(a) Phase-lock-loop for synchronization of the laser pulses (SpectraPhysics, Model3930 Lok-to-Clock).

(b) Nanosphere samples immobilized on regular microscopy glass slides (150 nm gold spheres, 40 nm and 170 nm fluorescent nanospheres).

(c) Sea of fluorescence (e.g., 10 mM calcein or fluorescein dissolved in buffered H_2O, pH 7.2).

(d) Power meters with sensitivity ranging from microwatts to 3 W.

6. Software
   (a) ImageJ (NIH).
   (b) Imspector (Abberior).
   (c) Agilis (Newport).
   (d) SpectraSuite (Ocean Optics).

2.2 Sample Preparation

1. Acute brain slices
   (a) Mice: $Thy1H^{+/eYFP}$ and $Thy1H^{+/eYFP}$ x $Cx3Cr1^{+/eGFP}$.
   (b) Preparation of sagittal brain slices (350 μm thick) using a vibratome (e.g., Leica VT1200S).
   (c) Artificial cerebrospinal fluid (ACSF): Sucrose-based ACSF (in mM): 210 sucrose, 10 glucose, 2 KCl, 26 $NaHCO_3$, 1.25 NaH_2PO_4, 0.1 $CaCl_2$ and 6 $MgCl_2$ (pH 7.4, osmolarity ~315 mOsm/L). Experiments are performed in carbogenated ACSF (in mM): 124 NaCl, 3 KCl, 26 $NaHCO_3$, 1.25 NaH_2PO_4, 10 glucose, 2 $CaCl_2$ and 1 $MgCl_2$ (pH 7.4, osmolarity ~305 mOsm/L). Perfusion (3–4 mL/min) and heated recording chamber (32–33 °C).
   (d) Dissection tools.
   (e) Stage for acute slices and electrophysiology (Motorized sample stage (e.g., 380FM, Luigs & Neumann)).
2. In vivo preparation
   (a) Mice: $Thy1H^{+/eYFP}$.
   (b) Anesthetics: Lidocaine 2% (Lurocaïne®)/Lidocaïne 2% (Xylocaïne®) and Isoflurane at 4% and 1.5% during induction and surgery/imaging, respectively.
   (c) Analgesics: Buprecare ten times diluted in PBS, 100 μL per 30 g bodyweight.
   (d) Other drugs: Dexamethasone sodium phosphate, 20 μL at 4 mg/mL and Ocry-gel.
   (e) Cortex buffer (in mM): 123 NaCl, 5 KCl, 10 glucose, 10 Hepes, 2 $CaCl_2$, and 2 $MgSO_4$.
   (f) Surgery tools.
   (g) Head-holder, in vivo platform and stage (customized from Luigs & Neumann).

3 Methods

3.1 Step-by-Step Procedure for Working with a 2P-STED Microscope

The achievable resolution and contrast of the images depends critically on the quality of the optical components, as well as on the precise alignment of the laser beams (in x, y, z, and t) and detectors (*see* **Note 1**). We have described the details of the design and construction of a 2P-STED microscope from scratch previously [15].

3.1.1 Description of the Laser Beam Paths

The 2P beam clears an electrical optical modulator (EOM, e.g., a Pockels cell), which enables fast and convenient power modulation of the transmitted intensity of the 2P light (*see* Fig. 1b). The 2P beam diameter is expanded with a telescope to properly overfill the back aperture of the objective. By slightly de-collimating the beam with the telescope, the z-focus of the 2P beam can be adjusted to coincide with the STED doughnut in the focal plane. In order to increase the depletion efficiency, the polarization of the 2P beam can be adjusted to the polarization of the STED beam by using a λ/2-wave plate. Eventually, the 2P beam is combined with the STED beam path via a dichroic mirror.

The STED light pulses are stretched in time by sending them through a 12 cm long SF6 glass rod before passing through an EOM, which allows for rapid power modulation (*see* Fig. 1b). Subsequently, the beam is launched into a 20 m long polarization maintaining single-mode optical fiber (SMF) to obtain a pulse length of ~70 ps. After the fiber, the re-collimated beam acquires its characteristic doughnut shape by passing through a helical 2π phase mask. The shape of the doughnut is fine-tuned by adjusting a λ/2-wave and λ/4-wave plate. Specifically, they serve to establish circular polarization of the STED beam (*see* **Note 2**). The STED and 2P beams are combined by a short-pass dichroic mirror. The foci of both beams are moved over the sample pixel-by-pixel in all three dimensions by the beam scanner and the piezo z-positioner. The emitted fluorescence is de-scanned by passing it through the scanner in reverse and sent toward the detectors by the way of a long-pass dichroic mirror. The emitted fluorescence of green-yellow fluorophores, such as GFP and YFP, can be spectrally separated via a dichroic mirror before getting focused into multi-mode fibers coupled to avalanche photodiode light detectors (APDs).

3.1.2 Preparation of the Microscope

1. Determine the power of the STED and 2P light and calibrate the transmission through the Pockels cells. Correct alignment of the Pockels cell can be checked by the transmission and extinction ratio, which should be >90% and >1000×.
2. Verify regularly that the laser beams are centered on the mirrors and optical elements. The installation of pinholes will facilitate the assessment of beam alignment and its optimization (*see* also **Note 3**). Long-term alignment stability is greatly aided by eliminating thermal gradients and fluctuations on and around the setup.
3. Coupling the STED beam into the single-mode optical fiber (*see* also **Note 4**): At low laser power, ensure that at least 50% of the light is transmitted. If necessary, align the beam ("beam walking") until maximal transmission of the light is reached (>65%, depending on fiber length).

4. STED doughnut construction and optimization: A smooth and symmetric shape in all three spatial dimensions featuring a low intensity minimum at the center (the "null") is crucial for obtaining a high gain in resolution and a good signal-to-noise ratio. The doughnut is visualized by scanning the reflection from 150 nm gold nanospheres immobilized on a glass slide. The reflected light is sent via a pellicle beam splitter to a PMT (*see* Fig. 2b). The quality of the doughnut can be optimized by adjusting the wavefront shaping optical elements. First, rotation of the wave plates can improve the symmetry and diameter of the central minimum. Second, repositioning of the helical phase mask can additionally adjust the position of the null. Preferably, the remaining intensity in the center of the doughnut should be less than 1% of the intensity measured on the rim of the doughnut. Third, the correction collar on the water-immersion objective should be adjusted to produce a tight z-profile. All these elements should be tweaked iteratively until the optimal doughnut shape is reached in x, y, and z (*see* Fig. 2a).
5. Co-alignment of the 2P and STED beam: Utilizing the reflection of the illuminated gold bead sample, verify that the 2P and the STED beam overlap in x, y, and z. If this is not the case, the 2P beam can be moved using a motorized mirror external to the common beam path (*see* Fig. 5). The telescope in the 2P beam path can serve to overcome misalignment in z by slight de-collimation (*see* Fig. 2a).
6. Detector alignment: Emitted fluorescence is de-scanned and spectrally detected by two APDs. For two-channel detector alignment, a highly concentrated sea of fluorescence is used (e.g., 10 mM calcein or fluorescein). Note, when using a water-dipping lens, the objective is in direct contact with the fluorescent solution and needs to be thoroughly rinsed afterward. Large field of view scans (e.g., 100 × 100 μm) can be used to assess whether signals are centered on both the detection channels (*see* Fig. 2c). If both the channels are not well aligned, the signal can be steered with the last common mirror in the detection path. Ultimately, both the YFP and GFP signals are centered in their dedicated channels by moving the beam with the respective mirrors in front of the multimode fibers connected to the APDs. To verify that both the channels are well superimposed, we check the overlay of the channels in ImageJ (*see* Fig. 2c).
7. Test of the quenching efficacy of the STED doughnut: While performing rapid z-scans across the interface of the sea of fluorescence and the glass slide, we make sure that the fluorescence quenching efficiency by the STED doughnut is higher than 80%. Applied laser powers should be similar to the

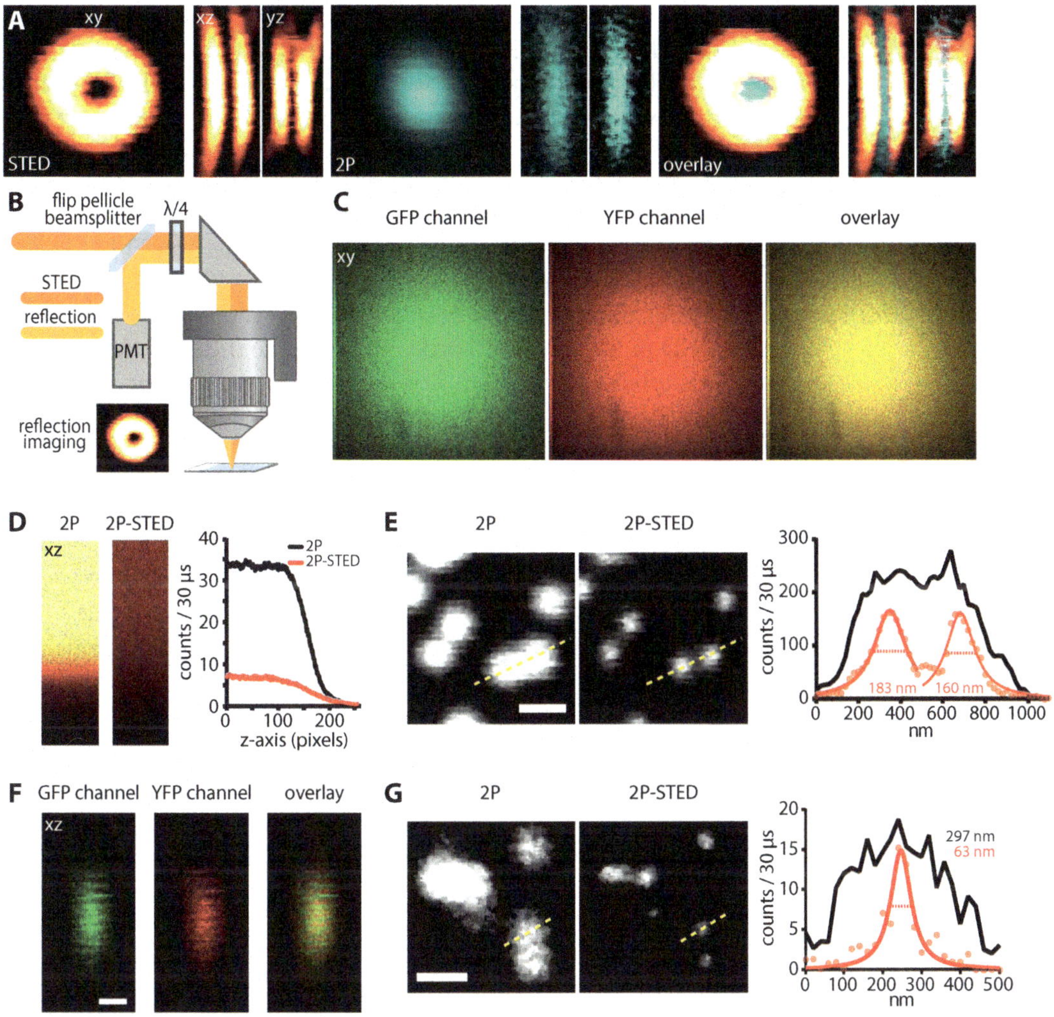

Fig. 2 Setting up the microscope and testing its resolution. (**a**) Lateral and axial views of point spread functions (PSFs) using gold beads illuminated by 2P and STED light. A symmetric STED doughnut with a zero-intensity center (*left*). Gold bead illuminated by 2P light showing a Gaussian beam profile (*middle*). For 2P-STED imaging, it is crucial to perfectly overlap the two beams in x, y, and z (*right*). (**b**) Gold bead PSF imaging is performed by using a flip pellicle beam splitter (55%/45% transmission/reflection) to send the reflected light to a photomultiplier tube (PMT). (**c**) Lateral scans of the sea of fluorescence in a large field of view (FOV) are performed to align the two detectors. (**d**) The quenching efficiency of the STED beam is assessed by scanning axially the sea of fluorescence across the coverslip interface in 2P and 2P-STED mode. Example z-scan images show approximately 80% quenching in the 2P-STED mode. (**e**) Two closely spaced fluorescent beads (with 170 nm diameters) can only be resolved in 2P-STED mode. (**f**) Axial alignment is verified using the fluorescent beads. (**g**) Consecutive imaging of 40 nm fluorescent beads in 2P and 2P-STED mode demonstrates a major gain in resolution in 2P-STED mode. The line profile and the FWHM of the Lorentzian fit provide a quantitative measure of the spatial resolution, which is around 60 nm. Scale bars: 500 nm (**e**–**g**)

situation when the biological sample is actually used. At this stage the co-alignment of the detectors in the z-dimension can be verified by exciting the sea of fluorescence with the 2P beam. An offset of the position of the edge of the interface between fluorophore and glass slide between both the channels is rectified by adjusting the distance of the collection lens in front of the multimode fibers. The synchronization system (Lok-to-Clock, SpectraPhysics) phase-locks the 2P to the STED beam and allows introducing a variable time delay. The use of a rough (12 ns) and fine (2 ns) delay allows for overlapping the 2P and STED pulses in time in order to maximize the quenching effect (i.e., quenching of >80% of the original fluorescent signal; *see* Fig. 2d). Low quenching efficiency can be improved by adjusting the relative intensities of the excitation and STED beams. While too little STED power will not efficiently quench the 2P fluorescence, excessive STED power can produce direct excitation of the fluorophore, thereby reducing quenching efficiency. If these manipulations do not yield sufficient fluorescence quenching, we recommend going back to improving the doughnut on the gold beads and checking the spatial and temporal registry of the 2P and STED laser PSFs, as low quenching efficacy can originate from a low-quality doughnut and/or from poor co-alignment of the 2P and STED beams.

8. Evaluation of the spatial resolution achieved by 2P-STED imaging using fluorescent nanospheres: First, while imaging 170 nm fluorescent beads in a small field of view (FOV, 10 × 10 μm) with both the 2P and STED beams, we can test whether the two beams are well aligned. Small corrections can be carried out by using the two motorized mirrors in the 2P and STED beam paths. Z-scans of single beads allow for the validation of the co-alignment of the detectors in the z-dimension (*see* Fig. 2f). As a result, the beads should be resolved at approximately 170 nm (measured by full-width half max (FWHM)) and the peak of the intensity should be centered (*see* Fig. 2e).

   To evaluate subdiffraction resolution, we recommend using 40 nm fluorescent nanospheres and image them with a pixel size of 10–20 nm in x-y. They are about ten times smaller than the resolution limit of regular 2P microscopy, yet still fairly bright and produce robust signals with a sufficient signal-to-noise ratio. The same beads are imaged by 2P and 2P-STED light and line profiles with their FWHM measurements are used to quantitatively assess the PSF. The resolution enhancement can be improved by finding the optimal power ratio between the 2P and STED light as mentioned above (*see* **step 7**). Using our 2P-STED microscope, we achieve a lateral resolution of around 60 nm of beads on a cover slip (*see* Fig. 2g).

3.2 Sample Preparation

The 2P-STED setup is designed for imaging acute brain slices as well as the intact brain through a "cranial window." Below, we shortly describe how to prepare the different samples for imaging.

3.2.1 Acute Brain Slice Preparation

Acute sagittal brain slices are prepared from P25 - P35 aged Thy1H$^{+/YFP}$ or Thy1H$^{+/YFP}$ x Cx3Cr1$^{+/eGFP}$ mice. In both transgenic mice a subset of pyramidal neurons is labeled with YFP, while in the latter in addition all microglia are labeled with GFP. Microglia are the resident immune cells of the brain and are very sensitive to the cellular damage and debris stemming from the slicing procedure, requiring quick and skillful execution. Slices are prepared as described previously [32]. In brief, after the slicing procedure slices are kept in carbogenated (95% O_2 and 5% CO_2) artificial cerebrospinal fluid (ACSF) until use. Experiments are to be performed in a submerged recording chamber at 33 °C with continuous perfusion of carbogenated ACSF (3–4 mL/min).

3.2.2 Preparation for In Vivo Brain Imaging

Standard procedures for in vivo brain imaging include thinned skull preparations, the implementation of a cranial (glass) window, or an open skull preparation. We utilize a cranial window (*see* Fig. 3a) in P55 – P75 aged Thy1H$^{+/YFP}$ male mice, which allows for keeping the cranial pressure constant, thereby reducing significantly movement artifacts of the brain compared to the open skull preparation. A 170 μm thick glass coverslip serves as a window and permits better optical access (for superior signal/noise contrast and spatial resolution) to structures deeper in the brain than the thinned skull preparation. Mice are chosen in this age-range based on their skull thickness and ease of surgery. At this age the skull is thin enough to easily penetrate it with the drill, but at the same time thick enough to resist breaking and possibly doing damage to the underlying cortical tissue.

In principle, a wide variety of mouse lines expressing different fluorescent proteins can be used for 2P-STED microscopy in vivo, as long as their spectral properties are matched to the microscope and the expression is strong but not too dense. In our case we used the classic Thy1H$^{+/YFP}$ mouse line [7], which transgenically expresses a robust level of YFP in a subset of pyramidal neurons, providing a nearly ideal label for volume-metric STED imaging.

As the YFP molecules can freely diffuse inside the cell's cytosol, the usual bleaching problem encountered when taking z-stacks or time-lapse series is greatly mitigated, because bleached molecules quickly get replaced by unbleached ones from areas outside of the imaging field of view.

Successful in vivo experiments critically depend on the mechanical stability of the sample. A headholder, which is glued to the brain via a helmet-like structure, allows for the fixation of the

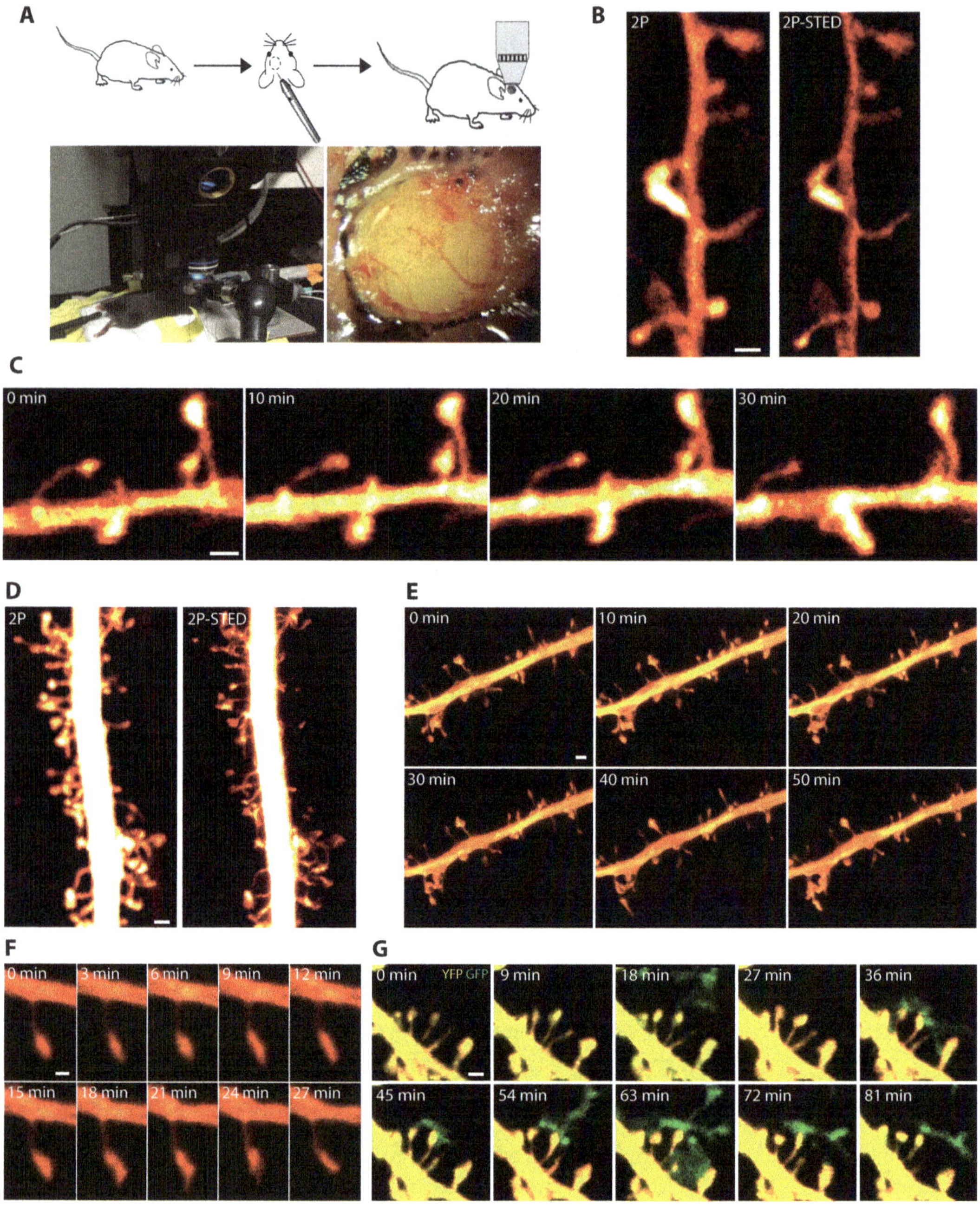

Fig. 3 2P-STED imaging in live brain tissue. (**a**) Outlining of the cranial window in Thy1H$^{+/YFP}$ transgenic mice (*top*), and positioning of the mouse below the objective (*bottom left*), and typical view through the cranial window of an operated mouse (*bottom right*). (**b**) 2P and 2P-STED image of YFP-labeled dendritic spines in layer 1 of the barrel cortex in vivo, showing the improvement in spatial resolution. (**c**) Time-lapse STED imaging of the morphology of dendritic spines in layer 1 of the barrel cortex in vivo at 40 μm depth. (**d**) Comparison between 2P and 2P-STED images (maximum intensity projections) of dendritic spines in an acute cortical brain slice of layer 5 (L5) obtained from a Thy1H$^{+/YFP}$ mouse. (**e**) Time-lapse imaging of spines on apical oblique dendrites in L5. Maximum intensity projections were acquired at a depth of approximately

mouse to the microscopy platform and ultimately for the physical stabilization of the skull. In addition, the sample stability and eventually the quality of the acquired images can be optimized by considering the following points. First, the head of the mouse is turned in such a way that breathing is not obstructed, but the region of interest can still be easily accessed by the objective. Second, the body posture of the mouse is adjusted so that the lungs can easily expand during breathing, while the front paws are still resting on the heating pad for a stable body posture. The anesthesia mask should cover the nostrils of the mouse and anesthesia is set to oxygen supply at 0.6 L/min and ~1.5% isoflurane. At this rate the breathing of the mouse is calm and abdominal, thereby reducing vibrations of the brain compared to diaphragmatic breathing. Having established a stable and optically accessible cranial window, one can image neuroanatomical structures like dendritic spines (*see* Fig. 3b).

3.3 Super-Resolution Image Acquisition

For STED imaging experiments high-quality samples are of outmost importance. Cellular debris on the surface of the tissue or an overall unhealthy state of the tissue will elevate the background fluorescence induced by the excitation and STED lasers. In general, laser intensities, field of view (FOV), pixel size, and dwell time depend on the purpose of the experiment, depth in the tissue, label intensity of the structure of interest and whether single images or time-lapse experiments are envisaged. For imaging dendritic spines we choose FOVs located at least 20 and up to 80–90 μm below the surface of the brain slice and intact brain. Imaging of microglia-spine interactions was performed at least 50 μm below the slice surface, where microglia are less likely to be affected by the slicing procedure. 2P-STED microscopy enables the observation of dynamic microglia-spine interactions at the nanoscale (*see* Fig. 3g). Slice health and moderate laser powers (2P and STED) are the key for successful super-resolution time-lapse image acquisitions.

During in vivo imaging, breathing (1 Hz) and blood pulsation (5–7.5 Hz) of the mouse can severely degrade the quality of the images. Clearly, motion artifacts are a serious limitation for obtaining high-quality images, especially in the case of super-resolution imaging, which imposes much stricter requirements for mechanical stability than regular 2P microscopy. To minimize the impact of

Fig. 3 (continued) 30 μm below the surface of the brain slice. (**f**) The morphology of dendritic spines can be tracked with a relatively high temporal resolution. Single z-section images were acquired approximately 35 μm below the surface in an acute brain slice. (**g**) Two-color 2P-STED time-lapse imaging can be used to record the dynamic morphological interactions between microglial processes and dendritic spines in acute brain slices (Thy1H$^{+/YFP}$ x Cx3Cr1$^{+/eGFP}$ mouse). The image is a maximum intensity projection based on two z-planes acquired 45 μm below tissue surface. Scale bars: 500 nm (**f**, **g**); 1 μm (**b**–**d**); 2 μm (**e**)

brain motion, the region of interest should be far from major blood vessels and the image acquisition parameters (pixel size and dwell time) need to be adjusted frequently. By continuously checking and adapting the anesthesia settings, a stable breathing pattern can be ensured. Scanning faster, i.e., the reduction of the FOV and/or the pixel dwell time, can overcome the effects of brain motion. However, scanning faster means higher 2P and STED laser intensities, which can in turn induce photodamage. The needed power depends both on the intensity and the stability of the label, on the imaging depth and whether the sample is an acute brain slice or the intact mouse brain. Typically, they range between 5 ~ 20 mW (2P) and 15 ~ 30 mW (STED) (measured at the back aperture of the objective). Image stacks are acquired with voxel sizes between 20 × 20 × 500 nm and 40 × 40 × 1000 nm (x, y, and z) and typical pixel dwell times are 10 ~ 50 μs. Time-lapse imaging can be performed at a temporal resolution of 5 ~ 10 min for a few hours on end. We advise acquiring small stacks of 3 μm in z for in vivo acquisitions in order to guard against drift of the sample out of the focus because of brain motion.

3.4 Analysis of Super-Resolved Images

Acquired 2P-STED images of dendritic stretches and spines are exported as 16-bit tiff files from the Imspector software to NIH ImageJ. For illustration purposes, brightness and contrast can be adjusted and a 1 pixel median filter applied to improve image quality. Z-image stacks can be visualized in maximal intensity projections (MIP, *see* Fig. 3b, d). For the visualization of time-lapse acquisitions (*see* Fig. 3c, e–g), we use the ImageJ plugin StackReg to correct post-hoc lateral sample drift.

To quantify the nanoscale morphology of dendritic spines (over time) (*see* Fig. 3c, e–g), it is possible to measure the full width at half maximum (FWHM) of spine neck width (*see* Fig. 4b), which is done on raw data of single z-sections. To automatically measure FWHM a custom-made image analysis program was used. At multiple positions on the spine neck, the program automatically estimates the best orthogonal fit and plots the intensity profile. Measuring across multiple parallel lines guards against noise in line profiles based on single pixels. The FWHM value can be extracted from the nonlinear Gaussian fits of the intensity profiles (*see* Fig. 4a).

4 Notes

1. Backward-forward alignment is not only mandatory when first setting up a STED microscope, but it is also regularly carried out to maintain a good performance of the system. To this end, a collimated beam from a 600 nm light source and an additional optical multimode fiber are used. The beam, which is coupled in the fiber, is sent backward through the beam path by clearing as centrally as possible the objective, the scanner, and

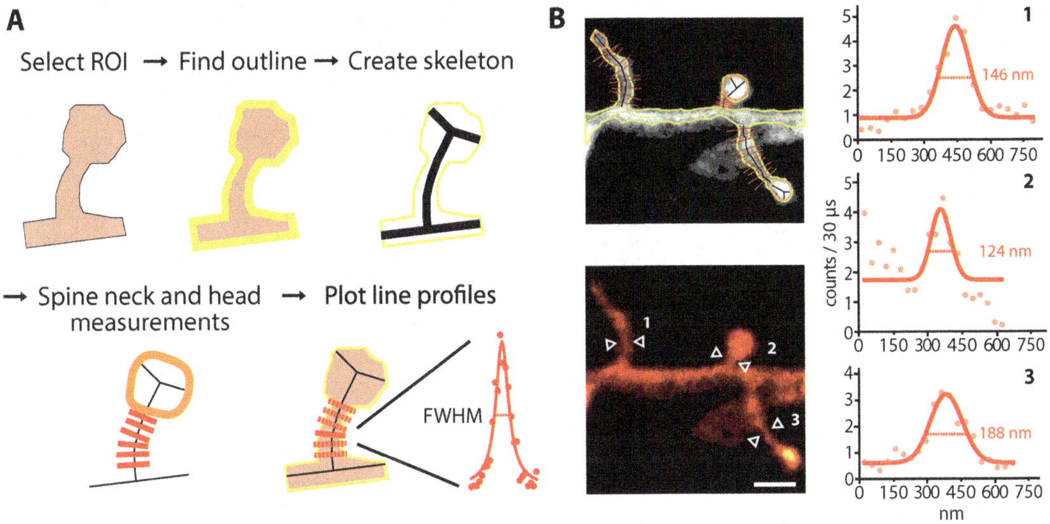

Fig. 4 Quantitative analysis of dendritic spine morphology. (**a**) A custom-made image analysis program facilitates analysis of morphometric parameters. After defining a region of interest, the program produces an outline of the region and creates a skeleton. Orthogonal lines are drawn on the skeleton and the spine head is fitted with an ellipse. The intensity distribution of the orthogonal lines is plotted and the FWHM is extracted from the Gaussian fit. (**b**) Example image showing the difference in FWHM of different spine neck widths on the same dendritic stretch. Scale bar: 1 μm

the detection and excitation beam paths. Ultimately, the backward propagating beam has to be aligned into the 20 m polarization-maintaining single-mode fiber. The two pinholes to be installed before the scanner in the common beam path are positioned accordingly (*see* Fig. 5a). They will serve as a reference for the forward alignment and all minor daily alignment routines.

Subsequently, the STED beam will be aligned forward to the reference pinholes (*see* Fig. 5b). The backward-forward alignment is repeated until both the beams clear the pinholes and couple correctly into the optical fibers. At this stage, the 2P beam is superimposed on the STED beam path. Subsequently, the overlap of the 2P and the STED beam is ensured on a large distance before continuing the regular beam alignment (see the preparation of the microscope).

2. In 2P-STED microscopy, the efficiency of the STED effect depends greatly on the ability to control the polarization of the light, which should be circular when it enters the objective. The STED beam must be circularly polarized so that the focal doughnut can exhibit a deep intensity minimum. This is particularly an issue for high NA objectives, where the marginal rays coming in at shallow angles otherwise do not interfere destructively. A circular polarization is achieved by the insertion of a rotatable λ/4-wave plate into the beam path. The circularity can be checked by measuring the intensity of the light after a

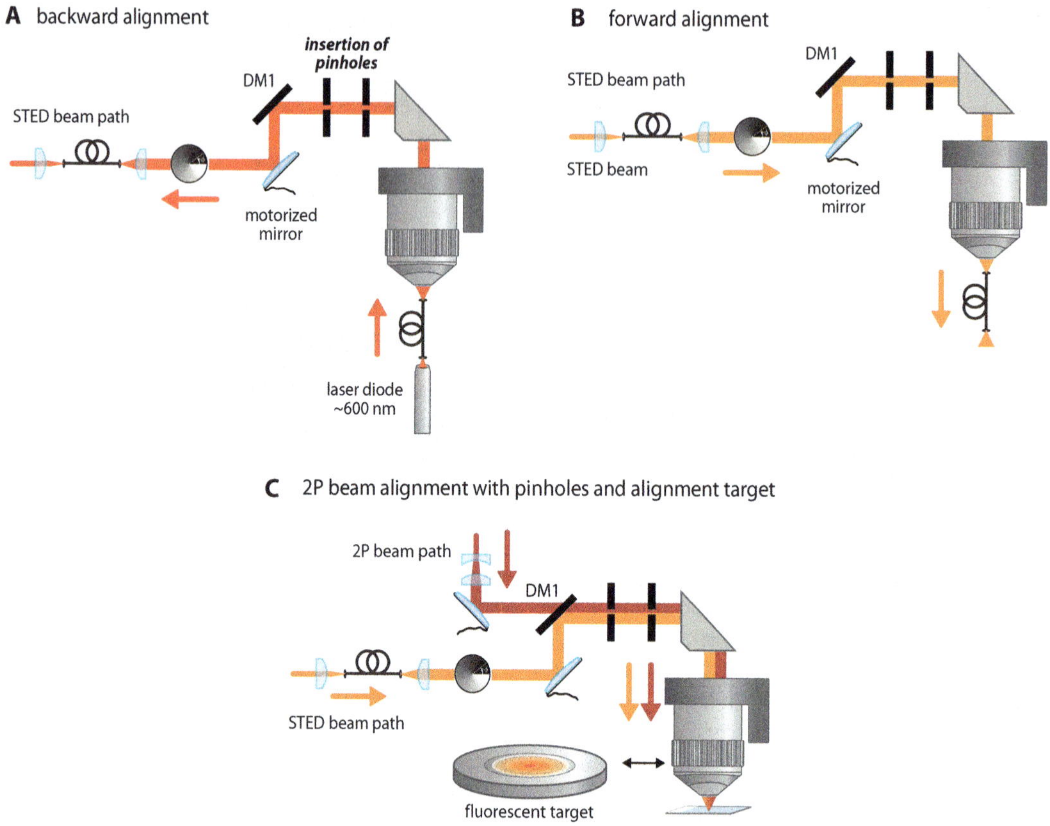

Fig. 5 Microscope alignment procedure. (**a**) Schematic explanation of the backward alignment procedure. A collimated beam from a laser diode is sent backward through the setup. The beam travels in a centered way through the objective, z-positioner, beam scanner, and the STED beam path. Pinholes are inserted based on the backward beam, and serve as reference for the forward beam. (**b**) Forward alignment: the STED beam is aligned to the reference pinholes and is guided through all optical elements. Back- and forward-alignment is repeated until the pinholes are perfectly cleared. (**c**) 2P beam alignment: the 2P beam is subsequently co-aligned to the STED beam by utilizing the reference pinholes and a fluorescent target to verify co-alignment is maintained until the sample

rotating polarizer. In case of perfect circularity, a flat line will appear on the oscilloscope, since all polarization directions are equally represented; otherwise, the signal will be modulated by a sine wave, indicating poor circularity.

3. In order to facilitate alignment procedures, we recommend inserting pinholes at strategic positions of the common (*see* Fig. 5a) and external beam paths. This allows verifying whether the laser beams are still aligned to the centers of the optical elements and, more importantly in the common beam path, whether the 2P and STED beam are still co-aligned. By iterative beam walking perfect clearing of the pinholes can be achieved (*see* Fig. 5a, b). Replacing the objective by a fluorescing alignment target makes it possible to check whether

the co-alignment of the two beams on the pinholes holds true until reaching the sample (*see* Fig. 5c).

4. Fiber-coupling can be facilitated by using a power meter suitable for the pulsed STED beam in order to measure the input power and the amount of transmitted light. In order to couple the light properly into the fiber, as well as in order to align to pinholes, one should walk the beam. Two different mirrors (each having three degrees of freedom) are used to control translation and rotation of the laser beam separately.

Acknowledgments

This work was supported by a PhD fellowship from the "European Neuroscience Campus" network (ENC) to MJTTV, a Boehringer Ingelheim PhD fellowship and a 4th-year extension grant from the "Fondation pour la Recherche Médicale" (FRM, FDT20160435677) to TP, and grants from the "Agence Nationale de la Recherche" (ANR) (ANR-13-BSV4-0007-01), France-BioImaging (ANR-10-INSB-04), and European Research Area Network NEURON II (ANR-12-NEUR-0007-03) to UVN. We thank Florian Levet (CNRS, University of Bordeaux) for the custom-made image analysis program.

References

1. Holtmaat A, Svoboda K (2009) Experience-dependent structural synaptic plasticity in the mammalian brain. Nat Rev Neurosci 10 (9):647–658. doi:10.1038/nrn2699. nrn2699 [pii]
2. Tonnesen J, Nägerl UV (2016) Dendritic spines as tunable regulators of synaptic signals. Front Psych 7:101. doi:10.3389/fpsyt.2016.00101
3. Kettenmann H, Kirchhoff F, Verkhratsky A (2013) Microglia: new roles for the synaptic stripper. Neuron 77(1):10–18. doi:10.1016/j.neuron.2012.12.023
4. Gray EG (1959) Electron microscopy of synaptic contacts on dendrite spines of the cerebral cortex. Nature 183(4675):1592–1593
5. Denk W, Strickler JH, Webb WW (1990) Two-photon laser scanning fluorescence microscopy. Science 248(4951):73–76
6. Svoboda K, Yasuda R (2006) Principles of two-photon excitation microscopy and its applications to neuroscience. Neuron 50(6):823–839. doi:10.1016/j.neuron.2006.05.019
7. Feng G, Mellor RH, Bernstein M, Keller-Peck C, Nguyen QT, Wallace M, Nerbonne JM, Lichtman JW, Sanes JR (2000) Imaging neuronal subsets in transgenic mice expressing multiple spectral variants of GFP. Neuron 28 (1):41–51
8. Holtmaat A, Bonhoeffer T, Chow DK, Chuckowree J, De Paola V, Hofer SB, Hubener M, Keck T, Knott G, Lee WC, Mostany R, Mrsic-Flogel TD, Nedivi E, Portera-Cailliau C, Svoboda K, Trachtenberg JT, Wilbrecht L (2009) Long-term, high-resolution imaging in the mouse neocortex through a chronic cranial window. Nat Protoc 4(8):1128–1144. doi:10.1038/nprot.2009.89
9. Hell SW (2007) Far-field optical nanoscopy. Science 316(5828):1153–1158. doi:10.1126/science.1137395
10. Hell SW, Wichmann J (1994) Breaking the diffraction resolution limit by stimulated emission: stimulated-emission-depletion fluorescence microscopy. Opt Lett 19(11):780–782
11. Nägerl UV, Bonhoeffer T (2010) Imaging living synapses at the nanoscale by STED microscopy. J Neurosci 30(28):9341–9346. 30/28/9341 [pii]. doi:10.1523/JNEUROSCI.0990-10.2010

12. Nägerl UV, Willig KI, Hein B, Hell SW, Bonhoeffer T (2008) Live-cell imaging of dendritic spines by STED microscopy. Proc Natl Acad Sci U S a 105(48):18982–18987. 0810028105 [pii]. doi:10.1073/pnas.0810028105
13. Berning S, Willig KI, Steffens H, Dibaj P, Hell SW (2012) Nanoscopy in a living mouse brain. Science 335(6068):551. doi:10.1126/science.1215369
14. Willig KI, Steffens H, Gregor C, Herholt A, Rossner MJ, Hell SW (2014) Nanoscopy of filamentous actin in cortical dendrites of a living mouse. Biophys J 106(1):L01–L03. doi:10.1016/j.bpj.2013.11.1119
15. Bethge P, Chereau R, Avignone E, Marsicano G, Nägerl UV (2013) Two-photon excitation STED microscopy in two colors in acute brain slices. Biophys J 104(4):778–786
16. Ding JB, Takasaki KT, Sabatini BL (2009) Supraresolution imaging in brain slices using stimulated-emission depletion two-photon laser scanning microscopy. Neuron 63(4): 429–437. doi:10.1016/j.neuron.2009.07.011
17. Takasaki KT, Ding JB, Sabatini BL (2013) Live-cell superresolution imaging by pulsed STED two-photon excitation microscopy. Biophys J 104(4):770–777. doi:10.1016/j.bpj.2012.12.053
18. Li Q, SS W, Chou KC (2009) Subdiffraction-limit two-photon fluorescence microscopy for GFP-tagged cell imaging. Biophys J 97 (12):3224–3228. doi:10.1016/j.bpj.2009.09.038
19. Moneron G, Hell SW (2009) Two-photon excitation STED microscopy. Opt Express 17 (17):14567–14573
20. Helmchen F, Denk W (2005) Deep tissue two-photon microscopy. Nat Methods 2 (12):932–940. doi:10.1038/nmeth818
21. Zipfel WR, Williams RM, Webb WW (2003) Nonlinear magic: multiphoton microscopy in the biosciences. Nat Biotechnol 21 (11):1369–1377. doi:10.1038/nbt899
22. Abbe E (1873) Beiträge zur Theorie des Mikroskops und der mikroskopischen Wahrnehmung. Arch Mikrosk Anat 9(1):413–418
23. Harke B, Keller J, Ullal CK, Westphal V, Schonle A, Hell SW (2008) Resolution scaling in STED microscopy. Opt Express 16 (6):4154–4162
24. Klar TA, Jakobs S, Dyba M, Egner A, Hell SW (2000) Fluorescence microscopy with diffraction resolution barrier broken by stimulated emission. Proc Natl Acad Sci U S A 97 (15):8206–8210. doi:97/15/8206 [pii]
25. Hell SW (2015) Nanoscopy with focused light (Nobel lecture). Angew Chem 54(28): 8054–8066. doi:10.1002/anie.201504181
26. Kittel RJ, Wichmann C, Rasse TM, Fouquet W, Schmidt M, Schmid A, Wagh DA, Pawlu C, Kellner RR, Willig KI, Hell SW, Buchner E, Heckmann M, Sigrist SJ (2006) Bruchpilot promotes active zone assembly, Ca2+ channel clustering, and vesicle release. Science 312 (5776):1051–1054. 1126308 [pii]. doi:10.1126/science.1126308
27. Sieber JJ, Willig KI, Kutzner C, Gerding-Reimers C, Harke B, Donnert G, Rammner B, Eggeling C, Hell SW, Grubmuller H, Lang T (2007) Anatomy and dynamics of a supramolecular membrane protein cluster. Science 317(5841):1072–1076. doi:10.1126/science.1141727. 317/5841/1072 [pii]
28. Willig KI, Rizzoli SO, Westphal V, Jahn R, Hell SW (2006) STED microscopy reveals that synaptotagmin remains clustered after synaptic vesicle exocytosis. Nature 440 (7086):935–939. nature04592 [pii]. doi:10.1038/nature04592
29. Westphal V, Rizzoli SO, Lauterbach MA, Kamin D, Jahn R, Hell SW (2008) Video-rate far-field optical nanoscopy dissects synaptic vesicle movement. Science 320(5873):246–249. doi:10.1126/science.1154228. 1154228 [pii]
30. Urban NT, Willig KI, Hell SW, Nägerl UV (2011) STED nanoscopy of actin dynamics in synapses deep inside living brain slices. Biophys J 101(5):1277–1284. doi:10.1016/j.bpj.2011.07.027. S0006-3495(11)00885-X [pii]
31. Tonnesen J, Katona G, Rozsa B, Nägerl UV (2014) Spine neck plasticity regulates compartmentalization of synapses. Nat Neurosci 17 (5):678–685. doi:10.1038/nn.3682. nn.3682 [pii]
32. Pfeiffer T, Avignone E, Nägerl UV (2016) Induction of hippocampal long-term potentiation increases the morphological dynamics of microglial processes and prolongs their contacts with dendritic spines. Sci Rep 6:32422. doi:10.1038/srep32422

Chapter 6

STED Imaging of Golgi Dynamics with Cer-SiR: A Two-Component, Photostable, High-Density Lipid Probe for Live Cells

Roman S. Erdmann, Derek Toomre, and Alanna Schepartz

Abstract

Long time-lapse super-resolution imaging in live cells requires a labeling strategy that combines a bright, photostable fluorophore with a high-density localization probe. Lipids are ideal high-density localization probes, as they are >100 times more abundant than most membrane-bound proteins and simultaneously demark the boundaries of cellular organelles. Here, we describe Cer-SiR, a two-component, high-density lipid probe that is exceptionally photostable. Cer-SiR is generated in cells via a bioorthogonal reaction of two components: a ceramide lipid tagged with *trans*-cyclooctene (Cer-TCO) and a reactive, photostable Si-rhodamine dye (SiR-Tz). These components assemble within the Golgi apparatus of live cells to form Cer-SiR. Cer-SiR is benign to cellular function, localizes within the Golgi at a high density, and is sufficiently photostable to enable visualization of Golgi structure and dynamics by 3D confocal or long time-lapse STED microscopy.

Key words Bioorthogonal chemistry, Click chemistry, Fluorophores, Membranes, Super-resolution microscopy, Inverse electron demand Diels-Alder reaction

1 Introduction

1.1 Visualizing the Boundaries of Cellular Organelles

Super-resolution microscopy can visualize cellular structure and dynamics at resolutions that simply cannot be attained using conventional microscopes [1–6]. However, increased spatial resolution extracts a price, as the concomitant high sampling rate and photon flux places enormous demands on both the photophysical properties of the fluorophore and the structural properties of the targeting probes. Nearly all super-resolution techniques [1–6] rely on fluorophores that cycle between ON and OFF states [1–7], either stochastically (as in PALM and STORM) [8, 9] or using point spread function engineering (as in STED) [10, 11]. In STED, cycling is achieved with highly powerful depletion and excitation lasers that switch the dye ON and OFF multiple times [1]. Fluorophore photostability is therefore paramount, with a clear

Holger Erfle (ed.), *Super-Resolution Microscopy: Methods and Protocols*, Methods in Molecular Biology, vol. 1663,
DOI 10.1007/978-1-4939-7265-4_6, © Springer Science+Business Media LLC 2017

preference for far red, bright, photostable dyes such as Atto647N [12] and SiR [13–15]. The structural properties of the targeting probes are also restrictive, as they must be simultaneously cell-permeable, benign to cell function, and possess the ability to localize at high density to achieve bright labeling.

Although proteins can now be tagged easily with fluorophores [16, 17], either by the expression of fusion proteins [18] or variants containing orthogonally reactive unnatural amino acids [19, 20], the labeling density achieved with most tagged proteins is generally insufficient for long time-lapse super-resolution imaging. The exceptions are protein polymers such as tubulin, actin, and clathrin - structures that are inherently dense and considered gold standards fore super-resolution imaging. By contrast, lipids represent ideal high-density probes, as they represent the primary constituent of an organelle membrane and are present at densities that are at least one hundred fold higher than the membrane-bound proteins they harbor [21–27]. Perhaps more importantly, it is the lipid—not the proteins held within—that defines the molecular outline of the organelle [28]. Although many lipid-dye conjugates, such as BODIPY FL C_5 and NBD C_6 ceramide have long been commercially available [29, 30], these molecules lack the requisite photostability to survive the powerful lasers required for super-resolution methods such as STED [1–5].

1.2 Labeling the Golgi with Cer-TCO and Sir-Tz

Herein, we describe a general approach to label organelle lipids at high density using fluorophores that are compatible with long time-lapse super-resolution imaging, and apply this approach to visualize the Golgi apparatus [31]. This approach relies on the well-characterized selective partitioning of lipids among cellular organelles [32, 33]; in the case of the Golgi apparatus, the enriched lipids are derived from ceramide [30, 32, 33]. We synthesized a ceramide derivative tagged with *trans*-cyclooctene (Cer-TCO), localized it to the Golgi apparatus in live cells, and reacted it in situ with a tetrazine-modified version of a photostable silicon-rhodamine dye (SiR-Tz) [13, 34–38] (Fig. 1). The resulting

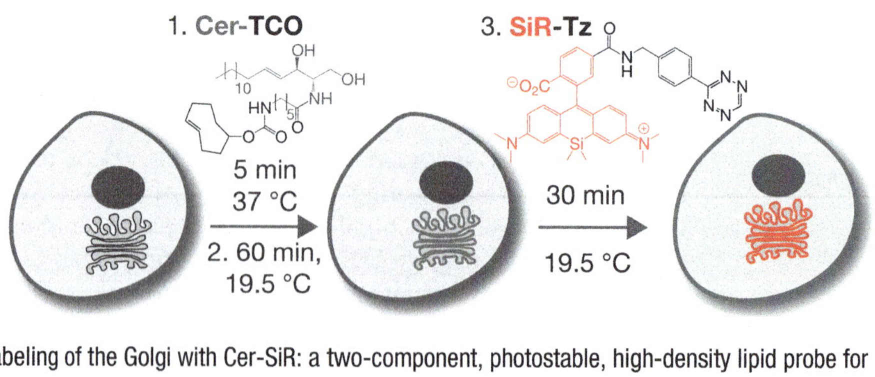

Fig. 1 Labeling of the Golgi with Cer-SiR: a two-component, photostable, high-density lipid probe for live cells. Hela cells are incubated with 2 μM Cer-TCO for 5 min, washed, and incubated at 19.5 °C for 60 min to localize the lipid to the Golgi. The cells are then incubated with 2 μM SiR-Tz for 30 min at 19.5 °C, washed, and imaged

product, Cer-SiR, is not cytotoxic and does not perturb protein trafficking within and through the Golgi: it is indeed benign to organelle function [31]. The two-component Cer-TCO/SiR-Tz labeling system we describe is suitable for STED imaging of the Golgi for over 900 frames and allows the visualization of vesicles budding and exiting the Golgi [31]. The one challenge is that the Golgi is a convoluted three dimensional object and typically STED only enhances the resolution in the xy axis, however new PSF engineering isotropic 3D STED microscopes, such as 4Pi STED should circumvent this challenge [39].

2 Materials

2.1 Microscopy Systems

2.1.1 STED Imaging (Commercial Microscope)

STED imaging was carried out on a commercial Leica TCS STED microscope. A picosecond pulsed laser diode (PicoQuant, LDH-P-F-640B) was used for excitation while depletion was achieved with a femtosecond pulsed mode-locked Ti:Sapphire laser (Spectra Physics, Mai Tai) tuned to 755 nm with the output pulses stretched to approximately 200 ps via propagation through a 120 m single mode polarization maintaining fiber. The excitation and depletion beams were directed into a 100× 1.4 NA oil immersion objective (Leica) which also collected fluorescence from the sample. Fluorescence was filtered from excitation light with a bandpass filter (Semrock, FF01–685/40) and detected with an avalanche photodiode.

2.1.2 STED Imaging (Custom Microscope)

Additional STED imaging was performed on a custom-built system with faster scanning capabilities compared to the Leica TCS STED microscope (Bewersdorf lab, Yale University). The system is centered around an 80 MHz mode-locked Ti:Sapphhire laser (Chameleon Ultra II, Coherent) tuned to 755 nm as the STED depletion beam. The output pulses from the depletion laser were stretched to several hundred picoseconds by first passing through a glass block made from a high dispersion material (SF6, Schott) and then coupled into a 100 m polarization-maintaining single mode optical fiber. After leaving the optical fiber, the depletion beam was collimated and directed onto an SLM, conjugate to the objective back pupil plane, where a 2π phase ramp was displayed. Excitation for fluorescence was accomplished with a 640 nm pulsed diode laser (PicoQuant) electronically synchronized to the depletion beam with an additional computer controlled electronic delay (Colby Instruments). The STED depletion and fluorescence excitation beams were combined with a dichroic mirror and directed into a 100 × 1.4 NA oil immersion objective lens (UPLSAPO 100XO/PSF, Olympus) where they were focused at the sample. A 16 kHz resonance scanning mirror, in combination with a galvanometer mirror, allowed imaging via beam scanning. The two mirrors were imaged into the objective pupil plane and allowed the beams to scan through the sample at a rate of 16 kHz along the fast axis. Fluorescence from the sample was collected by the objective lens, descanned by the scan mirrors, and separated from the excitation and depletion light using dichroic mirrors and band pass filters (FF01-685/40, Semrock). The fluorescence was then focused into a 105 μm core (~0.7 Airy units) multimode fiber connected to a single photon counting avalanche photodiode (ARQ-13-FC, Perkin Elmer). Counts from the APD were collected using an FPGA based data acquisition card (PCIe-7852R, National Instruments) and custom acquisition software (LabVIEW, National Instruments). Data collection was synchronized with the resonance mirror for uni-directional collection during the two-thirds of the half period where the motion of the mirror is most linear. Recorded pixel values were therefore linearized (on the DAQ card) to account for the sinusoidal velocity profile of the resonant mirror and normalized according to the pixel dwell times such that the center pixel was divided by unity.

2.1.3 Spinning Disk Confocal Microscopy

Spinning-disk confocal microscopy was performed using an Improvision UltraVIEW VoX system (Perkin-Elmer) built around a Nikon Ti-E inverted microscope, equipped with PlanApo objectives (60× 1.45-NA) and controlled by the Volocity software (Improvision). SiR was excited with a 640 nm laser and for the detection a 705 ± 45 nm filter was used. The microscope stage was surrounded by a box constantly held at 37 °C.

2.2 Synthesis of Cer-TCO

2.2.1 Equipment

1. 3 and 20 mL scintillation vials.
2. Erlenmeyer flasks.
3. Round-bottom flasks.
4. Filter funnels and filter paper.
5. Separatory funnel.
6. Chromatography column (*see* **Note 1**).
7. Spatulas.
8. Magnetic stir bar.
9. Rotary evaporator (*see* **Note 2**).
10. TLC developing chamber.
11. Fume hood.
12. Temperature controlled magnetic stir plate.
13. TLC plates.

2.2.2 Chemicals

1. TOC-PNB ester (((4*E*)-cyclooct-4-en-1-yl) (4-nitrophenyl) carbonate).
2. 6-aminohexanoic acid.

3. Sphingosine ((2*S*,3*R*)-2-aminooctadec-4-ene-1,3-diol).
4. DMF (*N*,*N*-Dimethylformamide).
5. Dichloromethane.
6. Methanol.
7. 1 M HCl.
8. Saturated aqueous $NaHCO_3$.
9. $MgSO_4$ (anhydrous).
10. HBTU (*N*,*N*,*N*′,*N*′-Tetramethyl-O-(1H-benzotriazol-1-yl) uronium hexafluorophosphate).
11. $^{i}Pr_2NEt$ (*N*,*N*-Diisopropylethylamine).
12. Silica gel.
13. Iodine chamber: Add 50 g of silica gel to a jar and add a few crystals of iodine.
14. $KMnO_4$ TLC stain: dissolve 3 g $KMnO_4$, 20 g K_2CO_3, 5 mL 5% NaOH (aq.) in 300 mL of deionized water (*see* **Note 3**).

2.3 Preparing Solutions for Cell Labeling

1. Prepare Cer-TCO working solution (2 μM): Add 1 μL of the 2 mM stock solution to 20 μL of 10% puronic F127 and mix thoroughly by drawing and releasing the solution into/from the pipet tip. Add 1 mL of 1% casein in PBS and mix again (*see* **Notes 4** and **5**).
2. Prepare SiR-Tz working solution (2 μM): Add 1 μL of the 2 mM stock solution to 1 mL of 1% casein in PBS and mix by pipetting up and down (*see* **Note 5**).

2.4 Imaging with Cer-TCO and SiR-Tz

All the solutions were prepared using ultrapure water (deionized water further purified to a sensitivity of 18 MΩ cm at 25 °C) and analytical grade reagents. All reagents and solutions were stored at −20 °C unless noted otherwise. HeLa cells were cultured in T75 flasks in Dulbecco's Modified Eagle Medium with phenol red (DMEM(ph+)) supplemented with 10% FBS and 1% Penicillin/Streptomycin solution (10,000 U/mL).

1. 2 mM SiR-Tz in DMSO (*see* **Notes 6** and **7**).
2. 2 mM Cer-TCO in DMSO (*see* **Notes 6** and **8**).
3. Dulbecco's Modified Eagle Medium without phenol red, supplemented with 10% FBS (DMEM (ph-)) (store at 4 °C).
4. 10% Pluronic F127 in H_2O (*see* **Note 9**).
5. 1% casein hydrolysate in 1x PBS (sterile filtered; store at 4 °C).
6. 35 mm glass-bottom dish (glass no. 1.5).
7. Parafilm™.
8. Waterbath with thermostat (*see* **Note 10**).

3 Methods

This section describes our protocols for the synthesis of Cer-TCO, maintenance of the cells to be labeled, preparation of all working solutions, labeling of the cells using Cer-TCO and SiR-Tz, as well as the imaging of labeled cells by 3D confocal and STED microscopy.

3.1 Synthesis of Cer-TCO

All the chemical steps should be performed in a well-vented fume hood, and appropriate personal protective equipment (googles, gloves, lab coat) should be worn throughout all the procedures. All waste should be disposed of in accord with appropriate regulations. None of the reactions described requires special measures to prevent contact with air or moisture.

3.1.1 Synthesis of TCO-C6 (rel-(1R-4E-pR)-Cyclooct-4-ene-1-yl-N-hexanoic Acid Carbamate)

1. Add 100 mg (343 μmol, 1.0 eq) of TCO-PNB ester to a 20 mL scintillation vial (*see* **Note 11**).
2. Add 135 mg of 6-aminohexanoic acid (1.03 mmol, 3.0 eq) to the vial.
3. Add a 5 mm magnetic stir bar and 3 mL of DMF.
4. Add 360 μL of $^{i}Pr_2NEt$ (2.06 mmol, 6.0 eq) and close the vial with its cap (*see* **Notes 12** and **13**).
5. Place an oil bath on a magnetic stir plate and set the temperature to 65 °C.
6. Mount the vial into a clamp and lower the closed vial into the oil bath until the level of the oil bath matches the level of the suspension in the vial.
7. Cover the reaction setup with aluminum foil to exclude light.
8. Stir the reaction for 3 days at 65 °C. Switch off the oil bath heater and remove the vial from the clamp.
9. Concentrate the reaction mixture under vacuum (*see* **Note 2**).
10. Transfer the reaction mixture to a separatory funnel together with 20 mL of dichloromethane.
11. Wash the organic layer 3× with 5 mL of 1 M aqueous HCl and combine the aqueous layers.
12. Extract the combined aqueous layers 2x with 5 mL of dichloromethane.
13. Combine the organic layers of **step 12** with the organic layer of **step 11**.
14. Dry combined organic layers over $MgSO_4$.
15. Concentrate the dried organic layers by rotary evaporation.
16. Prepare a chromatography column of silica gel (25 g) packed in dichloromethane (*see* **Note 1**).

17. Load the column with the crude product obtained in **step 15** and elute with a gradient from 100% dichloromethane to 5% methanol in dichloromethane (400 mL) and further elute with 5% methanol in dichloromethane (150 mL).
18. Identify the fractions containing the desired product by running silica gel TLC plates with 5% methanol in dichloromethane. The product can be visualized by developing the TLC place in an iodine chamber for 3 min. The product appears as a brown/yellow spot with a retention factor of 0.35 (*see* **Note 14**).
19. Combine the fractions containing the product and concentrate them by rotary evaporation to obtain the product as a pale yellow oil with a yield of approximately 40% (*see* **Note 15**).
20. Analyze the product by NMR and ESI-HRMS. The following analytical data is anticipated.

    ^{1}H NMR (400 MHz, MeOD-d_3) δ/ppm = 5.66–5.54 (m, 1H), 5.47 (ddd, J = 16.0, 10.7, 3.5 Hz, 1H), 4.39–4.25 (m, 1H), 3.06 (t, J = 7.0 Hz, 2H), 2.37–2.31 (m, 3H), 2.28 (t, J = 7.4 Hz, 2H), 2.04–1.86 (m, 4H), 1.79–1.65 (m, 2H), 1.65–1.54 (m, 3H), 1.54–1.44 (m, 2H), 1.39–1.27 (m, 2H).

    ^{13}C NMR (101 MHz, MeOD-d_3) δ 177.4, 158.6, 136.0, 133.7, 81.5, 42.2, 41.5, 39.6, 35.2, 34.8, 33.5, 32.1, 30.6, 27.3, 25.7.

    HRMS (ESI): m/z for $C_{15}H_{26}NO_4^+$: 284.1856.

3.1.2 Synthesis of Cer-TCO

1. Solution A: Dissolve 25.0 mg of sphingosine (83.5 μmol, 1.3 eq) in 2 mL of DMF.
2. Solution B: Dissolve 40.1 mg of HBTU (77.1 μmol, 1.2 eq) in 0.8 mL of DMF and add 33.7 μL of ^{i}Pr$_2$NEt (193 μmol, 3.0 eq).
3. Add 18.2 mg of TCO-C6 (from Subheading 3.2, **step 1**) (64.2 μmol, 1.0 eq) to solution A, add a magnetic stir bar, and stir for 2 min.
4. Combine solutions A and B and stir for 2.5 h at room temperature (*see* **Note 16**).
5. Concentrate the reaction mixture under reduced pressure to obtain a brownish-yellowish residue (*see* **Note 2**).
6. Dissolve the residue in 50 mL of dichloromethane and transfer into a separatory funnel.
7. Wash the organic layer 3× with 10 mL of sat. aqueous $NaHCO_3$ and wash 3× with 10 mL of 1 M aqueous HCl (*see* **Note 17**).

8. Combine the aqueous $NaHCO_3$ layers and wash 1× with 10 mL of dichloromethane.
9. Combine the aqueous HCl layers and wash 1× with 10 mL of dichloromethane.
10. Combine the organic layers from **steps** 7 to **9**, dry over $MgSO_4$, and concentrate by rotary evaporation.
11. Prepare a chromatography column of silica gel (12 g) packed in dichloromethane (*see* **Note 1**).
12. Load the column with the crude product obtained in **step 10** and elute with a gradient from 100% dichloromethane to 5% methanol in dichloromethane (200 mL) and further elute with 5% methanol in dichloromethane (100 mL).
13. Identify the fractions containing the product using silica gel TLC plates eluted with 5% methanol in dichloromethane. The product can be visualized by developing the TLC place in an iodine chamber for 3 min. It appears as a brown/yellow spot with a retention factor of 0.35. Product purity can be assessed by staining the TLC plate with $KMnO_4$. Impurities in the reaction mixture will appear as additional yellow spots on the TLC plate once the $KMnO_4$-stained plate is heated with a heat gun (*see* **Note 18**).
14. Combine those fractions that lack impurities as judged by TLC and concentrate them by rotary evaporation to obtain the product as a colorless wax with a yield of approximately 70% (*see* **Note 15**).
15. Analyze the product by NMR and ESI-HRMS. The following analytical data is anticipated.

    ^{1}H NMR (500 MHz, MeOD-d_3): δ/ppm = 5.72 (dt, J = 14.3, 6.7 Hz, 1H), 5.68–5.56 (m, 1H), 5.54–5.44 (m, 2H), 4.42–4.27 (m, 1H), 4.09 (t, J = 7.2 Hz, 1H), 3.89 (q, J = 5.8 Hz, 1H), 3.71 (d, J = 5.0 Hz, 2H), 3.08 (t, J=6.8 Hz, 2H), 2.42–2.29 (m, 3H), 2.23 (t, J=7.5 Hz, 2H), 2.06 (q, J = 7.1 Hz, 2H), 2.03–1.89 (m, 4H), 1.81–1.68 (m, 2H), 1.68- 1.56 (m, 3H), 1.50 (p, J = 7.3 Hz, 2H), 1.31 (s, 23H), 0.92 (t, J = 6.7 Hz, 3H).

    ^{13}C NMR (126 MHz, MeOD-d_3): δ/ppm = 176.5, 159.1, 136.5, 135.1, 134.2, 131.6, 81.9, 74.1, 62.7, 57.2, 42.7, 41.9, 40.1, 37.6, 35.6, 33.9, 33.8, 33.5, 32.6, 31.2, 31.2, 31.1, 30.9, 30.8, 30.8, 27.9, 27.1, 24.2, 14.9.

    HRMS (ESI): *m/z* calc. For $C_{33}H_{61}N_2O_5^+$: 565.4575.

3.2 Preparing Cells and Their Labeling with Cer-TCO and SiR-Tz

1. Seed 160,000 HeLa cells on a 35 mm glass-bottom dish in DMEM(ph+)and incubate for 24 h at 37 °C (*see* **Note 19**).
2. Wash the attached cells in the 35 mm glass-bottom dish 3× with 2 mL of DMEM (ph−).

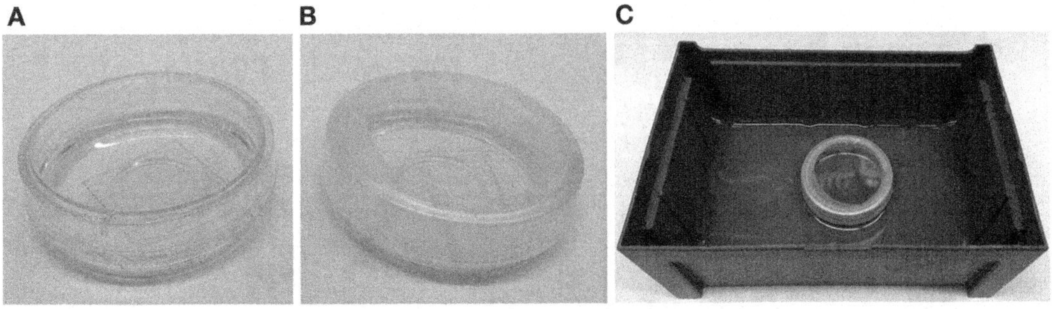

Fig. 2 Preparing cells for temperature block. (**a**) Hela cells in 35 mm glass bottom dish after incubation with Cer-TCO; (**b**) Glass bottom dish sealed with parafilm to prevent water from leaking into the dish; (**c**) Sealed glass bottom dish placed in a water bath which can be cooled to 19.5 °C

3. Add 500 μL of the Cer-TCO working solution (2 μM) to the dish and incubate for 5 min at 37 °C.
4. Wash the cells 3× with 2 mL of DMEM (ph−), add 2 mL of DMEM (ph−), and seal the dish with Parafilm™ (*see* Fig. 2).
5. Incubate the dish for 60 min in a water bath held at 19.5 °C (*see* **Notes 10** and **20**; Fig. 2).
6. Wash the cells 3× with 2 mL of DMEM (ph−) and add 500 μL of the SiR-Tz working solution.
7. Seal the dish again with Parafilm™ and incubate at 19.5 °C for 30 min (*see* **Note 10**).
8. Wash the cells 3× with 2 mL of DMEM (ph−) and add 2 mL of DMEM (ph−) (*see* **Note 21**).

3.3 Imaging Cells by 3D Confocal Microscopy

1. Place the dish on the stage of the microscope (*see* **Note 22**).
2. Find cells using the bright field mode of the microscope.
3. Turn on the 640 nm laser.
4. Adjust the exposure time to <300 ms and increase the camera sensitivity settings in order to decrease the laser intensity as low as possible while still obtaining an image with a good signal to noise ratio.
5. Find cells to image and adjust settings from **step 4** if necessary.
6. Set the z-slice distance to 0.5 μm and define the upper and lower z-boundary for imaging in the microscope settings (*see* **Note 23**).
7. Start imaging at the maximum rate (*see* **Note 24**).

3.4 Imaging Cells by STED Microscopy

1. Place the dish on the stage of the microscope (*see* **Note 22**).
2. Find cells using the eyepiece of the microscope.
3. Turn on the 640 nm laser.
4. Find cells to image.

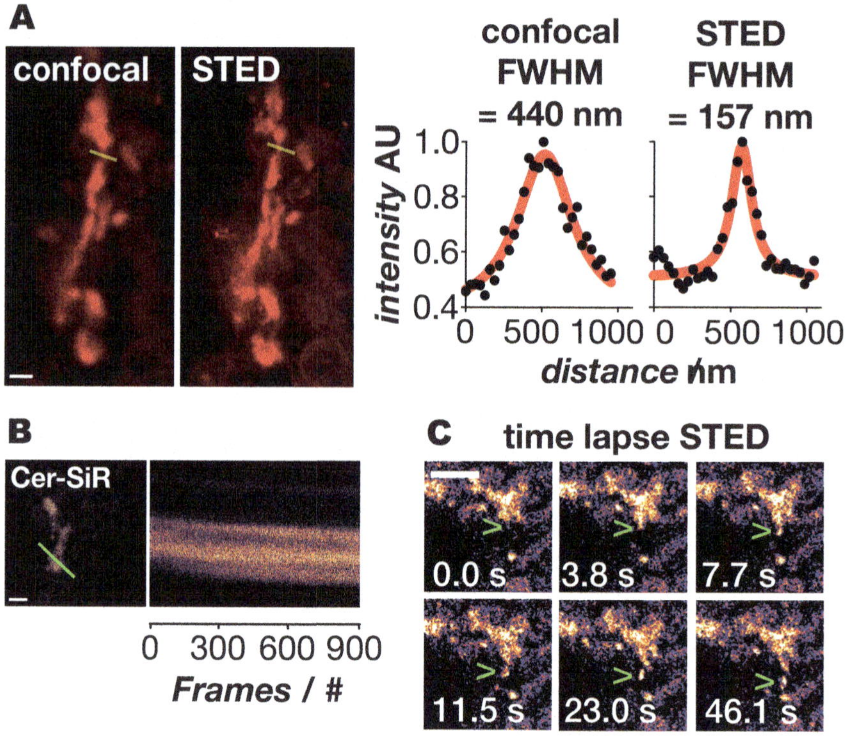

Fig. 3 Imaging of the Golgi with Cer-SiR. (**a**) Comparison of images obtained in confocal (*left*) and STED (*right*) mode; scale bar: 800 nm. The line profiles through the Golgi (*yellow*) illustrate the improvement in resolution. (**b**) Kymograph along the *yellow lines* illustrating the photostability of the probe; scale bar: 2 μm. (**c**) Time-lapse STED images of vesicles budding and exiting the Golgi; scale bar: 1 μm

5. Set field of view to 512 × 512 pixels with a pixel size of 30–40 nm.
6. Image with STED laser set to 755 nm with 60–120 accumulations per line (*see* **Note 25**, Fig. 3).

4 Notes

1. Alternatively, separation can be achieved with an automated chromatography system (such as a Teledyne Isco CombiFlash Rf with columns prepacked with RediSep Rf Silica (40–60 μm) or RediSep Rf Gold Silica (20–40 μm, spherical)). Use of the CombiFlash allows for better gradient control and better separation.
2. Rotary evaporation of DMF from the reaction mixture requires careful heating with a heat gun or use of a high vacuum pump to assist the rotary evaporator. Alternatively, a small-volume evaporation system can be used (e.g., Biotage V-10 evaporator).

3. The $KMnO_4$ solution should be stored in the dark. To minimize light-induced decomposition and contamination with silica particles from previously stained plates, a jar should be filled with only the amount of solution required to develop the TLC plates and the remainder of the solution should be stored separately.
4. The order of the addition is important. Dissolving the lipid in Pluronic F127 prevents the lipid from aggregating.
5. This solution should be prepared right before its use to avoid degradation.
6. Aliquots of the 2 mM stock solutions of Cer-TCO and SiR-Tz in DMSO can be stored at −20 °C for up to 9 and 18 months, respectively.
7. Stock solutions of SiR-Tz in DMSO are blue in its frozen from and become colorless to slightly pink when thawed.
8. Newly synthesized TCO-Cer should be stored as solid under argon/nitrogen at −20 °C, as the solid form is more stable than the stock solution. After more than 12 months of storage, Cer-TCO purity should be evaluated by both TLC **and** NMR. If impurities are detected, repeat purification **steps 11–13** (Subheading 3.1.2).
9. Pluronic F-127 dissolves very slowly. The dissolution rate can be increased by shaking the mixture in a thermoshaker at 37 °C for 2 h. The solution must be stored at RT (not 4 °C) to avoid precipitation.
10. Alternatively, the dish can also be incubated at 19.5 °C in an incubator with cooling capacities. The dish temperature must remain <20 °C to ensure that Cer-TCO accumulates in the Golgi. Variations of 1–2 °C below 19.5 °C are acceptable.
11. Alternatively, the reaction can be carried out at half the scale.
12. After the addition of $^{i}Pr_2NEt$ the reaction mixture might start to turn yellowish to deep yellow after 3 days.
13. The reaction mixture will not fully dissolve, even after 3 days.
14. If two spots are evident by TLC then the material should be chromatographed again using a shallower gradient to obtain pure material. Alternatively, the impure product can be used in the next synthetic step as the impurity does not interfere and can be removed during the workup of the next reaction.
15. Examples including images of the ^{1}H and ^{13}C NMR spectra of TCO-C6 and Cer-TCO can be found in the supporting information of reference [31].
16. The reaction can be stirred for up to 24 h without negatively impacting purity or yield.

17. The aqueous layer of the final wash with sat. $NaHCO_3$ (aq) should be (almost) colorless. If it is not, then continue to wash with sat. $NaHCO_3$ (aq) until the aqueous layer is clear.
18. If multiple yellow spots are visible on the TLC after staining with $KMnO_4$ the chromatography step should be repeated with a shallower gradient.
19. The confluency of the cells affects the labeling results. Ideal results are obtained with cells at 60–85% confluency.
20. The incubation time can be extended up to 2.5 h and will lead to brighter labeling of the Golgi. However, the morphology of the Golgi will become more compact over an extended incubation time.
21. Fixing the cells with 10% PFA is possible after three additional washing steps with PBS. However, the cells should still be imaged within one day since Golgi signal will blur over longer storage time, potentially as the probe is non- or weakly fixable.
22. For optimal cell health an incubator chamber around the microscope stage is useful.
23. Imaging of the whole Golgi usually requires 18–25 z-slices.
24. If the experiment requires imaging over a prolonged time (30 min and more), slower imaging speeds are recommended to minimize bleaching and phototoxicity.
25. Note that the depletion wavelength differs from the depletion wavelength commonly used to deplete the fluorescence of SiR attached to proteins.

Acknowledgments

This study was supported by the Wellcome Trust Foundation and by the National Institutes of Health (GM83257 to A.S.). R. S.E. was supported by an Advanced Postdoc. Mobility fellowship from the Swiss National Science Foundation.

References

1. Hell SW (2007) Far-field optical nanoscopy. Science 316:1153–1158
2. Huang B, Bates M, Zhuang XW (2009) Super-resolution fluorescence microscopy. Annu Rev Biochem 78:993–1016
3. Toomre D, Bewersdorf J (2010) A new wave of cellular imaging. Annu Rev Cell Dev Biol 26:285–314
4. Van De Linde S, Heilemann M, Sauer M (2012) Live-cell super-resolution imaging with synthetic fluorophores. Annu Rev Phys Chem 63:519–540
5. Fornasiero EF, Opazo F (2015) Super-resolution imaging for cell biologists. BioEssays 37:436–451
6. Uno SN, Tiwari DK, Kamiya M et al (2015) A guide to use photocontrollable fluorescent proteins and synthetic smart fluorophores for nanoscopy. Microscopy 64:263–277

7. Chozinski TJ, Gagnon LA, Vaughan JC (2014) Twinkle, twinkle little star: photoswitchable fluorophores for super-resolution imaging. FEBS Lett 588:3603–3612
8. Betzig E, Patterson GH, Sougrat R et al (2006) Imaging intracellular fluorescent proteins at nanometer resolution. Science 313:1642–1645
9. Rust MJ, Bates M, Zhuang XW (2006) Sub-diffraction-limit imaging by stochastic optical reconstruction microscopy (STORM). Nat Methods 3:793–795
10. Hell SW, Wichmann J (1994) Breaking the diffraction resolution limit by stimulated-emission – stimulated-emission-depletion fluorescence microscopy. Opt Lett 19:780–782
11. Vicidomini G, Moneron G, Han KY et al (2011) Sharper low-power STED nanoscopy by time gating. Nat Methods 8:571–U575
12. Westphal V, Rizzoli SO, Lauterbach MA et al (2008) Video-rate far-field optical nanoscopy dissects synaptic vesicle movement. Science 320:246–249
13. Lukinavicius G, Umezawa K, Olivier N et al (2013) A near-infrared fluorophore for live-cell super-resolution microscopy of cellular proteins. Nat Chem 5:132–139
14. Bottanelli F, Kromann EB, Allgeyer ES et al (2016) Two-colour live-cell nanoscale imaging of intracellular targets. Nat Commun 7:10778
15. Bottanelli F, Kilian N, Ernst AM et al (2017) A novel physiological role for ARF1 in the formation of bi-directional tubules from the Golgi. Mol Biol Cell 28(12):1676–1687. doi:10.1091/mbc.E16-12-0863
16. Fernandez-Suarez M, Ting AY (2008) Fluorescent probes for super-resolution imaging in living cells. Nat Rev Mol Cell Biol 9:929–943
17. Heilemann M, Van De Linde S, Mukherjee A et al (2009) Super-resolution imaging with small organic fluorophores. Angew Chem Int Ed 48:6903–6908
18. Hinner MJ, Johnsson K (2010) How to obtain labeled proteins and what to do with them. Curr Opin Biotechnol 21:766–776
19. Nikic I, Plass T, Schraidt O et al (2014) Minimal tags for rapid dual-color live-cell labeling and super-resolution microscopy. Angew Chem Int Ed 53:2245–2249
20. Uttamapinant C, Howe JD, Lang K et al (2015) Genetic code expansion enables live-cell and super-resolution imaging of site-specifically labeled cellular proteins. J Am Chem Soc 137:4602–4605
21. Guidotti G (1972) Membrane proteins. Annu Rev Biochem 41:731–752
22. Quinn P, Griffiths G, Warren G (1984) Density of newly synthesized plasma-membrane proteins in intracellular membranes. 2. Biochemical-studies. J Cell Biol 98:2142–2147
23. Neef AB, Schultz C (2009) Selective fluorescence labeling of lipids in living cells. Angew Chem Int Ed 48:1498–1500
24. Jao CY, Roth M, Welti R et al (2009) Metabolic labeling and direct imaging of choline phospholipids in vivo. Proc Natl Acad Sci U S A 106:15332–15337
25. Yang J, Seckute J, Cole CM et al (2012) Live-cell imaging of cyclopropene tags with fluorogenic tetrazine cycloadditions. Angew Chem Int Ed 51:7476–7479
26. Hang HC, Wilson JP, Charron G (2011) Bioorthogonal chemical reporters for analyzing protein lipidation and lipid trafficking. Acc Chem Res 44:699–708
27. Thiele C, Papan C, Hoelper D et al (2012) tracing fatty acid metabolism by click chemistry. ACS Chem Biol 7:2004–2011
28. Shroff H, Galbraith CG, Galbraith JA et al (2008) Live-cell photoactivated localization microscopy of nanoscale adhesion dynamics. Nat Methods 5:417–423
29. Pagano RE, Martin OC, Kang HC et al (1991) a novel fluorescent ceramide analog for studying membrane traffic in animal-cells – accumulation at the golgi-apparatus results in altered spectral properties of the sphingolipid precursor. J Cell Biol 113:1267–1279
30. Marks DL, Bittman R, Pagano RE (2008) Use of bodipy-labeled sphingolipid and cholesterol analogs to examine membrane microdomains in cells. Histochem Cell Biol 130:819–832
31. Erdmann RS, Takakura H, Thompson AD et al (2014) Super-resolution imaging of the Golgi in live cells with a bioorthogonal ceramide probe. Angew Chem Int Ed 53:10242 10246
32. Van Meer G, Voelker DR, Feigenson GW (2008) Membrane lipids: where they are and how they behave. Nat Rev Mol Cell Biol 9:112–124
33. Van Meer G, De Kroon AIPM (2011) Lipid map of the mammalian cell. J Cell Sci 124:5–8
34. Blackman ML, Royzen M, Fox JM (2008) Tetrazine ligation: fast bioconjugation based on inverse-electron-demand Diels-Alder reactivity. J Am Chem Soc 130:13518–13519
35. Devaraj NK, Hilderbrand S, Upadhyay R et al (2010) Bioorthogonal turn-on probes for imaging small molecules inside living cells. Angew Chem Int Ed 49:2869–2872

36. Karver MR, Weissleder R, Hilderbrand SA (2011) Synthesis and evaluation of a series of 1,2,4,5-tetrazines for bioorthogonal conjugation. Bioconjug Chem 22:2263–2270
37. Yang J, Karver MR, Li WL et al (2012) Metal-catalyzed one-pot synthesis of tetrazines directly from aliphatic nitriles and hydrazine. Angew Chem Int Ed 51:5222–5225
38. Carlson JCT, Meimetis LG, Hilderbrand SA et al (2013) BODIPY-tetrazine derivatives as superbright bioorthogonal turn-on probes. Angew Chem Int Ed 52:6917–6920
39. Dyba M, Keller J, Hell SW (2005) Phase filter enhanced STED-4Pi fluorescence microscopy: theory and experiment. New J Phys 7:134

Chapter 7

Four-Channel Super-Resolution Imaging by 3-D Structured Illumination

Ulrike Engel

Abstract

Multichannel imaging is used as a readout of relative localization of two or more components and is often the first step in investigating functional ensembles in cells. However, the localization volume of diffraction-limited light microscopy (approx. 200 nm by 500 nm) can accommodate hundred of proteins, calling for increased resolution for these types of analyses. Here, we present a protocol for 4-channel imaging using structured illumination microscopy (SIM), which increases resolution by a factor of two. We use adherent, fixed cells to identify the localization of adhesion proteins using immunofluorescence and fluorescent proteins. We discuss how labeling with the necessary brightness is achieved and how data has to be processed for colocalization analysis.

Key words Cytoskeleton, Actin, Focal adhesion, Structured-illumination microscopy, Super-resolution, Colocalization, Chromatic aberration, Fluorescent protein

Abbreviations

3D	Three-dimensional
3D–SIM	Three-dimensional structured illumination microscopy
FA	Focal adhesion
FP	Fluorescent protein
GFP	Green fluorescent protein
PALM	Photo activation localization microscopy
STED	Stimulated emission depletion
STORM	Stochastic optical reconstruction microscopy

1 Introduction

Resolution beyond the diffraction limit of light (approx. 200 nm) opens new routes in cell biological research such as the analysis of macromolecular complexes to determine their composition and

Holger Erfle (ed.), *Super-Resolution Microscopy: Methods and Protocols*, Methods in Molecular Biology, vol. 1663, DOI 10.1007/978-1-4939-7265-4_7,

localization inside cells. Super-resolution microscopy comprises many different technical solutions, each with its own advantages and trade-offs. When choosing a particular method for super-resolution imaging, researchers should carefully evaluate techniques available in order to choose the one that suits their applications best (*see* refs. 1, 2). Here, we present a protocol for structured illumination microscopy (SIM) achieving 4-channel imaging in fixed adherent cells. We use a commercial available instrument (Nikon N-SIM) to assay the differential localization of adhesion proteins in focal adhesions (FAs) and at cell-cell contacts.

Structured illumination microscopy for super-resolution was developed by Mats Gustafsson. In his landmark paper in 2000 [3] he demonstrated doubling of resolution in *xz* and *z*. The system we use achieves a resolution of about 120 nm laterally and 300 nm axially. Using a high numerical aperture (NA) objective is the basis for obtaining super resolution images, as the pattern is focused into the specimen by the objective (Figs. 1 and 2a) and also captures the resulting fluorescence. While many patterns of structural illumination are possible [4], Gustafsson used a sinusoidal stripe pattern (Fig. 2b), which is sequentially moved across the specimen (Fig. 2c). Importantly, axial resolution improvement depends on 3D-structured illumination, generated in the specimen plane [3, 5], *see* Fig. 1a and 2a). To produce a single plane super-resolution image, the pattern is shifted 5× in three different orientations (Fig. 2c).

Multichannel imaging is often the only way to verify that a protein is part of a particular organelle. This is done by correlating the signal representing a protein of interest in one channel with that of "marker" proteins in additional channels. In light microscopy, the specific readout of four channels is easily achieved by careful selection of fluorophores and filters, allowing for localization of four components. Additional channels can be separated using computational techniques based on spectral signatures (spectral unmixing). Higher localization accuracy by super-resolution is of course an additional advantage, however, of the current super-resolution techniques, only structured illumination microscopy offers 4-channel imaging (*see* ref. 2). This restriction in STORM, PALM, and STED is owed to the fact that already a single channel needs more than one spectral input, one for imaging and the other for switching of fluorophores (localization techniques) or emission depletion (STED). While STED and localization techniques allow for higher resolution than SIM, we find that the relative ease of multichannel imaging in SIM is compensating in many applications for the limited gain in resolution. The big field of view and rapid acquisition provides richness in information and the capability to evaluate many samples in a limited time.

The scope of this article is to provide guidance in sample preparation, microscope setting, and postprocessing of images to produce

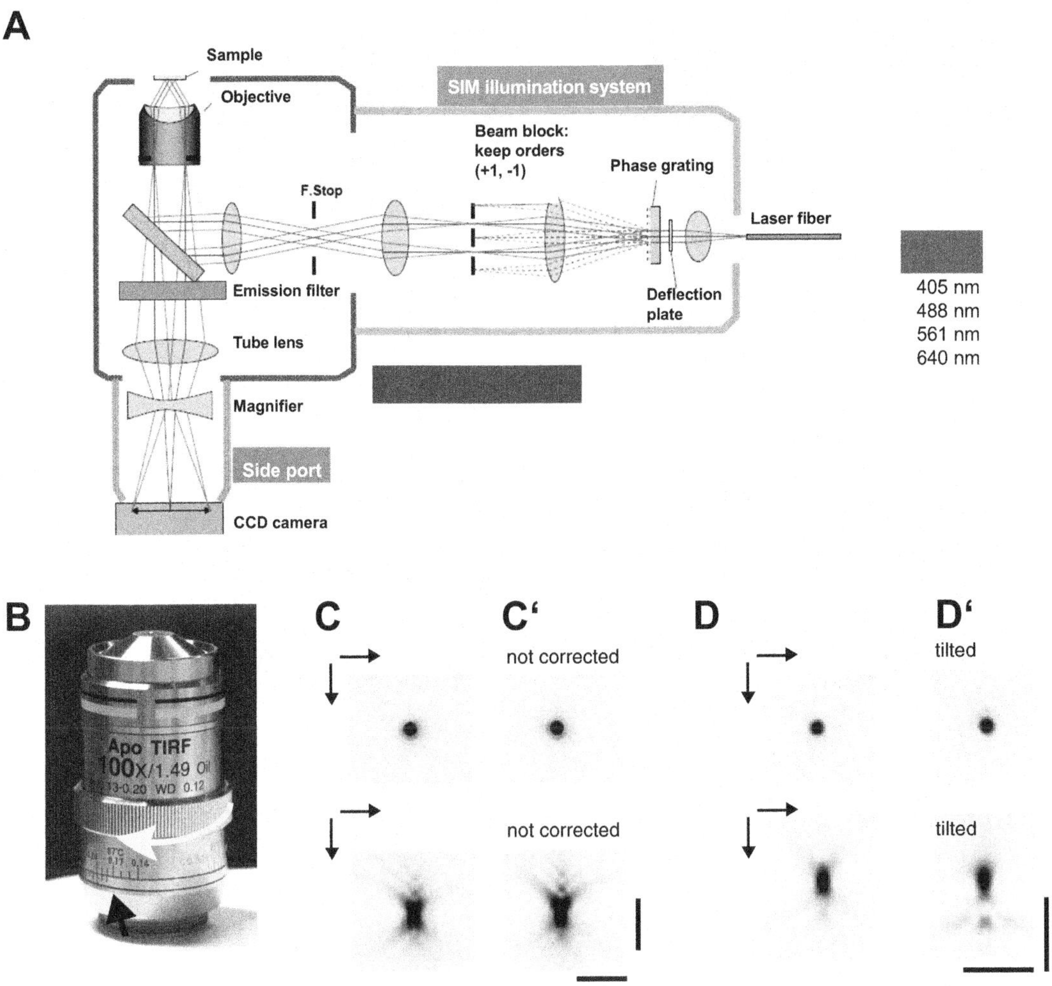

Fig. 1 N-SIM setup with high resolution objective. (**a**) Schematic of pattern generation in the specimen plane for SIM microscopy. For 3D–SIM, the 0-order is used in addition to −1 and +1 laser orders (*see* Fig. 2a). Emission captured as in wide field fluorescence onto a sensitive camera (EM-CCD or sCMOS). (**b**) TIRF Apochromat NA 1.49 with adjustable coverslip correction/temperature correction ring (*arrow*). (**c**) 3D–intensity distribution of bead image with well-adjusted coverslip correction and (**c'**) 3 units off optimum (corresponding to 30 μm coverslip thickness). This difference corresponds to approx. 5 °C temperature shift with constant coverslip thickness. Note asymmetry in axial distribution in c′. (**d–d′**) Distortion might also result from the coverslip not being oriented 90° in respect to the optical axis (in **d** sample is orthogonal, in **d′** is tilted in respect to optical axis)

robust results when analyzing colocalization. This protocol puts a lot of attention to details that are not specific to super-resolution imaging. True, most points would also benefit multichannel light microscopy in general. However, successful super-resolution imaging requires attention to the "nitty-gritty" detail to avoid artifacts and to provide meaningful results [6, 7]. Measurements of beads and known structures should be considered the "optical controls"

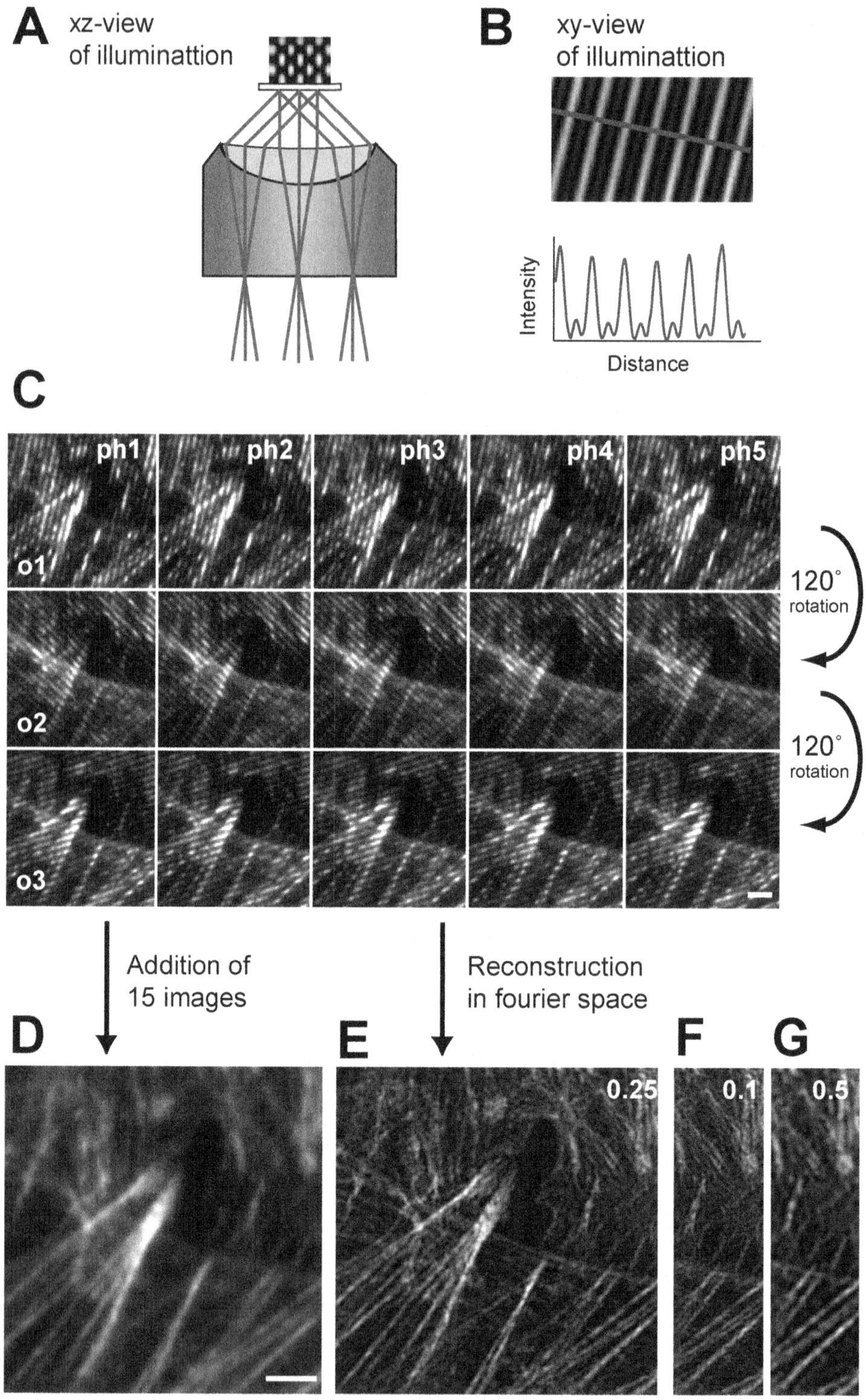

Fig. 2 Illumination pattern and image reconstruction in 3D–SIM. (**a**) Interference of −1, 0, +1 laser orders creates an axial pattern. (**b**) In the plane of focus a stripe pattern with major and minor stripes results. (**c**)

needed to generate data we can trust. Taking these controls into account, SIM imaging produces highly informative and quantitative data about nanoscale position of up to four components in the cell.

2 Materials

2.1 Cell Culture and Transfection

Cell culture and transfection protocols need to be optimized for the cell line of interest. In general flat, well-spread cells produce better results. In Figs. 4 and 5, BS-C1 cells were used.

1. Precision coverglass, thickness No. 1.5H, 170 ± 5 μm (Paul Marienfeld GmbH). Alternatively, use coverslip dishes (*see* **Note 1**).
2. Transfection reagent: Lipofectamine 3000.
3. Plasmid: Zyxin EGFP N1 plasmid [8].

2.2 Primary and Secondary Antibodies

1. Anti-paxillin monoclonal mouse antibody: 1:200.
2. Anti-β-catenin rabbit serum: 1:1000.
3. Examples for anti-mouse and anti-rabbit secondary antibodies, selected for bright signal and high photo-stability are listed in Table 1.

2.3 Mounting and Index Matching

1. ProLong Gold or ProLong Diamond antifade mountant, let it cure 24 h or more after mounting.
2. TDE for index matching (Abberior GmbH, Göttingen, Germany).

2.4 Beads for Calibration and Correction

1. Micro beads: FluoSpheres 0.1 μm, red fluorescent (F-8801) or yellow-green fluorescent (F-8803) (Thermo Fisher Scientific).
2. Multicolor beads: TetraSpeck 0.5 μm (ThermoFisher).
3. Poly-D Lysine for coating.

2.5 Microscope System

Nikon N-SIM equipped with 405 nm, 488, 561, and 640 nm laser on Nikon Ti-E using a (SR) TIRF Apochromat 100× NA 1.49 objective.

Emission filters (band pass):

Blue: 470/50 nm.

Green: 520/45 nm.

Fig. 2 (continued) 3D–SIM requires five phases (ph1–ph5) for each of three orientations (o1–o3). (**d**) Addition of either five phases, or all 15 raw images results in a wide field image. (**e**) Reconstruction results in an image with increased resolution. In (**e**), the parameter which controls suppression of out of plane artifacts (3D apodization) was set to 0.25. In (**f**, **g**), this parameter was varied as indicated, resulting in artifacts in (**f**) and loss of resolution in (**g**). Note that instead of full field of view, only a crop is shown. Scale bar 2 μm

Table 1
Choices made for Figs. 4 and 5 are highlighted by dark gray shading

Channel 1–4		Excitation 405 nm	Excitation 488 nm	Excitation 561 nm	Excitation 640 nm
1.	Fluorescent proteins	–	EGFP, TagGFP, mneongreen, mclover	mCherry TagRFP, mscarlet, tandem tomato	–
2.	Detection of rabbit IgG	Biotium CF405 goat anti rabbit	Alexa Fluor® 488 goat anti-rabbit IgG	Atto 550 goat anti-rabbit IgG, Alexa Fluor® 568 goat anti-rabbit IgG	Atto 655 goat anti-Rabbit IgG, Biotium CF640 goat anti-rabbit IgG
3.	Detetion of mouse IgG	Brilliant Violet 421 anti-mouse	Alexa Fluor® 488 goat anti mouse IgG, goat anti mouse DyLight®488	Alexa Fluor® 568 goat anti-mouse IgG, Anti-mouse (Cy3)	Atto 647 N–goat anti-Mouse-IgG, Biotium CF640 goat anti-mouse IgG
4.	Direct labeling	4′,6-Diamidino-2-phenylindole dihydrochloride (DAPI)		Phalloidin-Atto 565	Phalloidin-Atto 655

An additional combination is indicated by light grey shading

Red: 610/60 nm.

Far-red 700/70 nm.

Camera: Andor iXon EM-CCD camera.

3 Methods

3.1 Choice of Immuno-Labeling and Direct Stains

The quality of the staining is extremely crucial for the outcome of the SIM reconstructions (*see* also refs. 7, 9). The considerations in the detection are all aimed at high signal, while preserving specificity. When imaging up to four channels, spectral overlap and cross-reactivity of antibodies have to be optimized in addition to specificity and brightness of the individual markers, be it fluorescent proteins, immunofluorescence or direct stain. Table 1 shows possible and tested fluorophore combinations, choose one label per column (channel) and one per row. This is by no means a comprehensive list of possibilities, and guidelines for modification are given below.

1. For Figs. 4 and 5, EGFP-zyxin-expressing cells were stained with rabbit anti-β-catenin antibody detected with sheep anti-rabbit CF405, a mouse anti-paxillin antibody detected by anti-

mouse CF640 and a phalloidin Atto 565 conjugate (*see* Table 1, highlighted in dark gray). We used cells fixed with 4% paraformaldehyde in PBS with 0.05% glutaraldehyde for 15 min. Procedures for transfection, fixation, blocking, and incubation are cell and antibody specific, different conditions have to be checked beforehand.

2. Spectral separation: As all the channels are acquired sequentially with single band pass filter sets, excitation of two dyes by the same laser line (cross excitation) can be tolerated as long as their emission is separated by the emission filters (2.5). This condition is not strictly true for td-tomato, which we sometimes chose for its exceptional brightness. It shows a very weak signal in the GFP channel, which can be balanced by combining a weakly expressing td-tomato with a strong EGFP expression or an immunostaining in the green channel. Orange fluorophores occupy both the green and the red channel, and are therefore not useful for 4-channel experiments. For 3-channel experiments they can be used in combination with a far-red label omitting the red channel. Yellow FPs (YFP, venus etc.) can be used instead of EGFP in most cases, but level of contamination of the red channel needs to be checked.
3. Fluorescent proteins: Proteins can be detected by expression of fusion proteins with either green or red fluorescent proteins exited by 488 nm or 561 nm, respectively. The advantage of labeling with fluorescent proteins (FPs) is low unspecific background as long as expression is kept low. Overexpression in wild-type background might lead to a large cytoplasmic pool yielding high background, due to binding sites occupied by the endogenous protein. To preserve specificity, expression should be kept near the physiological levels of the protein of interest. Therefore, FPs of high brightness and high photostability need to be chosen for tagging. Alternatively, FP signal can be enhanced by immunodetection of FPs (*see* **Note 2**). Despite the wide range of colors, bright FPs are restricted to the green to red emission range [10]. In addition to being less bright, cyan fluorescent proteins show spectral overlap with green fluorophores. To avoid multimerization due to FP aggregation, monomeric, aggregation reduced species are preferred [11].
4. Immunodetection: In addition to expressed fusion proteins, choose primary antibodies of different species (*see* **Note 3**), e.g., a mouse and a rabbit antibody. While cross-reactions of secondary antibodies limit the number of primary antibodies, primary antibodies conjugated to a fluorophore can only be used for abundant antigens (*see* **Note 4**). The choice of secondary antibodies depends on: the available lasers, e.g., 405, 488, 561, and 640 laser line, and the fluorescent proteins present. If a green fluorescent and red fluorescent protein are present,

640 nm excitable dyes are the first choice, 405 nm excitable dyes are very poor in performance and bleach rapidly. Therefore, they should only be used for primary antibodies that produce ample labeling (e.g., polyclonal antibodies or serum, *see* Table 1).

5. Additional labeling with direct small molecules that are fluorescent by themselves (e.g., DAPI) or coupled to organic fluorophores (e.g., fluorescent phalloidin) can be used. Labeling of (repetitive) biopolymers such as chromatin or actin produces high signal.

3.2 Sample Mounting and Test Specimens

For imaging of adhesions, cells can be imaged in coverslip dishes covered with PBS after staining (*see* **Note 1**). For imaging structures more than 3 μm away from the coverslip, we recommend index matching in TDE (use according to the manufacturer, *see* **Note 5**) to avoid spherical aberration [12, 13]. For cells grown on coverslips that have to be mounted on slides, mounting in Prolong Gold or ProLong Diamond antifade hardening medium produces good results (*see* **Note 6**).

Bead samples are mounted on poly-lysine coated coverslips:

1. Add 200 μl of 0.5 mg/ml poly-D-lysine to coverslips for 30 min and wash 2× with water, let it dry.
2. Add a drop of approx. 200 μl of a 1:10,000 dilution of Fluospheres or a 1:200 of a Tetraspeck solution. Incubate for 30 min.
3. Remove bead solution and mount in hardening medium (ProLong Gold antifade).

3.3 Image Acquisition on N-SIM Microscope

1. Test for spherical aberration in the system by taking a z-stack of beads mounted on same coverslips/coverslip dishes than used in the experiment. Create a yz or xz slice view (orthogonal view) to visualize asymmetry along the z-axis in the image of the bead. When creating the views, the cursor needs to be set in the center of the bead in the xy-view. Then adjust the objective correction ring (Fig. 1b, white arrow) if necessary, and repeat the procedure until the z-slice of the bead image shows a symmetric intensity distribution (Fig. 1c compare to c′, lower panel, *see* **Note 7**). The in-focus xy image of the bead is much less affected by the coverslip correction, and cannot be used to assay the quality of the setting (Fig. 1c compare to c′, upper panel).
2. Mount your sample, taking great care that the coverslip is level (exactly perpendicular to the optical axis). If the sample is tilted, this will result in distortion of pattern projection and imaging (Fig. 1d' compared to Fig. 1d).
3. Use epifluorescence illumination provided by mercury or metal halide lamp to find a cell of interest by eye. Focus the sample on the camera.

4. View sample in 3D–SIM mode where the pattern is moved sinusoidal across the specimen (Fig. 2c).
5. Optimize (a) exposure time, (b) EM-gain of camera, and (c) laser intensity for each channel. In 3D–SIM every channel requires 15 raw images per plane (Fig. 2c). (a) To avoid any drift during the 4-channel experiment, exposure time should be kept below 1 second. Depending on the temperature control of the room and the thermal stability of the system, long acquisition times may lead to misalignment of z planes or channels. (b) We use the iXon EM-CCD, camera readout time at 14 bit 3 Mhz readout mode. When optimizing raw images, we start with 100 ms exposure time (which equals the readout time). EM gain is kept at or below 80 for well-stained sample, but can be used up to 130 for critical ones, e.g., 405 excited dyes. But generally image quality of constructed images is poor at gain settings >100 (*see* **Note 8**). (c) The laser is then adjusted to result in maximum levels of 5000 counts. If this cannot be reached with a short exposure time, exposure time is increased up to 300 ms.
6. Set a z-stack with 0.12–0.15 μm spacing. Image acquisition captures 15 images per plane and channel.
7. Reconstruct raw images in acquisition software. The reconstruction filters need to be carefully adjusted (*see* Fig. 2, compare e, f, and g) but for a given sample or experimental series they are kept constant as they affect intensity distribution.

3.4 Chromatic Correction of Image Stacks

The objectives used for SIM imaging are specified to be completely color corrected. However, these specifications are based on diffraction-limited light and the resolution achieved in SIM surpasses these specifications. Currently accuracy of color correction is not documented numerically for the objectives or the microscope system; therefore, the best solution is to measure chromatic shifts in xy and z experimentally with multicolor fluorescent beads. As these measurements characterize the chromatic alignment of the very system used (*see* **Note 9**), they can be used to correct image stacks after acquisition (*see* **Note 10**). Here, the procedure is outlined using 0.5 μm Tetraspeck beads with four colors.

1. Beads should be mounted as similar as possible to the sample, or inside the sample itself. For adhesions we use beads adsorbed to the coverslip surface, for other samples we mix the beads in mountant medium to achieve a 3D distribution.
2. Beads are much brighter than samples and laser and exposure time need to be adjusted accordingly. A 3-D stack is acquired with many beads across the field of view in SIM mode and reconstructed (Fig. 3a, *see* **Note 11**).

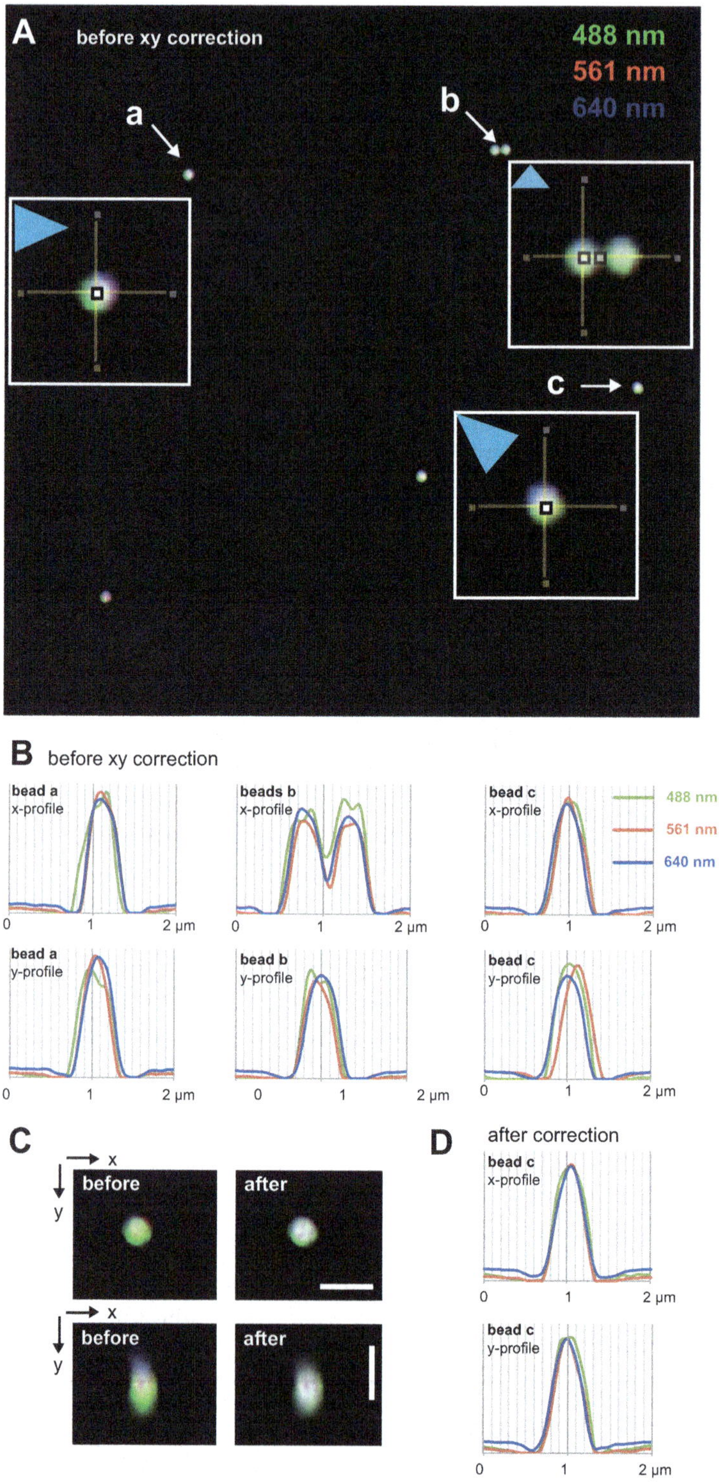

Fig. 3 Chromatic alignment of channels. (**A**) Full field of view (32 × 32 μm) SIM image of tetraspeck beads at five positions. Magnified *inserts* of positions *a–c* indicated by *arrow*. Cross hairs for profiles in **B** are 2 μm in

3. Correction of multichannel data: The correction of sample images is done post-acquisition, where you define the shifts on the bead images, and than apply them to your sample images. The image correction is a geometrical transformation, such that the bead images coincide for all the channels in xyz (Fig. 3c). In the simplest of all cases, a common shift vector for all pixels would be enough, but a tool that can perform sub-pixel shifts (interpolation) is required as the misalignment of channels in xy in your system is not systematic across the field of view, such as shown in Fig. 3a, b.
4. Axial shifts can usually be corrected by simple subpixel shifts. In a z-view (Fig. 3c) they can be measured, and corrected by shifting one image stack against the reference channel (in NIS-Elements this translation of one channel versus the other is called alignment). Despite the high color correction of the TIRF Apochromat, far-red emission (around 700 nm) is not included in the specification of this correction. Therefore, while axial shifts for blue to red are below 100 nm, the far-red channel might be further off (*see* **Note 12**).
5. In xy we resort to registration of the images, which uses magnification and rotation of the channels to map them on each other. This is determined by marking the center of 3 beads of the image (NIS-Elements 4.3 image registration) in respect to the 488 channel. For each channel a separate template is stored, which can than be applied to images. As a test for robustness, the whole procedure should be tested on bead images, which were not used to create the template.

3.5 Displaying Multichannel SIM Images

Multicolor SIM images are very rich in information. For the images showing adhesions in BS-C1 cells, we use four channels to show components of focal adhesions (paxillin, zyxin [8, 14]) and proteins of the adherens junction (β-catenin and zyxin [15, 16]), in respect to the actin cytoskeleton (Fig. 4a). In these not-yet confluent cells, we can distinguish nascent adherens junctions (positive for β-catenin) from well-developed focal adhesions (FAs). The two cells that are in contact have their leading edges intertwined. As only the left cell is expressing GFP-zyxin (Fig. 4c), it is possible to trace its outline (Fig. 4b–f, white line). Where cells overlap, it is hard to

Fig. 3 (continued) length. The direction and approximate magnitude of the lateral shift from green to far red is indicated by *blue arrow heads*, pointing in different directions. Alignment of channels in *xy* is best for position *b*. (**B**) Profiles of normalized intensity for three channels as indicated in *inserts* for position *a–c*. For **A** and **B**, the axial shift was corrected already. (**C**) Lateral and axial view of bead image in position *c* before (*left panels*) and after axial and lateral correction (*right panels*), with templates created from reference bead images. (**C**) Intensity profiles after correction show excellent alignment in *x* and good alignment in *y* for three channels (remaining lateral misalignment <50 nm)

decide which cell protrusion is closer to the substrate. Using a color-coding for the z-position in the phallodin channel (*see* **Note 13**), we can demonstrate, which protrusion of the zyxin-positive cells reaches below the neighboring cell (arrows). A volume of only 300 nm is sufficient to capture the entire height of the cell peripheries. The actin protrusions (orange) are 150–300 nm beneath the stress fibers (blue-green) of the cell on the top (Fig. 4b, arrow heads). The lamellipodial protrusions of both the cells contain a relative diffuse β-catenin signal and in addition elongated patches of localized β-catenin. These patches are emerging cell-cell contacts, which contain β-catenin, zyxin, and f-actin, but not paxillin (arrow head in Fig. 4g).

In an enlarged region of the zyxin-positive cell (Fig. 5), we see both paxillin and zyxin localized to FAs (arrows). Actin stress fibers are anchored to the substrate at FAs. In the color-coded z-projection of f-actin, they stand out as orange structures (<100 nm from substrate, Fig. 4b and 5c, arrows) indicating their juxtaposition to the membrane. Both zyxin and paxillin localize to FAs (Fig. 5d, e, arrows) and the pattern of localization looks qualitatively the same. However, a merge in contrasting colors (magenta and green) demonstrates that zyxin is enriched closer to the actin stress fibers (proximal to the arrows), giving them a kind of Italian flag look (Fig. 5f) This is true for both large FAs (Fig. 5f arrows) and smaller adhesions at the cell periphery (Fig. 5f, insert). Moreover, in actin bundles positive for β-catenin at cell-cell contacts, zyxin is present, but not paxillin (insert Fig. 5, magnified in g–j) as expected for adherens junctions. Note that showing single channels or quantification is absolutely necessary for these conclusions, as the visibility of colocalization in merges depends strongly on colors chosen and on their intensity scaling: the relatively weak zyxin signal at nascent adherens junctions (Fig. 5d arrows) does not show in the merge in Fig. 5b, but is visible in h.

4 Notes

1. If SIM is combined with live imaging we use either MatTek glass-bottom dishes, 35 mm diameter, P35G-1.5-14C (MatTek cooperation, Ashland, MA, USA) or chambered cover glass (Lab-Tek II, Thermo Fisher Scientific).
2. Signal of FPs can be busted by immuno-detection, especially useful are single chain camel antibodies [17]. However, antibodies cannot discriminate between spectral variants of GFP.
3. For multiplexing of antibodies, it is important that secondary antibodies do not cross-react between species of primary antibodies. This is naturally the case between mouse and rabbit,

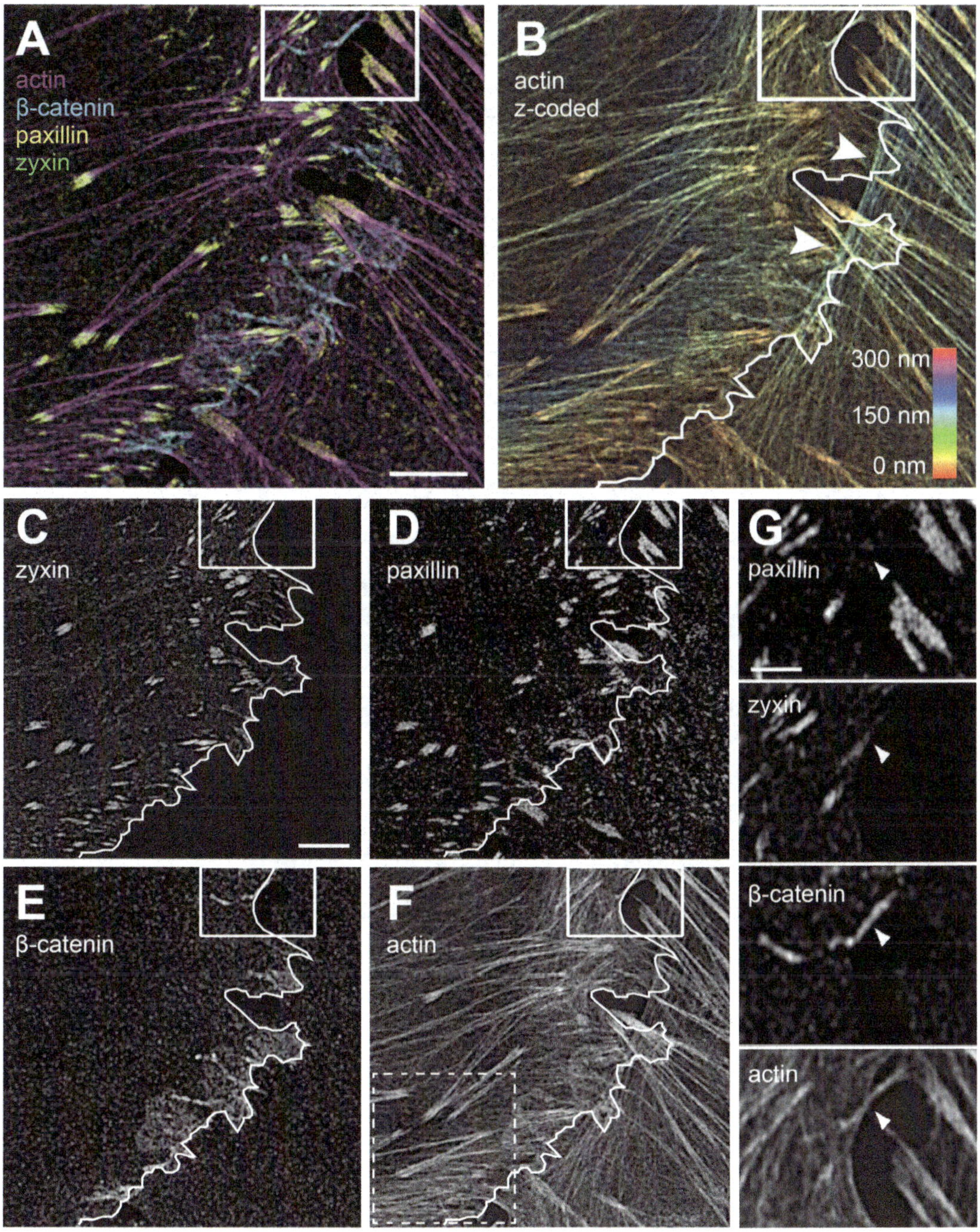

Fig. 4 Example of 4-channel imaging of BS-C1 cells reaching confluence. (**a**) Merge of four channels as indicated in slice closest to substrate. (b) F-Actin in color-coded z-projection of three slices spaced 150 nm. (**c–f**) Single channels as indicated in slice closest to substrate. The *white line* highlights the cell border of the left cell. (**g**) *Inserts* with *arrow* indicating cell-cell contact. *Dotted frame* indicates region magnified in Fig. 5. Scale bars: **a–f** 5 μm; **g** 2 μm

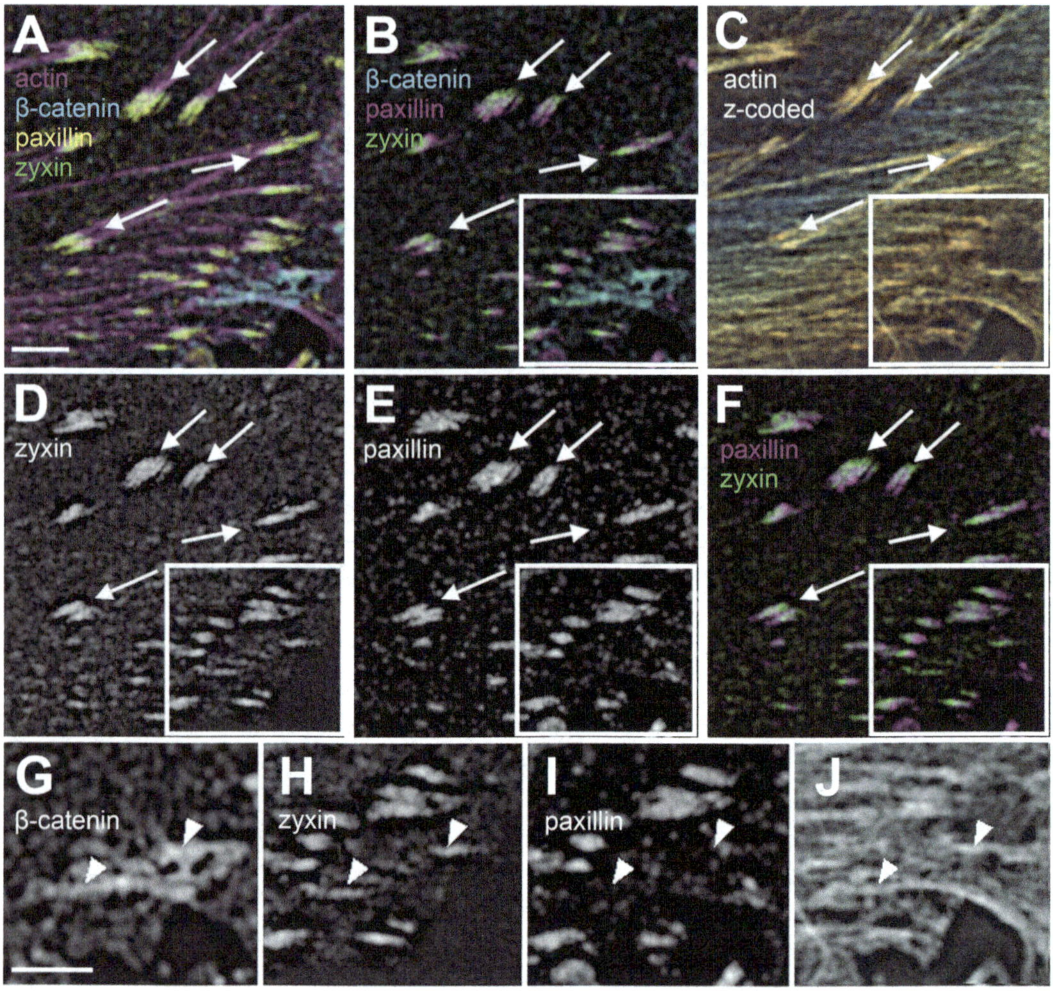

Fig. 5 Magnified region showing components of FAs (*arrows*) and markers for adherens junctions. (**a**) Merge of four channels as indicated. (**b**) Merge of adhesion components. (**c**) Z-projection of f-actin as in Fig. 5b. (**d–f**) Distribution of zyxin and paxillin in FAs, *arrows* point to FAs indicating direction of stress fibers (*see* **c**). (**g–j**) Single channels in regions of cell-cell contact. *Arrow heads* point at actin rich cell-cell contacts positive for zyxin and β-catenin. Scale bar: 2 μm

but for rat and mouse cross-adsorbed secondary antibodies have to be used.

4. Indirect immunofluorescence is a means to amplify signal, as several secondary antibodies can bind to the same primary antibody.
5. If the final concentration of TDE is above 70% fluorescent proteins are quenched and phalloidin staining is lost at even lower concentration [12].

6. Antifade reagents need to be chosen depending on which fluorophores they protect (see the documentation of the manufacturer).
7. When using the same coverslip thickness, the setting of the correction ring is only dependent on the temperature. Adjustment is only required if the temperature at the microscope changes by several degrees.
8. A wide range of different labeling intensities can be captured by SIM. Our experience is that if we have to use a gain >100 to achieve 4000 counts, image reconstruction will be suboptimal, as the high frequencies are dominated by noise. If any variation in sample preparation allows for a better signal, it is worth preparing a new sample.
9. All parts of the microscope in the emission pathway can contribute to chromatic aberration. Our discussion focuses on the objective, because this is the limiting factor in our system. Filters can also contribute, but zero pixel shift filter sets are used to minimize this effect. In systems, like the OMX (Applied Precision) the biggest contribution is the alignment of the different cameras (one per channel).
10. Both optical performance of objectives and the validity of color correction templates should be tested regularly with bead samples to avoid correcting images by nonvalid templates and distorting them further, especially after changes in the setup were made (objectives, camera alignment, immersion oil).
11. Color correction can also be done from stacks in wide field mode (metal halide lamp, 0-order of laser). However, to obtain valuable parameters/templates, the image needs to be resized by a factor 2 in xy (512×512–1024×1024), as the SIM images are resized during reconstruction.
12. Because of the large z discrepancy between visible light and the far-red channel, we use a shift of the z-channel by −200 nm during acquisition.

    Nikon is now pre-selecting TIRF Apochromats with superb performance for super-resolution (SR-objectives), and in objectives we tested, z-shifts were around 100 nm from blue to far-red.
13. In FIJI/ImageJ use temporal color-code plugin within FIJI to project plane 01–03 (corresponding to 0–300 nm) with the look up table "spectrum," in NIS-Elements use 3D-rendering depth-encoded maximum projection mode.

Acknowledgments

I am grateful to Tim Mitchison and Matthias Krause for BS-C1 cell and the zyxin plasmid, respectively and Kees van der Oord and Pablo Hernandez Varas at Nikon BV for discussions on 3D-SIM technology. I want to thank all the users of the Nikon Imaging Center at the University of Heidelberg, who helped in improving our SIM routines, especially Diana Rüthnick for testing FPs and reading the manuscript. This work was supported by the CellNetworks Excellence Cluster and the SFB 873.

References

1. Schermelleh L, Heintzmann R, Leonhardt H (2010) A guide to super-resolution fluorescence microscopy. J Cell Biol 190:165–175
2. Lidke DS, Lidke KA (2012) Advances in high-resolution imaging–techniques for three-dimensional imaging of cellular structures. J Cell Sci 125:2571–2580
3. Gustafsson MG (2000) Surpassing the lateral resolution limit by a factor of two using structured illumination microscopy. J Microsc 198:82–87
4. Heintzmann R (2006) Structured illumination methods. In: Pawley JB (ed) Handbook of biological confocal microscopy. Springer, New York, NY, pp 265–277
5. Gustafsson MG, Shao L, Carlton PM, Wang CJ, Golubovskaya IN, Cande WZ, Agard DA, Sedat JW (2008) Three-dimensional resolution doubling in wide-field fluorescence microscopy by structured illumination. Biophys J 94:4957–4970
6. Burgert A, Letschert S, Doose S, Sauer M (2015) Artifacts in single-molecule localization microscopy. Histochem Cell Biol 144:123–131
7. Smeets D, Neumann J, Schermelleh L (2014) Application of three-dimensional structured illumination microscopy in cell biology – pitfalls and practical considerations. In: Fornasiero E, Rizzoli S (eds) Super-resolution microscopy techniques in the neurosciences, neuromethods, vol 86. Springer, New Yor, NY, pp 167–188
8. Rottner K, Krause M, Gimona M, Small J V, Wehland J (2001) Zyxin is not colocalized with vasodilator-stimulated phosphoprotein (VASP) at lamellipodial tips and exhibits different dynamics to vinculin, paxillin, and VASP in focal adhesions. Mol Biol Cell 12:3103–3113
9. Hubner B, Cremer T, Neumann J (2013) Correlative microscopy of individual cells: sequential application of microscopic systems with increasing resolution to study the nuclear landscape. Methods Mol Biol 1042:299–336
10. Chudakov DM, Matz MV, Lukyanov S, Lukyanov KA (2010) Fluorescent proteins and their applications in imaging living cells and tissues. Physiol Rev 90:1103–1163
11. Cranfill PJ, Sell BR, Baird MA, Allen JR, Lavagnino Z, de Gruiter HM, Kremers GJ, Davidson MW, Ustione A, Piston DW (2016) Quantitative assessment of fluorescent proteins. Nat Methods 13:557–562
12. Staudt T, Lang MC, Medda R, Engelhardt J, Hell SW (2007) 2,2′-thiodiethanol: a new water soluble mounting medium for high resolution optical microscopy. Microsc Res Tech 70:1–9
13. Engel U (2014) Structured illumination superresolution imaging of the cytoskeleton. Methods Cell Biol 123:315–333
14. Case LB, Waterman CM (2015) Integration of actin dynamics and cell adhesion by a three-dimensional, mechanosensitive molecular clutch. Nat Cell Biol 17:955–963
15. Crawford AW, Beckerle MC (1991) Purification and characterization of zyxin, an 82,000-dalton component of adherens junctions. J Biol Chem 266:5847–5853
16. Takeichi M (2014) Dynamic contacts: rearranging adherens junctions to drive epithelial remodelling. Nat Rev Mol Cell Biol 15:397–410
17. Ries J, Kaplan C, Platonova E, Eghlidi H, Ewers H (2012) A simple, versatile method for GFP-based super-resolution microscopy via nanobodies. Nat Methods 9:582–584

Chapter 8

Correlative SIM-STORM Microscopy

O. Burri, T. Laroche, R. Guiet, and A. Seitz

Abstract

The ability to specifically label subcellular structures or even proteins of interest in combination with the ability to look at live specimens turned fluorescence light microscopy into an invaluable tool. However, conventional light microscopy is diffraction limited, which restricts the lateral resolution to around 200 nm laterally and 600–800 nm axially. In 2014, the Nobel Prize in Chemistry was awarded to Eric Betzig, Stefan W. Hell, and William E. Moerner for the development of super-resolved fluorescent microscopy techniques. Since then, it has become evident that imaging techniques that enable the visualization of structures below the diffraction limit are essential for the field of life sciences. However, each one of these approaches has inherent advantages and limitations. Here, we describe an imaging workflow suitable for combining structured illumination microscopy (SIM) with direct stochastic optical reconstruction microscopy (dSTORM) data. This is invaluable, since it allows us to put highly resolved dSTORM data into its cellular context.

Key words Nanoscopy, Structured illumination microscopy, Stochastic optical reconstruction microscopy, Correlative microscopy

1 Introduction

Technologies that enable the precise labeling of genetically encoded proteins such as GFP have been available for over 30 years and have revolutionized life sciences in general and cell biology in particular. Virtually every protein in the genome of an organism can now be fluorescently tagged, thus allowing the visualization of its expression pattern at different time-points of the cell cycle or of development. In order to obtain better mechanistic insight into cellular processes, it is often needed to achieve precise localization of the labeled proteins. Until a few years ago, the diffraction of light was the major limitation to achieve this goal. Due to the so-called diffraction limit, it was impossible to laterally resolve structures smaller than 200 nm. Unfortunately, this is nearly two orders of magnitude larger than the size of a typical protein and 5–10 times larger than important biological structures (e.g., vesicles). In recent years, this limitation has been overcome using

Holger Erfle (ed.), *Super-Resolution Microscopy: Methods and Protocols*, Methods in Molecular Biology, vol. 1663, DOI 10.1007/978-1-4939-7265-4_8, © Springer Science+Business Media LLC 2017

several different approaches, which have been grouped under the name "super-resolution" optical light microscopy [1, 2]. The existing super-resolution approaches can be classified according to their underlying working principle. As an example, Stimulated emission depletion (STED) [3] and Structured illumination microscopy (SIM) [4, 5] are based on patterning the excitation light whereas stochastic optical reconstruction microscopy (STORM) [6] makes use of single molecule localization. In order to overcome the inherent limitations of the individual techniques, combining them seems to be a suitable and recommended approach. In the field of correlative light and electron microscopy this has already turned out to be a very promising and successful methodology [7, 8]. As SIM and dSTORM are based on classical wide-field microscopy setups (in contrast to STED that is using the concept of beam scanning) combining them is a rather straightforward approach and can be easily done in the same instrument [9, 10]. However, as different optical elements need to be used (e.g., a cylindrical lens has to be inserted into the beam path in the case of 3D STORM) a registration procedure is necessary in order to achieve a high quality correlation between the two modalities. We describe a detailed workflow in order to correlate 3D SIM images with 3D dSTORM images (Fig. 1). This was performed on a commercial Nikon N-SIM/N-STORM setup. However, it should be noted that it can be done on any other commercial or home-built system. For the

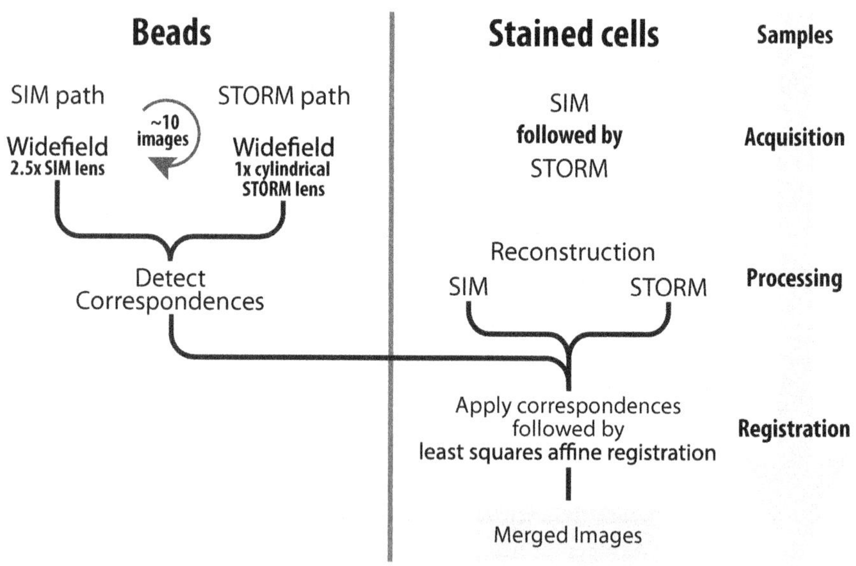

Fig. 1 The experimental workflow comprises the acquisition of the sample itself as well as a series of bead images in widefield mode using the SIM and STORM optical paths. In our case, there is a 2.5× lens when performing SIM acquisitions and a 1.0× astigmatism lens when performing STORM acquisitions. The beads can be used to build a "warping map" of the field of view

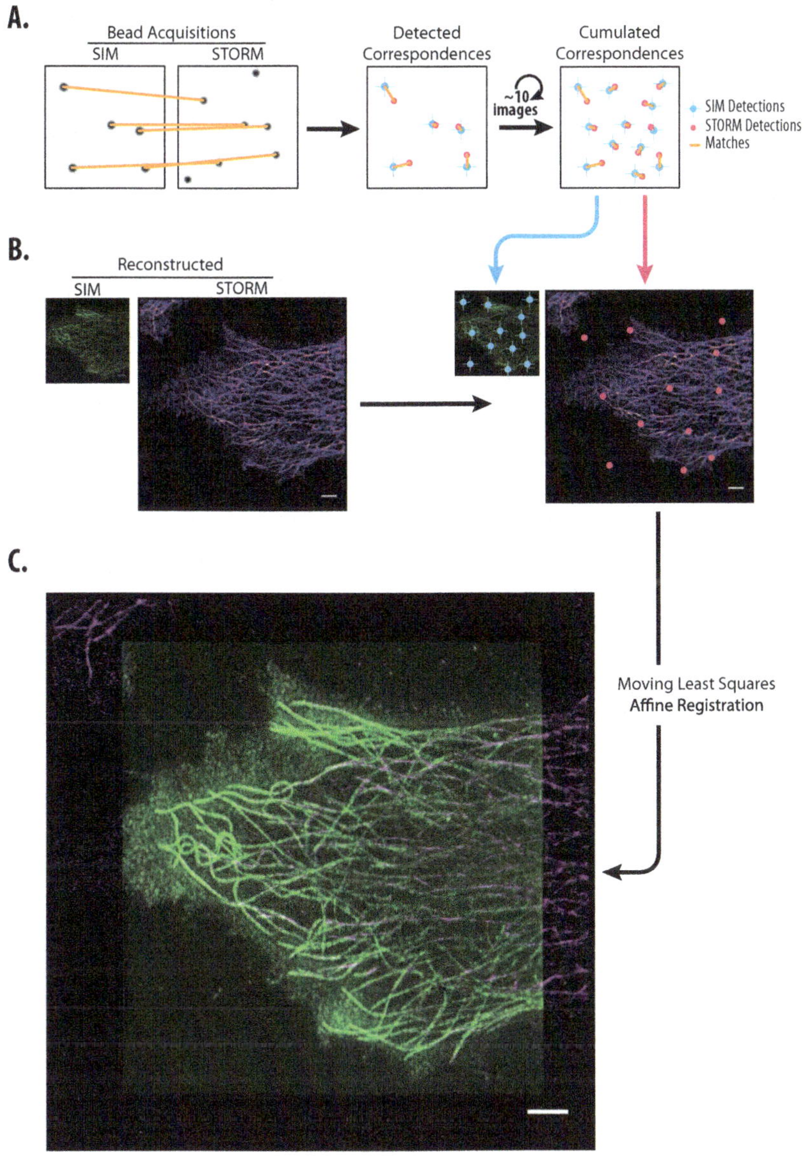

Fig. 2 (**a**) In practice, the SIM-like and STORM-like bead images are analyzed and the detected beads are matched between images. This is repeated until a large enough proportion of the field of view is covered by correspondences. (**b**) To warp the images, we reapply the correspondences on the SIM and STORM images and run the "Moving Least Squares" plugin using an affine registration. The resulting warped image is then overlayed as a new channel, shown in (**c**)

reconstruction of the point localization data [11] as well as for the registration of the images [12], freely available ImageJ plugins are used in order to guarantee maximal freedom in instrument choice. We also included a protocol to prepare test samples that can be used to validate an existing setup (Fig. 2).

2 Materials

1. Coverslips, Diam. 25 mm.
2. Sample holder (Attofluor® cell chamber for microscopy, ThermoFisher A7816).
3. Poly-L-Lysine solution: 0.1% w/v in H_2O.
4. Beads: TetraSpeck™ Microspheres, 0.1 μm.
5. CB buffer: 10 mM MES pH 6.1/138 mM KCl/3 mM MgCl/ 2 mM EDTA.
6. Phosphate-Buffered Saline (PBS Buffer) 137 mM NaCl, 2.7 mM KCl, 10 mM Na_2HPO_4, 2 mM KH_2PO_4, pH 7.4 .
7. Fixation buffer: CB buffer + 3% (v/v) Paraformaldehyde + 0.1% (v/v) Glutaraldehyde.
8. Permeabilization buffer: PBS + 0.1% (w/v) Saponin.
9. Blocking buffer: PBS + 3% (w/v) Bovine Serum Albumin (BSA).
10. Staining solutions (*see* **Note 1**) PBS + 1% (w/v) BSA:

    Tubulin: Monoclonal Anti-β-Tubulin antibody produced in mouse (Sigma-T4026).
11. Mitochondria: Tom20 antibody polyclonal (Santacruz-SC-11415)

    Secondary antibodies: Life technologies, F(ab′)2-Goat anti-mouse/or rabbit conjugated to Alexa555 and Alexa 647.
12. F-Actin: AlexaFluor 488-phalloidin.
13. The STORM buffer contains 50 mM Tris–HCl (pH 8.0), 10 mM NaCl, and 10% (w/v) Glucose. Immediately before imaging, the buffer was complemented with 2-Mercaptoethanol (final concentration 0.14 mol/L), glucose oxidase (final concentration 0.56 mg/ml), and catalase (final concentration 0.03 mg/ml) [9].

3 Methods

All the steps are carried out at room temperature unless otherwise stated.

3.1 Bead Sample

1. Clean coverslips with 70% (v/v) ethanol solution, rinse with deionized water, and let it dry.
2. Prepare a sheet (6 × 4 cm) of plastic paraffin film and put it on the bench.
3. Pipette 50 μl Poly-L-Lysine solution onto the film.
4. Lay the coverslip down on the Poly-L-Lysine drop.

5. Incubate for 5 min.
6. Remove the coverslip from the drop and let it dry.
7. Prepare a new sheet (6 × 4 cm) of plastic paraffin film and put it on the bench.
8. Pipette 50 μl beads solution onto the film (*see* **Note 2**).
9. Incubate for 5 min.
10. Remove the coverslip from the drop and let it dry.
11. Mount coverslip into the sample holder and fill it with PBS.
    Or
    Dispose in a plate, and expose to U.V. to "sterilize" before adding cells.

3.2 Cells Fixation and Staining

1. Plate Hela cells on coverslips 24 h before the staining procedure.
2. Fix cells using fixation buffer.
3. Permeabilize cells by incubation with Permeabilization buffer for 5 min (*see* **Note 3**).
4. Remove the solution, wash three times with PBS.
5. Incubate with blocking buffer for 30 min.
6. Remove the buffer and wash three times with PBS.
7. Prepare a sheet (6 × 4 cm) of plastic paraffin film on the bench.
8. Pipette 50 μl of Primary Antibodies Solution.
9. Incubate for 60 min and wash three times with PBS.
10. Prepare a sheet (6 × 4 cm) of plastic paraffin film and put it on the bench.
11. Pipette 50 μl of secondary antibodies and phalloïdin solution.
12. Incubate for 30 min and wash three times with PBS.
13. Mount coverslip into the sample holder.

3.3 Image Acquisition

1. Preheat the microscope for at least 4 h prior to starting image acquisition.
2. All images were acquired with a NIKON SR Apochromat TIRF 100× 1.49 N.A.
3. Images were taken at 28 °C, and the correction collar was set to 0.22 (*see* **Note 4**).

3.4 SIM/STORM Acquisition of Beads for Distortion Map Estimation

1. Acquire a series of image of beads (10–20 beads per field, at least ten fields) in widefield mode with the beam path for the different imaging modalities (e.g., 2.5× lens for SIM, 1× cylindrical lens for STORM).

3.5 SIM Image

1. Set "Perfect Focus System" (PFS) ON.
2. Define the center of the z-stack with a range of 3 μm corresponding to 51 images with a step size of 120 nm. Beads coated onto the coverslips can be used as fiducial markers and facilitate finding the same focal plane in SIM and dSTORM modes.
3. Acquire multi-channel SIM stack.
4. Switch to N-STORM mode.

3.6 STORM Image

1. Set "Perfect Focus System" (PFS) ON.
2. The focus position should be the same than the SIM. If you observe a drift and that you have beads coated on the coverslips, you can use them as fiducials.
3. Acquire STORM time-stack.

3.7 SIM Reconstruction

1. Reconstruct the acquired SIM images (*see* **Note 4**).

3.8 Astigmatic Lens Calibration (Protocol for Beads Acquisition?)

1. Open bead stack dataset in Fiji (*see* **Note 5**).
2. Start ThunderStorm "Cylindrical Lens Calibration."
3. Under Camera Setup, ensure pixel size, and base-level A/D Counts are correct.
4. In this protocol, pixel size = 157 nm and base-level A/D Counts = 100.
5. Set Z Stage Step and Z Range Limits in relation to your stack.
6. In this protocol, 10 nm Z step and 400 nm Z Range Limit.
7. Under "Save to file," define where to store the resulting calibration.
8. Use Preview to check quality of detections and modify Fitting Radius and Initial Sigma as needed.
9. In this protocol, Fitting Radius = 3 px, and initial sigma of 1.6 px.
10. Click OK, the stack is analyzed and a fitting result is presented.
11. The calibration is saved automatically.

3.9 STORM Reconstruction Using ThunderSTORM

1. Open the raw STORM stack in Fiji.
2. Start ThunderSTORM "Run Analysis."
3. Under Sub-pixel localization of molecules.
4. Choose Method: "PSF: Elliptical Gaussian (3D Astigmatism)."
5. Provide the calibration file path.
6. Set fitting radius and initial sigma as for the "Cylindrical Lens Calibration."
7. Test detection on some frames using "Preview."
8. Use default visualization options.

3.10 SIM/STORM Image Overlap

1. Run BIOP SIM STORM Tools ActionBar.
2. Create "Field Distortion" Map.
3. Under Working Parameters, set Filter radius and noise tolerance for SIM/STORM bead detection.
   (a) In this protocol, Gaussian Filter radius = 3 px.
   (b) Maxima Noise Tolerance = 50.
4. Points Distance Threshold = 5.Perform initial "Rough Calibration."
   (a) Load both a SIM and STORM image in Fiji.
   (b) Click on "Rough Calibration."
   (c) Define at least three points on the SIM image and click OK.
   (d) Define the same number of points, in the same order on the storm image and click OK.
   (e) Rough calibration settings are saved.
5. Find the registration points.
   (a) Load both a SIM and STORM Bead Image.
   (b) Click on "Find Registration Points."
   (c) A ROI set is generated with the detected points on each image in the ImageJ ROI Manager.
   (d) Repeat for at least ten images, so that the "Ordered SIM Points" and "Ordered STORM Points" cover most of the image area.
6. Save the ROI set from the ROI Manager.
   (a) Click "More" > Save and save the ZIP File somewhere.
7. Register a reconstructed SIM and STORM Image.
   (a) Open both images and the ROI set saved in the step above.
   (b) Click on Register SIM-STORM.
   (c) Make sure that the right images are selected.
8. Click OK, the result will be computed, overlaid, and saved.

4 Notes

1. For an optimal "Microtubules STORM imaging" staining, cells should be permeabilized prior to fixation using CB—0.2%NP-40 for 15 s, to "remove" free tubulin from cytoplasm. The protocol described above is the best compromise we found to image Actin, Mitochondria, and Microtubules.
2. The stock bead solution needed to be diluted in order to obtain 10–20 beads per field.

3. Optimal results when imaging Mitochondria were achieved by using saponin as detergent [13].
4. It turned out that a stable temperature is the key for successful image acquisition especially for SIM. Therefore, we decided to actively heat the box which is typically used as wind shield to 28 °C. The temperature was constantly measured throughout the experiments.
5. We used NIS-Elements (NIKON) for the reconstruction of the images. Typically, the following parameters were used: Illumination Modulation Contrast 0.80; High Resolution Noise Suppression 1.00; and Section out of Focus Blur Suppression 0.50. The reconstruction parameters for the Z-stack 3D SIM were: Illumination Modulation Contrast 0.80; and High Resolution Noise Suppression 0.80. An open source alternative for reconstruction is fairSIM (https://github.com/fairSIM) [14].
6. For the calibration of the cylindrical lens in order to determine the z-position of the fluorophore, the following protocol was used: Dark Red Fluospheres F8789 (Invitrogen) were mounted onto a coverslip (*see* Subheading 3.1, **step 1**). A stack of 200 images was acquired starting with 20 frames at the actual focus position followed by a z-stack of 160 images ranging from −400 nm to +400 nm around the focal position (step size 5 nm) and finally 20 images at the focus position.
7. For image acquisition coverslips were mounted in imaging chamber in STORM acquisition Buffer.

Acknowledgments

We are grateful to Joao Firmino and Jens Rietdorf for critically reading the manuscript.

References

1. Eggeling C et al (2015) Lens-based fluorescence nanoscopy. Q Rev Biophys 48:178–243
2. Turkowyd B, Virant D, Endesfelder U (2016) From single molecules to life: microscopy at the nanoscale. Anal Bioanal Chem 408:6885–6911
3. Hell SW (2007) Far-field optical nanoscopy. Science 316:1153–1158
4. Gustafsson MG (2000) Surpassing the lateral resolution limit by a factor of two using structured illumination microscopy. J Microsc 198:82–87
5. Gustafsson MG, Agard DA, Sedat JW (1999) I5M: 3D widefield light microscopy with better than 100 nm axial resolution. J Microsc 195:10–16
6. Huang B et al (2008) Three-dimensional super-resolution imaging by stochastic optical reconstruction microscopy. Science 319:810–813
7. Muller-Reichert T, Verkade P (2014) Preface. Correlative light and electron microscopy II. Methods Cell Biol 124:xvii–xviii
8. Muller-Reichert T, Verkade P (2012) Introduction to correlative light and electron microscopy. Methods Cell Biol 111:xvii–xxix
9. Hamel V et al (2014) Correlative multicolor 3D SIM and STORM microscopy. Biomed Opt Express 5:3326–3336
10. Rossberger SB, Best G, Baddeley D, Heintzmann R, Birk U, Dithmar S, Cremer C (2013)

Combination of structured illumination and single molecule localization microscopy in one setup. J Opt 15. http://iopscience.iop.org/article/10.1088/2040-8978/15/9/094003/meta;jsessionid=50E70D08A947F1419CE808327F9B9174.c4.iopscience.cld.iop.org

11. Ovesny M et al (2014) ThunderSTORM: a comprehensive ImageJ plug-in for PALM and STORM data analysis and super-resolution imaging. Bioinformatics 30:2389–2390

12. Thevenaz P, Ruttimann UE, Unser M (1998) A pyramid approach to subpixel registration based on intensity. IEEE Trans Image Process 7:27–41

13. Kunz WS, Kuznetsov AV, Gellerich FN (1993) Mitochondrial oxidative phosphorylation in saponin-skinned human muscle fibers is stimulated by caffeine. FEBS Lett 323:188–190

14. Muller M et al (2016) Open-source image reconstruction of super-resolution structured illumination microscopy data in ImageJ. Nat Commun 7:10980

Chapter 9

Correlative Super-Resolution Fluorescence Imaging and Atomic Force Microscopy for the Characterization of Biological Samples

Patricia Bondia, Santiago Casado, and Cristina Flors

Abstract

Recent advances in imaging tools have greatly improved our ability to analyze the structure and molecular components of a wide range of biological systems at the nanoscale. High resolution imaging can be performed with a handful of techniques, each of them revealing particular features of the sample. A more comprehensive picture of a biological system can be achieved by combining the information provided by complementary imaging methods. Specifically, the correlation between super-resolution fluorescence imaging and atomic force microscopy (AFM) provides high resolution topography as well as specific chemical information, the latter with a spatial resolution that approaches that of AFM. We present a detailed protocol and discuss the requirements and challenges in terms of sample preparation, instrumentation, and image alignment to combine these two powerful techniques. This hybrid nanoscale imaging tool has the potential to provide robust validation for super-resolution methods as well as new insight into biological samples.

Key words Correlative microscopy, Super-resolution fluorescence microscopy, Single-molecule localization microscopy (SMLM), Atomic force microscopy (AFM)

1 Introduction

Advances in high resolution imaging techniques have led to an increasingly detailed insight into structures and mechanisms in cells and materials. Hybrid approaches that exploit the synergies between two (or more) of these techniques can provide an even more comprehensive understanding of these systems. Herein, we focus on the combination of two remarkably successful imaging techniques for the characterization of biological structures at the nanoscale: super-resolution fluorescence microscopy and AFM. The correlation of standard fluorescence microscopy and AFM has been popular to characterize biological systems due to the complementarities between both the techniques [1, 2]. AFM stands out for its ability to produce high-resolution surface

Holger Erfle (ed.), *Super-Resolution Microscopy: Methods and Protocols*, Methods in Molecular Biology, vol. 1663, DOI 10.1007/978-1-4939-7265-4_9,

topography images as well as to test local mechanical properties or intermolecular interactions. It works well in air or in liquid without special sample treatment, which allows acquiring nanometer scale images of biological systems in near-physiological conditions. In parallel, fluorescence microscopy takes advantage of selective molecular labels, the possibility to image large sample areas in aqueous environments, low perturbation of the system, and fast acquisition. The development of super-resolution techniques has now extended these advantages to a range of about tens of nm, approaching the resolution of AFM.

We describe here a detailed protocol to combine super-resolution fluorescence microscopy based on single molecule localization microscopy (SMLM) coupled in situ with AFM [3]. Specific requirements for the optimal combination of both techniques in terms of sample and substrate preparation, instrumentation, vibration isolation, and image alignment are discussed. We consider a typical experiment in which AFM and SMLM are performed sequentially, rather than simultaneously, which simplifies the procedure and is adequate for most applications. As an example, we describe correlative imaging of stretched DNA labeled with the fluorophore YOYO-1, a small intercalating dye that does not significantly alter the 3D structure of the sample (an important consideration for AFM imaging). We discuss in the Notes section some alternative procedures for other types of samples.

One of the possible applications of correlative AFM and super-resolution microscopy is the identification of artifacts in the latter due to poor labeling, photobleaching, or image reconstruction issues, as described below. The AFM image can be used as "ground truth" to validate and improve novel super-resolution methods [3]. Similar correlative experiments have been used to gain complementary information about biological systems at the nanoscale [4, 5], and their scope could be greatly extended taking advantage of the nanomanipulation capabilities of AFM [6].

2 Materials

2.1 Sample

1. λ-DNA.
2. *YOYO-1 stock solution: 1 μM YOYO-1 in DMSO.* Fluorescent probe YOYO-1, a green DNA intercalating cyanine dye with relatively high brightness ($\Phi_F = 0.52, \varepsilon_{491} = 98{,}900\ M^{-1}\ cm^{-1}$), which has been previously reported for super-resolution microscopy applications in the presence of a "switching buffer" [7] (*see* **Note 1**).
3. Switching buffer: 0.5 mg/mL glucose oxidase, 40 μg/mL catalase, 10% w/v glucose, and 50 mM β-mercaptoethylamine in phosphate buffer saline (PBS) pH 7.4.

2.2 Surface

A thin layer of mica is detached with tape and carefully glued onto a 24 mm diameter #2 glass coverslip with a small drop of UV-curable adhesive (NOA63, Norland). The glue is cured by exposing it to UV light for 20 min. The tape is then peeled off resulting in an even thinner layer of mica on the surface. Afterward, the surface is functionalized by immersion of the coverslips in a solution of polylysine (3.5 mL 0.1% w/v, 2 mL PBS, and 15 mL ultra pure water) for 1 h (*see* **Note 2**) [3].

2.3 Microscopy

We use an AFM (JPK Nanowizard II) mounted on the top of an inverted optical microscope (Nikon Eclipse Ti-U) which has been adapted for SMLM (Fig. 1) [3]. The instrument is placed on an active vibration isolation table (Thorlabs) (*see* **Note 3**). Fluorophore excitation is provided by a 488 nm laser (0.75 kW/cm^2 at the sample, Luxx, Omicron). The laser beam is expanded, collimated, and focused onto the back focal plane of the objective (TIRF, 60×, 1.49 NA, oil immersion, Nikon). Fluorophore emission passes through a dichroic mirror (z488rdc, Chroma Technology) and additional spectral filters (HQ500 LP, HQ530/50, Chroma Technology) and is detected by a EMCCD camera (iXon Ultra 897, Andor Technology). Additional lenses provide a final magnification of 219×, equivalent to a pixel size of about 73 nm.

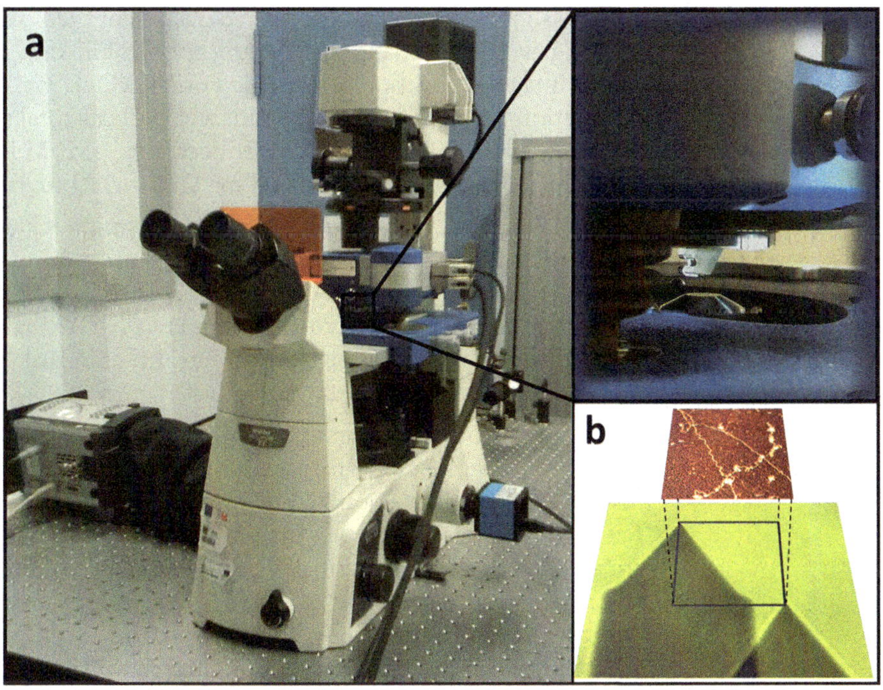

Fig. 1 (**a**) Correlative AFM/SMLM setup. (**b**) Alignment of the images is achieved by optically imaging the AFM tip at several scan positions. Adapted from [3] with permission

For AFM, a Top Visual cantilever (force constant ~50 N/m, NT-MDT) is used, which is specifically designed for using in combination with optical microscopy because it allows easy brightfield illumination and precise optical focusing of the scanning probe. The AFM instrument is equipped with 850 nm laser, which is convenient to avoid unwanted fluorophore excitation (*see* **Note 1**).

3 Methods

1. 1 μg of λ-DNA is mixed with YOYO-1 stock solution to obtain 1:17 dye/bp in PBS. After 15–30 min of incubation, 60–80 μL of the sample is deposited onto a polylysine-coated substrate. The sample is allowed to bind for 20 s and then spin-coated at 5000 rpm. To achieve DNA stretching and remove unbound material, 4 mL of water are pipetted onto the surface during spinning [7].
2. The sample is placed on the microscope and a region of interest is first selected by standard fluorescence microscopy using low excitation power. Then, the objective is slightly retracted and AFM imaging in air is performed using dynamic mode (*see* **Notes 4** and **5**).
3. After AFM imaging, the objective is carefully repositioned for fluorescence imaging. The switching buffer is gently added to avoid sample displacement and detachment. Fluorescence imaging is initiated, and when fluorophore blinking is optimum, image acquisition is started. For YOYO-1, we use and integration time per frame of 100 ms and collect a total number of about 3000–4000 frames [7]. If necessary, SMLM imaging could be performed first followed by AFM (*see* **Note 6**).
4. Data analysis: raw images from both the techniques should be processed and analyzed following the same procedures as with the individual techniques. A wide range of software tools are available either open-source or commercially. We have used *Gwyddion* [8] and *Localizer* [9] to perform standard analyses of AFM and SMLM images, respectively.
5. Image alignment or registration: since fluorescence microscopy and AFM imaging are performed in situ, topography and fluorescence images are localized within the same approximate region of interest (Fig. 1). Image alignment that (partially) compensates distortions and aberrations arising from the individual techniques may be performed with software such as DirectOverlay™ (JPK). This software uses a set of 25 known positions of the cantilever to calibrate the optical image at each point in relation to the AFM lateral piezo coordinates. This allows automatic image registration and distortion correction

[10]. Alternatively, if separate microscopes are used or higher precision is required, other alignment methods can be applied, as discussed below (*see* **Notes 7** and **8**).

6. Correlative image inspection: one possible application of correlative imaging is the comparison of different image reconstruction parameters in SMLM, using the AFM image as "ground truth" [3]. To illustrate this, Fig. 2 shows how having an AFM image as a reference can be helpful to optimize analysis parameters such as threshold sensitivity or point spread function (PSF) width [9]. For example, the two triangles within the green circle in the AFM image (panel a) can be recognized in panels b and c (which only differ in PSF width), although the images are rather sparse (especially in panel b). Lowering the detection threshold of the image reconstruction software to increase localization density (panel d) blurs the smaller triangle. On the other hand, the lower threshold in panel d does make

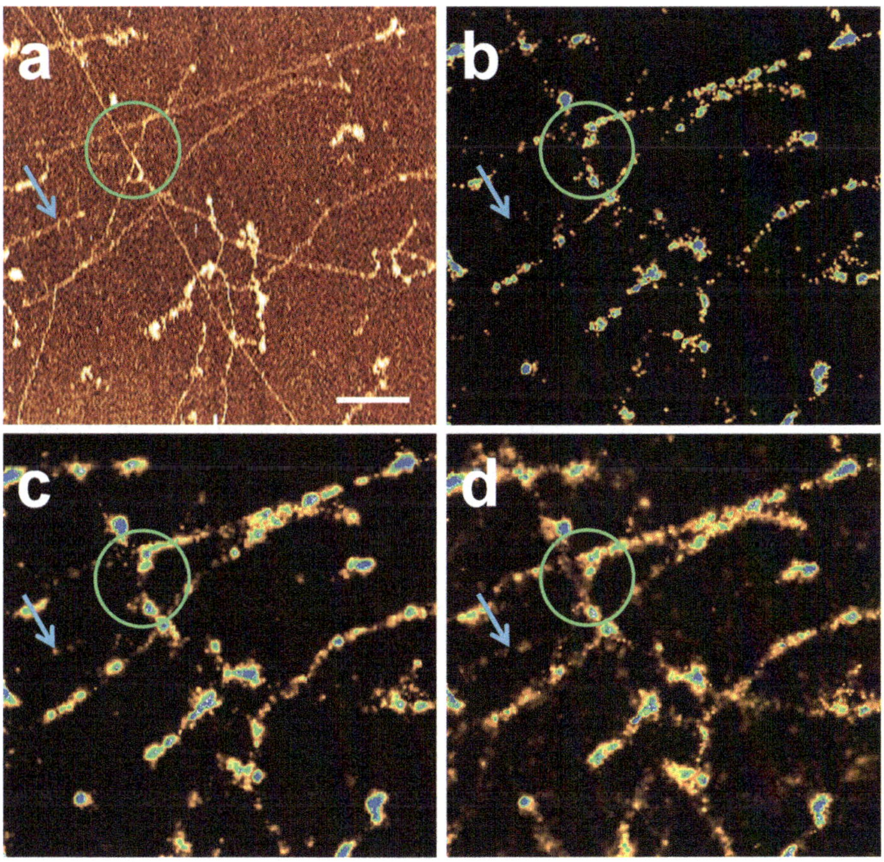

Fig. 2 Correlative microscopy is a useful tool for comparing different parameters for super-resolution image reconstruction of λ-DNA labeled with YOYO-1. (**a**) AFM image; Super-resolution reconstructed image [9] with (**b**) detection sensitivity 25 and PSF fixed to 2.5 pixels; (**c**) detection sensitivity 25 and PSF 3.5 pixels; (**d**) detection sensitivity 10 and PSF 2.5 pixels. Scale bar is 1 μm. Reproduced from [3] with permission

visible some structures that do not appear in panel b, for example that marked by the blue arrow. The AFM image provides verification that it is indeed a DNA fragment and not background due to, e.g., unspecific attachment of YOYO-1 on the surface. Therefore, correlative imaging helps in finding the best balance between spatial resolution and localization density.

4 Notes

1. If the AFM instrument uses a red detection laser, the fluorophore should be carefully selected since it could bleach during AFM imaging. This problem has been described for example for Alexa Fluor 647 [5]. As an alternative, SMLM could be performed before AFM, although it could be detrimental to AFM imaging as discussed in **Note 6**.
2. Choosing a surface can be challenging for correlative microscopy since AFM requires a flat and smooth substrate and SMLM good optical properties for efficient photon detection and precise localization of the fluorophores. Glass or quartz are usually preferred for super-resolution microscopy; however, they have a higher average roughness than mica (0.5 nm for glass [11] or quartz [12] vs 0.05 nm for mica [11]*)*. For small objects such as stretched DNA, which has a height of 2 nm, glass is not flat enough for AFM imaging, and therefore the use mica is recommended. Although mica has poor optical properties (light absorption and birefringence) their impact can be minimized if a very thin layer is used, providing a good compromise for correlative microscopy [3, 11]. For larger structures such as protein fibers [13, 14] or cells [4, 5], glass is suitable for AFM. The use of thicker glass coverslips (#2 instead of the more common #1.5 used in SMLM) reduces fluctuations on the sample induced by the tip scanning. A common problem for AFM imaging in liquid environment is that the sample is easily detached from the surface. Functionalization of the surface can improve adsorption, typically by electrostatic interaction and a wide variety of options can be found in the literature. A common method is to coat the surface with polylysine, whose positive charge is advantageous for adhesion of negatively charged structures such as bare DNA or cells [3, 5]. This is also very helpful for cell-imaging since they are able to adhere and grow on the functionalized glass [15]. Another possibility is to functionalize with cross-linking groups such as 3-aminopropyltriethoxysilane (APTES) which protonates at neutral pH [5]. However, surface functionalization could affect the native structure of the molecules of interest, and it may decrease the flatness and optical properties of the surface.

3. Vibration could be a critical problem as it blurs the AFM image. In addition to an active vibration isolation table, other issues need to be considered when the AFM and the optical microscope are integrated on the same system. For example, it is helpful to slightly retract the objective during AFM scanning to reduce coupled vibration. The EMCCD camera should be off to avoid fan vibration, or if simultaneous measurement is required, a water-cooling system should be used. The lasers used for SMLM should have a heat sink instead of a fan (or the fan should be turned off during AFM imaging). Acoustic noise, air flow (like air conditioning), and large variations in room temperature should be avoided. Vibrations could be minimized by placing the instrument inside an acoustic isolation box, and additional isolation of the AFM by holding it on a mechanical support without contacting the microscope body could be helpful [5].
4. AFM can be performed in dry conditions or in liquid environment. Measurement in air is more convenient since the sample has better adhesion to the surface and the cantilever tuning is simpler. However, to address biological questions, measurements near to physiological environment provide more relevant conclusions since drying of biological samples could modify their structure [16].
5. Different AFM measuring modes are available. Briefly, contact mode is based on measuring changes in topography while the tip of the cantilever remains in contact with the sample. In this mode, the sample experiences high lateral forces that produce damage. For biological samples, softer methods such as dynamic or jumping mode should be applied. In our experiment we use dynamic mode, in which the cantilever tip oscillates near its resonance frequency (generally about 200–400 kHz) lightly tapping the surface [17]. Jumping mode, which achieves a careful control of the forces applied to the sample by bringing the tip in and out of contact, would also be suitable [18]. It is also critical to select a proper cantilever that should be adapted for each measurement depending on the sample, measuring medium and imaging mode. In general terms, softer cantilevers are usually used in contact mode, with spring constants generally lower than 0.5 N/m. In dynamic mode, stiffer cantilevers with a spring constant near 10 N/m are more stable for measuring in air. For jumping mode, soft cantilevers would be recommended to avoid damage of the sample during acquisition [19].
6. The order of imaging can be modified and must be carefully evaluated for each correlative experiment. Generally, it is advisable to perform AFM before SMLM imaging since the switching buffer components (especially enzymes) could interact with

the AFM tip and be adsorbed on the sample surface. Additionally, some structures like actin become damaged by laser exposition during SMLM image acquisition, and therefore AFM must be performed in advance, although with low peak forces to avoid mechanical damage of the sample [5]. If for some reason SMLM has to be performed first, exhaustive washing should be done before AFM. However, this very probably leads to sample disturbance [5, 13, 20].

7. A critical step for correlative microscopy is image alignment or registration. While DirectOverlay™ software allows coarse alignment, which is enough for many applications, other methods are necessary if nanometer precision is required. Fiduciary markers such as fluorescent nanoparticles that give contrast in both topographic and optical images are used as references to determine a correction matrix, and in turn obtain an accurate overlay [11, 14, 21, 22]. Similarly, nano-patterned slides with gold fiduciary features fabricated by electron beam lithography could facilitate the identification of the same imaged region and be used as reference to overlay both images [13, 20, 23].
8. The strategies discussed in **Note 7** are also useful to find the same sample area if separate AFM and fluorescence microscopes are used [5, 13, 20].

Acknowledgments

This work has been financed by the Spanish *Ministerio de Economía y Competitividad* (RyC2011-07637, MAT2012-34487, MAT2015-66605-P) and the European Commission Marie Curie Actions (FP7-PEOPLE-2011-CIG n° 303620). S.C. thanks the European Union structural funds and the *Comunidad de Madrid* MAD2D-CM Program (S2013/MIT-3007). We thank Dr. Peter Dedecker (Katholieke Universiteit Leuven, Belgium) for the Localizer software.

References

1. Oreopoulos J, Yip CM (2008) Combined scanning probe and total internal reflection fluorescence microscopy. Methods 46(1):2–10
2. Caplan J, Niethammer M, Taylor RM 2nd, Czymmek KJ (2011) The power of correlative microscopy: multi-modal, multi-scale, multi-dimensional. Curr Opin Struct Biol 21 (5):686–693
3. Monserrate A, Casado S, Flors C (2014) Correlative atomic force microscopy and localization-based super-resolution microscopy: revealing labelling and image reconstruction artefacts. ChemPhysChem 15 (4):647–650
4. Chacko JV, Zanacchi FC, Diaspro A (2013) Probing cytoskeletal structures by coupling optical superresolution and AFM techniques for a correlative approach. Cytoskeleton 70 (11):729–740
5. Odermatt PD, Shivanandan A, Deschout H, Jankele R, Nievergelt AP, Feletti L, Davidson MW, Radenovic A, Fantner GE (2015) High-resolution correlative microscopy: bridging the gap between single molecule localization

microscopy and atomic force microscopy. Nano Lett 15(8):4896–4904

6. Chacko JV, Canale C, Harke B, Diaspro A (2013) Sub-diffraction nano manipulation using STED AFM. PLoS One 8(6):e66608
7. Flors C, Ravarani CN, Dryden DT (2009) Super-resolution imaging of DNA labelled with intercalating dyes. ChemPhysChem 10 (13):2201–2204
8. Nečas D, Klapetek P (2012) Gwyddion: an open-source software for SPM data analysis. Open Phys 10(1):181–188
9. Dedecker P, Duwé S, Neely RK, Zhang J (2012) Localizer: fast, accurate, open-source, and modular software package for superresolution microscopy. J Biomed Opt 17 (12):126008–126008
10. Harke B, Chacko JV, Haschke H, Canale C, Diaspro A (2012) A novel nanoscopic tool by combining AFM with STED microscopy. Opt Nanoscopy 1(3):10.1186
11. Fronczek D, Quammen C, Wang H, Kisker C, Superfine R, Taylor R, Erie D, Tessmer I (2011) High accuracy FIONA–AFM hybrid imaging. Ultramicroscopy 111(5):350–355
12. Saraji S, Piri M, Goual L (2014) The effects of SO_2 contamination, brine salinity, pressure, and temperature on dynamic contact angles and interfacial tension of supercritical CO_2/brine/quartz systems. Int J Greenh Gas Control 28:147–155
13. Pinotsi D, Schierle GSK, Rees E, Kaminski CF (2013) Localization microscopy for the study of amyloid fibril formation. In: SPIE NanoScience+engineering. International Society for Optics and Photonics, p 88150G, 20 September 2013
14. Bondia P, Jurado R, Casado S, Domínguez-Vera JM, Gálvez N, Flors C (2017) Hybrid nanoscopy of hybrid nanomaterials. Small 13 (7):1603784
15. Zhang WI, Röhse H, Rizzoli SO, Opazo F (2014) Fluorescent in situ hybridization of synaptic proteins imaged with super-resolution STED microscopy. Microsc Res Tech 77 (7):517–527
16. Dorobantu LS, Goss GG, Burrell RE (2012) Atomic force microscopy: a nanoscopic view of microbial cell surfaces. Micron 43 (12):1312–1322
17. Hansma P, Cleveland J, Radmacher M, Walters D, Hillner P, Bezanilla M, Fritz M, Vie D, Hansma H, Prater C (1994) Tapping mode atomic force microscopy in liquids. Appl Phys Lett 64(13):1738–1740
18. De Pablo P, Colchero J, Gomez-Herrero J, Baro A (1998) Jumping mode scanning force microscopy. Appl Phys Lett 73 (22):3300–3302
19. Moreno-Herrero F, Colchero J, Gomez-Herrero J, Baro A (2004) Atomic force microscopy contact, tapping, and jumping modes for imaging biological samples in liquids. Phys Rev E Stat Nonlinear Soft Matter Phys 69 (3):031915
20. Duim WC, Chen B, Frydman J, Moerner W (2011) Sub-diffraction imaging of Huntingtin protein aggregates by fluorescence blink-microscopy and atomic force microscopy. ChemPhysChem 12(13):2387–2390
21. Sanchez H, Kertokalio A, van Rossum-Fikkert S, Kanaar R, Wyman C (2013) Combined optical and topographic imaging reveals different arrangements of human RAD54 with presynaptic and postsynaptic RAD51–DNA filaments. Proc Natl Acad Sci U S A 110 (28):11385–11390
22. Sanchez H, Kanaar R, Wyman C (2010) Molecular recognition of DNA–protein complexes: a straightforward method combining scanning force and fluorescence microscopy. Ultramicroscopy 110(7):844–851
23. Lee JK, Jäckel F, Moerner W, Bao Z (2009) Micrometer-sized DNA–single-fluorophore–DNA supramolecule: synthesis and single-molecule characterization. Small 5 (21):2418–2423

Chapter 10

Quantitative Single-Molecule Localization Microscopy (qSMLM) of Membrane Proteins Based on Kinetic Analysis of Fluorophore Blinking Cycles

Franziska Fricke, Joel Beaudouin, Sebastian Malkusch, Roland Eils, and Mike Heilemann

Abstract

Photoswitchable or photoactivatable fluorophores are the key in single-molecule localization microscopy. Next to providing fluorescence images with subdiffraction spatial resolution, additional information is available from observing single fluorophores over time. This includes the characteristic photophysical phenomenon of "blinking" that is exhibited by single fluorescent proteins or fluorophores and follows well-defined kinetic laws. Analyzing the kinetics of "blinking" allows determining the number of fluorophores in a multi-molecular complex. As such, quantitative information at the molecular level can be extracted, representing a tremendously useful extension of single-molecule super-resolution microscopy. This concept is in particular useful to study homo- and heterooligomeric signaling protein complexes in the plasma membrane of an intact cell with molecular resolution. Here, we provide an experimental framework for deciphering the stoichiometry of membrane proteins on the basis of SMLM and photoswitching statistics.

Key words SMLM, Membrane proteins, Molecular counting, Cluster analysis

1 Introduction

Single molecule localization microscopy (SMLM) belongs to the toolbox of "super-resolution" fluorescence microscopy techniques and provides a spatial resolution below the diffraction limit of light microscopy [1, 2]. Subdiffraction spatial resolution is achieved by (1) employing photoswitchable fluorophores or fluorescent proteins that are stochastically activated (i.e., "switched on" to be fluorescent) at very low densities using light and/or chemical buffers, by (2) determining the position of each detected single fluorophore with high precision of a few nanometers, and by (3) reconstructing "pointillistic" images from the ensemble of positions of all detected fluorophores. SMLM has been used to visualize

Holger Erfle (ed.), *Super-Resolution Microscopy: Methods and Protocols*, Methods in Molecular Biology, vol. 1663, DOI 10.1007/978-1-4939-7265-4_10, © Springer Science+Business Media LLC 2017

cellular features smaller than the diffraction limit (~200 nm), for example the orientation of proteins in the nuclear pore complex [3] or the nanoscale organization of membrane proteins in the vicinity to budding HI viruses [4].

1.1 Quantitative Single-Molecule Localization Microscopy (qSMLM)

Since SMLM initially determines coordinates of each detected fluorophore, the raw data can also be used for quantitative evaluation, including spatial distribution analysis and molecular counting. This unique feature turned SMLM into a super-resolution microscopy method that at the same time can provide quantitative information, e.g., copy numbers of proteins in membranes [5, 6], endosomes [7], and bacteria [8, 9]. In contrast to other single-molecule fluorescence methods that address biomolecular quantification [10], SMLM is not limited to low protein surface densities owing to its intrinsic capability for super-resolution imaging.

Reliable protein quantification using photoswitchable probes is, however, hampered: The blinking behavior of photoswitchable probes can introduce over-counting artifacts [11]. This can be overcome by introducing an empirical global dark time that combines fluorophore emissions closely clustered in time [7, 11]. Under-counting can arise when not every target is fluorescently tagged or blinking cycles of separate fluorophores overlap in time caused by high protein densities, fast photoactivation schemes, or extended blinking of the fluorescent probe. In this context, the maturation time of fluorescent proteins is a key parameter. For example, reported photodetection efficiencies of the photoactivatable protein PAmCherry span a range between 4% in *E. coli* [12], 45% in *Xenopus* oocytes [13], and 77% in BHK21 cells [6].

Today, there are few strategies that truly overcome the technical difficulties of molecular counting with SMLM. The pair-correlation approach [14] is useful for analyzing structures that are larger than the effective resolution of the measurement. By comparing the pair-correlation function to a model, various parameters including cluster size and the number of proteins per cluster can be derived. The single-molecule counting method developed by Lee et al. and Avilov et al. [9, 15] corrects for over-counting of fluorescent proteins, but requires detailed knowledge of photokinetic rates to estimate the optimal global dark time. A recent improvement [16] based on an aggregated Markov model is independent of a temporal threshold and a priori knowledge of photokinetic parameters. The method is sophisticated, but computationally advanced and, so far, mostly synthetic data sets were analyzed. Moreover, it is rarely the case that all photoswitchable probes are photodetected [13], though frequently ignored in the past. Undetectable fractions of fluorescent proteins are largely attributed to incomplete maturation, misfolding, protonation states, or premature photobleaching of the fluorescent probe [17–19].

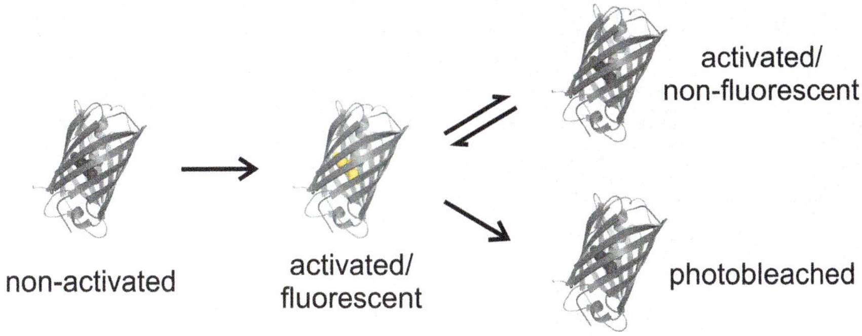

Fig. 1 Four-state kinetic model for the photoactivatable fluorescent protein mEos2. The model accounts for photoactivation, reversible photoswitching ("blinking"), and photobleaching

1.2 Counting Single Molecules by Analyzing Photoswitching Statistics

We recently developed a simple approach for extracting the stoichiometry of fluorophore-labeled membrane proteins by analyzing the fluorescence blinking kinetics [5]. This method directly relates the number of fluorophore "blinking" events of the photoswitchable fluorescent protein mEos2 to the number of underlying molecules, and takes into account the stochastic nature of fluorophore blinking to overcome impediments of dark time thresholding. It builds on a four-state photokinetic model that accounts for photoactivation, reversible photoswitching, and photobleaching (Fig. 1) [9, 20].

Once activated, the fluorescent protein (or any photoswitchable fluorophore that is described by this model) can "switch" between a non-fluorescent ("off") and a fluorescent ("on") state, commonly referred to as "blinking," before irreversible photobleaching occurs. The model predicts that the distribution of the number of times a single fluorescent protein i blinks upon intial photoactivation ($N_{\text{blinks},i}$) is described by a geometric distribution [9].

$$P_p(N_{\text{blinks},i}) = p(1-p)^{N_{\text{blinks},i}}. \tag{1}$$

Here, p is the probability to observe no blinking of the fluorescent probe. We verified this model by imaging single-molecule surfaces of purified mEos2 with SMLM (Fig. 2a and b). We extracted intensity time traces of single mEos2 (Fig. 2c), determined the number of blinking events per mEos2 protein, and generated a distribution of $N_{\text{blinks},i}$ that was approximated by a geometric distribution with $p = 0.30 \pm 0.01$ (Fig. 2d), demonstrating that this model is valid to describe the blinking statistics of mEos2 [5].

If more than one fluorescent protein is present in a protein complex (e.g., in case of oligomerization as found for membrane receptors), the number of emitting molecules N is larger than one. The distribution of blinking events now is described by a negative binomial distribution shifted by $N - 1$,

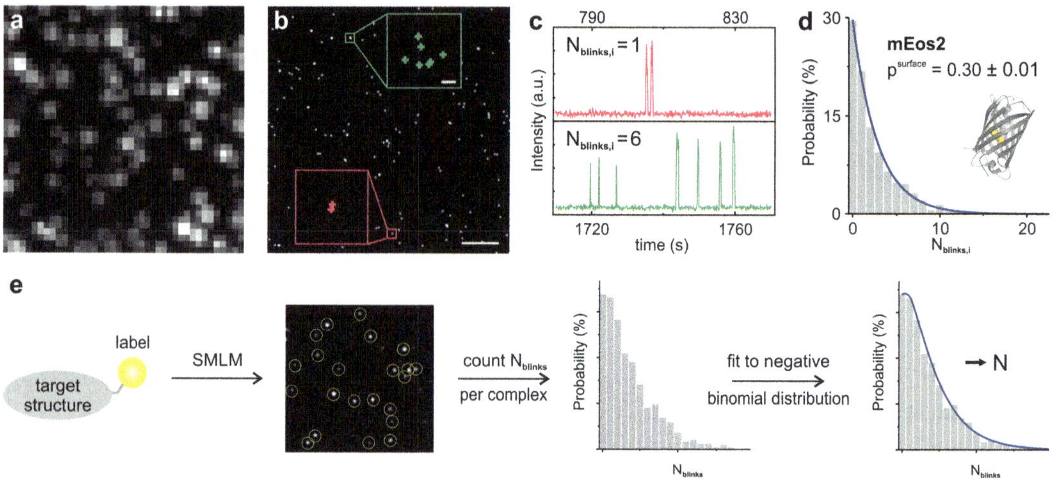

Fig. 2 Experimental analysis of mEos2 blinking kinetics. From an image stack generated by TIRF microscopy (**a**), the positions of single fluorophores were determined and used to reconstruct an SMLM image (**b**). Because of reversible photoswitching, a single mEos2 fluorophore might be detected multiple times (**c**) (the parameter $N_{\text{blinks,i}}$ is defined as the number of re-occurrences of fluorescence emission). The distribution of blinking events for a single fluorophore is described by a geometric distribution and returns a value for the blinking probability, $p = 0.30 \pm 0.01$. Applied to protein complexes of more than one fluorophore (**e**), the blinking distribution is approximated by a negative binomial distribution and returns the average number of mEos2 fluorophores per complex (scale bars 1 μm, 50 nm (*inset*), adapted from [5])

$$P_{N,p}(N_{\text{blinks}}) = \binom{N_{\text{blinks}}}{N-1} p^N (1-p)^{N_{\text{blinks}}-N+1}, \tag{2}$$

which represents a generalization of the geometric distribution (the latter is obtained for $N = 1$) [21]. Again, N_{blinks} is defined as the number of fluorescence blinks/reocurrences upon initial photoactivation of the FP-labeled complex. The shift of the distribution corrects for the number of fluorescence appearances expected from the stoichiometry of the complex. In the actual experiment, individual protein complexes are determined from SMLM data sets to extract the number of blinking events per complex, which relates to N_{blinks}. The distribution of N_{blinks} is then approximated by a negative binomial distribution yielding the number of underlying molecules N as free fit parameter (Fig. 2e). It should be noted that the photodetection efficiency of the fluorescent probe has to be considered for deciphering the true stoichiometry of the complex. For the exact approximation of the blinks distribution, it may be useful to use additional terms that take into account “artificial” lower oligomeric states (Eq. 2), which arise from incomplete photodetection of the FPs [18].

1.3 Quantifying the Stoichiometry of Membrane Protein Complexes

We validated this strategy for protein counting by investigating the stoichiometry of known monomeric, dimeric, and trimeric membrane proteins and revealed the oligomeric states of these in intact cells [5] (Fig. 3). On the basis of these proteins, a photodetectable fraction of mEos2 of ~78% could be determined. Note that a simplified approach that does not account for a shift of the distribution still delivers the correct oligomeric state, yet overestimates the detection efficiency of mEos2 (90%) [5]. These findings demonstrated reliable molecular counting and the capability to distinguish between monomers, dimers, and higher-order oligomers. The method is further capable of reporting if a protein is found in multiple oligomeric states, e.g., as monomer and dimer. The counting strategy is intuitive and easily implemented and puts quantification of small protein complexes in immediate reach of interested users. So far, this approach was only demonstrated for one fluorescent protein, mEos2, and as a single-color application. We now aim to establish multi-color, quantitative SMLM (qSMLM) using photoswitchable/photoactivatable fluorescent proteins and organic fluorophores. This approach will build on a

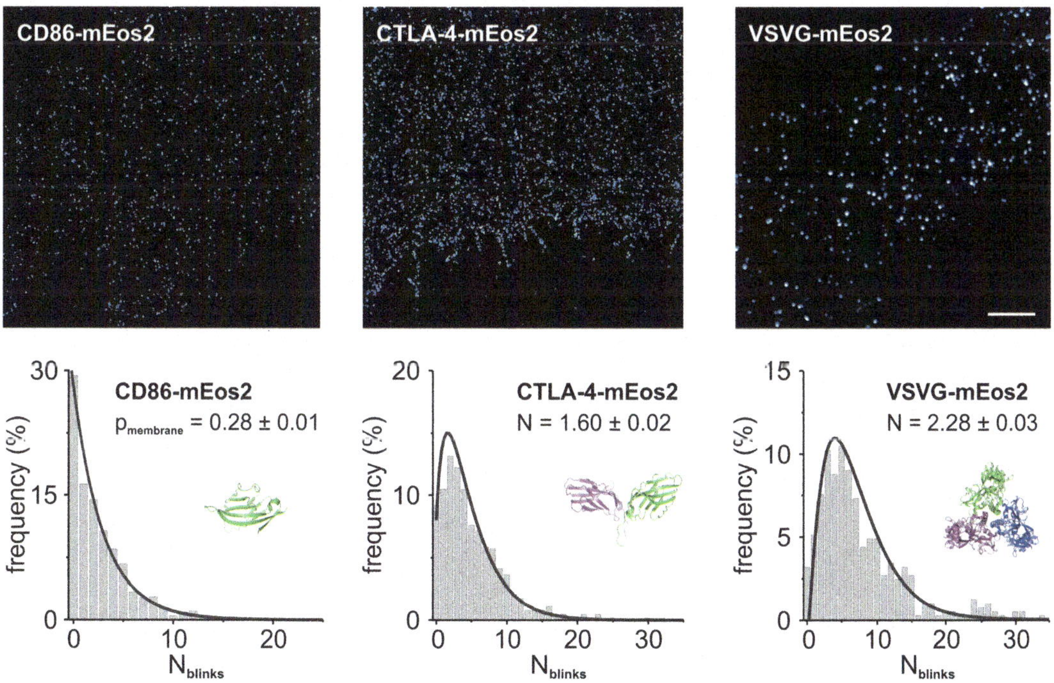

Fig. 3 Applying quantitative SMLM to investigate copy numbers in membrane protein complexes. Monomeric (CD86), dimeric (cytotoxic T-lymphocyte-associated protein 4, CTLA-4), and trimeric (vesicular stomatitis virus glycoprotein, VSVG) proteins conjugated to mEos2 were expressed in Hela cells, imaged with SMLM (*top row*), and the blinking kinetics extracted. Applying a shifted negative binomial distribution as fit function of the blinking distribution revealed the average protein copy number per complex (*bottom row*) (scale bar 2 μm, adapted from [5])

detailed investigation of fluorescence "blinking" properties of fluorescent probes, and will be corroborated with nanostructures of defined stoichiometry for calibration purposes.

1.4 Extracting Single Membrane Protein Complexes from SMLM

For the quantification of membrane proteins, single protein complexes have to be determined from the SMLM data on the basis of segmentation and cluster analysis. A very simple strategy for image-based cluster identification can be realized with the "Analyze Particles" plugin for ImageJ. A more sophisticated analysis builds on the fact that the initial raw data derived in an SMLM experiment are positions of single fluorophores, which can serve for a number of coordinate-based algorithms. Among others, possible segmentation algorithms that turn useful in this context include the "density-based spatial clustering of applications with noise" (DBSCAN) [22]. DBSCAN belongs to the class of hierarchical cluster analyses and categorizes single-molecule localizations by the number of neighbors and their distances to define clusters (Fig. 4) (*see* Subheading 3 for details).

An extension of DBSCAN was introduced with an algorithm termed "ordering points to identify the clustering structure" (OPTICS) [23]. A different approach suitable for this task is Voronoï tessellation [24].

Molecular counting with SMLM is intuitive, easily implemented and does not require advance knowledge of fluorophore photokinetics. Well-defined stoichiometric standards, e.g., protein monomers and dimers, are useful for determining the photodetectable fraction and the blinking probability of the fluorescent probe.

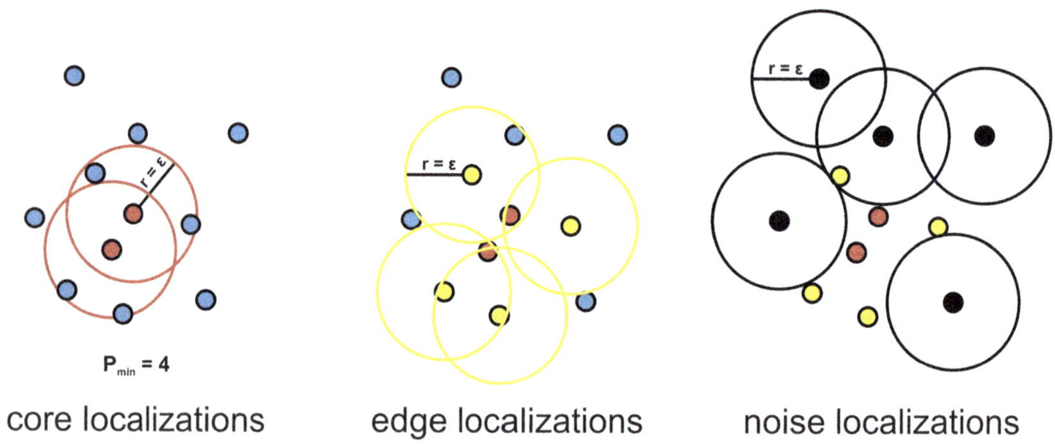

Fig. 4 The principle of hierarchical clustering. Localizations of single molecules are sorted into three categories: core, edge, and noise localizations. Core localizations (red) are surrounded by at least P_{min}-1 neighbors within a distance of ε. Edge localizations (yellow) do not have the sufficient number of neighbors to be a core localization, but are found within a distance ε from another core localization. Noise localizations (black) do not have enough neighbors to be a core localization and are spaced by a larger distance than ε from any core localization

In the context of membrane proteins, a possible selection is the monomeric protein CD86 and the dimeric protein CTLA-4. In this chapter, we provide a framework for SMLM-based molecular counting via photoswitching statistics on the basis of these proteins.

2 Materials

2.1 Reagents and Consumables

1. Hela cells.
2. Full DMEM medium: DMEM, 10% fetal bovine serum, 2 mM L-Glutamin, Penicillin-Streptomycin.
3. Phosphate buffered saline (PBS).
4. Trypsin.
5. Transfection solution: DNA plasmid, X-tremeGENE HP, sheared salmon sperm DNA, Opti-MEM (Gibco) prepared according to the manufacturer's instruction (X-tremeGENE HP, Roche; it is usually appropriate to use 50–150 ng of DNA plasmid plus 450–350 ng salmon sperm DNA adding up to 500 ng DNA in total per well of a 24-well plate).
6. Fibronectin solution: 25 μg/mL fibronectin in PBS.
7. Washing solution: 0.4 M sucrose in PBS.
8. Fixation solution: 4% formaldehyde, 0.1% glutaraldehyde, 0.4 M sucrose in PBS.
9. 24-Well plate.
10. 8-Well chambered microscopy glass.

2.2 Microscope and Equipment

Imaging experiments can essentially be carried out on any widefield microscope operated in total internal reflection fluorescence (TIRF) mode and suitable for SMLM. In our hands, experiments were carried out on a custom-built microscope setup operated in TIRF mode. The following list comprises essential components of the setup.

1. Inverted microscope (e.g., Olympus IX71).
2. High NA ($NA \geq 1.45$) 100× oil immersion TIRF objective, immersion oil.
3. Nosepiece stage for mechanical stabilization (IX2-NPS, Olympus).
4. Appropriate continuous wave lasers for fluorophore photoactivation and excitation (≥50 mW) such as diode lasers or solid state lasers (for mEos2, e.g., Sapphire 568 LP, Sapphire 488 LP, CUBE 405-100C Coherent), continuously variable neutral density filter for photoactivation laser.

5. Spectral filters, e.g., ET 525/50 for unconverted, green form of mEos2, BrightLine HC 590/20 for photoconverted, red form of mEos2 (AHF).
6. Dichroic mirror, e.g., HC Quad 410/504/582/669 (AHF).
7. EMCCD camera (e.g., Andor), camera acquisition software (e.g., Andor, μManager).
8. Single-molecule localization software (e.g., rapi*d*STORM, GraspJ), postprocessing software (Lama, ImageJ, OriginPro 9.1G).

3 Methods

3.1 Sample Preparation

1. Day 1: Plates 50,000 cells per well in a 24-well plate. Place cells in an incubator at 37 °C in 5% CO_2.
2. Day 2: Transfection. Prepare the transfection solution and add it to the wells in a dropwise manner. Gently shake the wells and place cells back in an incubator.
3. Day 4: Glass coating with fibronectin. Add 250 μL fibronectin solution per well to the 8-well chambered glass slide and incubate for 1 h at 37 °C. Remove the solution and place the glass slide for 1 h in laminar flow hood. Coated glass slides can be stored in the incubator until the next day.
4. Day 5: Trypsinate cells and plates 30,000–40,000 cells per well in the 8-well chambered, coated glass slide. Place cells back in an incubator until cell adhesion is restored (3–5 h). Prepare washing solution and fixation solution. Wash cells three times with washing solution and incubate cells for 20 min in fixation solution. Rinse cells three times with PBS. Cells are now ready for microscopy.

3.2 SMLM Imaging

1. Turn on all the components of the microscope. Cool the EMCCD camera to −70 °C and activate the EM gain (200–300). Prepare camera for image sequence acquisition with 100 ms integration time. Select an appropriate saving option for the image sequences. Many localization softwares are compatible with TIFF.
2. Place a drop of fresh immersion oil on the objective. Mount the sample on the microscope stage. Focus on the cells using the brightfield lamp. Check whether cells are visible on the camera chip.
3. Use appropriate laser (low power) and spectral filter to verify that transfection was successful (mEos2: 488 nm laser and ET 525/50). Most fluorescence signal should be localized to the plasma membrane of transfected cells. Focus on the bottom

plasma membrane of a transfected cell and change to SMLM-suitable filters and lasers. For mEos2, this would be the 405 and 568 nm lasers and the HC 590/20 filter. Turn on the 568 nm laser (0.5 kW/cm^2) and slowly increase 405 nm laser illumination (0–10 W/cm^2) by means of the adjustable neutral density filter until single-molecule fluorescence emissions of mEos2 appear. Correctly adjust the TIRF angle. Now, all the parameters are adjusted for SMLM imaging.

4. For quantitative SMLM, select a new cell using the brightfield lamp. This prevents premature photobleaching of fluorescence signal. Start camera acquisition and turn on the excitation laser (mEos2: 568 nm laser, 0.5 kW/cm^2). Slowly increase 405 nm laser power (0–10 W/cm^2) via the adjustable neutral density filter to photoactivate sparse subsets of mEos2. Pay attention that the density of active, fluorescent mEos2 molecules is moderate so that PSFs do not overlap. To ensure that most molecules are imaged, acquisition is maintained until no more mEos2 fluorescence appears. Depending on the protein expression level, image acquisition can last up to 100,000 frames (*see* **Note 1**).

3.3 SMLM Data Analysis

1. Detect and localize single molecules in the acquired image sequence using localization software (e.g., rapi*d*STORM) and appropriate parameters (camera pixel size, intensity threshold, etc.). Group fluorescence emissions that appear in consecutive camera frames and within a spatial distance provided by the localization uncertainty of the fluorophores (*see* **Note 2**). Many localization softwares offer the option for spatio-temporal grouping of single-molecule peaks. To give an example, in our hands the majority of mEos2 fluorescence peaks were localized with a spatial uncertainty <35 nm (determined according to [25]). For mEos2, we grouped peaks appearing in consecutive frames and within a spatial separation of 90 nm using rapi*d*STORM. These peaks were registered as single localizations with averaged spatial positions. The distance threshold was chosen so that 99% of peaks (corresponding to $\approx 2.58\sigma$) with a localization precision of 35 nm were grouped successfully. Exclude peaks that only appear in one frame.

2. Coordinate-based segmentation of SMLM data can be realized using the DBSCAN analysis. DBSCAN requires two input parameters: the distance parameter ε and a minimal number of localizations $P_{\min}$. All single-molecule localizations are grouped into "core" (all points that have at least $P_{\min} - 1$ neighbors within a distance of ε), "edge" (all points that have less than $P_{\min} - 1$ neighbors but at least one core localization within a distance of ε), or "noise" localizations (all other

points) (*see* also Fig. 4). The distance parameter ε is often expressed as multiples of the localization precision [8], whereas the parameter P_{min} needs to be matched with the density of localizations (a lower number may be used for fluorescent proteins with fewer blinking cycles) [8].

3. Obtained data from cluster analysis can be used for molecular counting. The important output is the number of localizations found for single protein complexes/clusters. Load the number of localizations per cluster for a set of identified clusters into OriginPro 9.1G or a similar program of your choice. It is important to realize that the presented counting approach requires the number of fluorescence reappearances after initial photoactivation/-conversion of the fluorescent probe. Hence, localization counts per cluster have to be reduced by one count. Generate the relative frequency distribution (start: 0; bin size: 1) and approximate the obtained distribution by Eq. 2 via the nonlinear curve fitting option of the program to obtain fit parameters p and N. It may be useful to express the binomial coefficient via gamma functions using the following expression: $\binom{n}{k} = \frac{\Gamma(n+1)}{\Gamma(k+1)\,\Gamma(n-k+1)}$. The obtained fit parameter N directly informs about the average number of photodetected subunits of the protein complex (*see* **Note 3–5**).

4 Notes

1. Pay attention to sample drift. When lateral drift is a problem, fiducial markers (e.g., small gold beads) may serve as drift control and allow for drift correction post image acquisition. Axial focus drift may be prevented by employing focus lock systems (e.g., Nikon, MadCityLabs) or a nosepiece stage for mechanical stabilization (e.g., IX2-NPS, Olympus).
2. The analysis of blinking cycles is compatible with almost any localization software that has a function implemented for grouping fluorescence bursts in consecutive camera frames (e.g., Rapi*d*STORM [26], ThunderSTORM [27], GraspJ [28]; all of them freely available). Different spatial segmentation techniques may be useful for grouping localizations into clusters.
3. It is important to evaluate the blinking characteristics and essentially the parameter p (Eqs. 1 and 2) of the fluorescent probe employed. A single-molecule surface of the fluorescent probe as monomeric reference may be helpful in this context. However, the target fusion molecule and microenvironment of the fluorescent probe can influence photokinetic properties so that, e.g., purely monomeric proteins, such as CD86, may

serve as a cellular calibration [29, 30]. Obtained blinks distributions of monomeric probes are approximated by the geometric distribution (Eq. 1) to extract the blink parameter p. The obtained p value may serve as reference for the analysis of unknown protein stoichiometry.

4. The covalent dimer CTLA-4 is a possible candidate for assessing the photodetection efficiency of the fluorescent probe. Upon blinking cycle analysis, the photodetectable fraction s is given by $s = N/2$, where N is the obtained fit parameter (Eq. 2). s serves as an important calibration parameter for deciphering the true stoichiometry complex $N_{true} = N/s$.
5. Although originally designed for probing the stoichiometry of membrane proteins, the concept can be extended to other oligomeric structures, such as mixed populations of monomers and dimers.

References

1. Klein T, Proppert S, Sauer M (2014) Eight years of single-molecule localization microscopy. Histochem Cell Biol 141(6):561–575
2. Furstenberg A, Heilemann M (2013) Single-molecule localization microscopy-near-molecular spatial resolution in light microscopy with photoswitchable fluorophores. Phys Chem Chem Phys 15(36):14919–14930
3. Szymborska A et al (2013) Nuclear pore scaffold structure analyzed by super-resolution microscopy and particle averaging. Science 341(6146):655–658
4. Muranyi W et al (2013) Super-resolution microscopy reveals specific recruitment of HIV-1 envelope proteins to viral assembly sites dependent on the envelope C terminal tail. PLoS Pathog 9(2):e1003198
5. Fricke F et al (2015) One, two or three? Probing the stoichiometry of membrane proteins by single-molecule localization microscopy. Sci Rep 5:14072
6. Nan X et al (2013) Single-molecule superresolution imaging allows quantitative analysis of RAF multimer formation and signaling. Proc Natl Acad Sci U S A 110(46):18519–18524
7. Puchner EM et al (2013) Counting molecules in single organelles with superresolution microscopy allows tracking of the endosome maturation trajectory. Proc Natl Acad Sci U S A 110(40):16015–16020
8. Endesfelder U et al (2013) Multiscale spatial organization of RNA polymerase in Escherichia coli. Biophys J 105(1):172–181
9. Lee SH et al (2012) Counting single photoactivatable fluorescent molecules by photoactivated localization microscopy (PALM). Proc Natl Acad Sci U S A 109(43):17436–17441
10. Fricke F, Dietz MS, Heilemann M (2015) Single-molecule methods to study membrane receptor oligomerization. ChemPhysChem 16 (4):713–721
11. Annibale P et al (2011) Identification of clustering artifacts in photoactivated localization microscopy. Nat Methods 8(7):527–528
12. Wang S et al (2014) Characterization and development of photoactivatable fluorescent proteins for single-molecule-based superresolution imaging. Proc Natl Acad Sci U S A 111 (23):8452–8457
13. Durisic N et al (2014) Single-molecule evaluation of fluorescent protein photoactivation efficiency using an in vivo nanotemplate. Nat Methods 11(2):156–162
14. Sengupta P et al (2011) Probing protein heterogeneity in the plasma membrane using PALM and pair correlation analysis. Nat Methods 8(11):969–975
15. Avilov S et al (2014) In cellulo evaluation of phototransformation quantum yields in fluorescent proteins used as markers for single-molecule localization microscopy. PLoS One 9(6):e98362
16. Rollins GC et al (2015) Stochastic approach to the molecular counting problem in superresolution microscopy. Proc Natl Acad Sci U S A 112(2):E110–E118

17. Adam V et al (2009) Structural basis of enhanced photoconversion yield in green fluorescent protein-like protein Dendra2. Biochemistry 48(22):4905–4915
18. Ulbrich MH, Isacoff EY (2007) Subunit counting in membrane-bound proteins. Nat Methods 4(4):319–321
19. Wiedenmann J et al (2004) EosFP, a fluorescent marker protein with UV-inducible green-to-red fluorescence conversion. Proc Natl Acad Sci U S A 101(45):15905–15910
20. Coltharp C, Kessler RP, Xiao J (2012) Accurate construction of photoactivated localization microscopy (PALM) images for quantitative measurements. PLoS One 7(12):e51725
21. Nelson R (1995) Probability, stochastic processes, and queueing theory: the mathematics of computer performance modeling. Springer, New York
22. Kriegel HP et al (2011) Density-based clustering. Wiley Interdiscip Rev Data Min Knowl Discov 1(3):231–240
23. Ankerst M et al (1999) OPTICS: ordering points to identify the clustering structure. Sigmod Rec 28(2):49–60
24. Levet F et al (2015) SR-Tesseler: a method to segment and quantify localization-based super-resolution microscopy data. Nat Methods 12 (11):1065–1071
25. Mortensen KI et al (2010) Optimized localization analysis for single-molecule tracking and super-resolution microscopy. Nat Methods 7 (5):377–381
26. Wolter S et al (2010) Real-time computation of subdiffraction-resolution fluorescence images. J Microsc 237(1):12–22
27. Ovesny M et al (2014) ThunderSTORM: a comprehensive ImageJ plug-in for PALM and STORM data analysis and super-resolution imaging. Bioinformatics 30(16):2389–2390
28. Brede N, Lakadamyali M (2012) GraspJ: an open source, real-time analysis package for super-resolution imaging. Opt Nanoscopy 1:11
29. Endesfelder U et al (2011) Chemically induced photoswitching of fluorescent probes—a general concept for super-resolution microscopy. Molecules 16(4):3106–3118
30. Nanguneri S et al (2014) Single-molecule super-resolution imaging by tryptophan-quenching-induced photoswitching of phalloidin-fluorophore conjugates. Microsc Res Tech 77(7):510–516

Chapter 11

Two-Color Single-Molecule Tracking in Live Cells

Siegfried Hänselmann and Dirk-Peter Herten

Abstract

Measuring the kinetics of protein-protein interactions between molecules in the plasma membrane of live cells provides valuable information for understanding dynamic processes, like cellular signaling, on a molecular scale. Two-color single-molecule tracking is a fluorescence microscopy-based method to detect and quantify specific protein-protein interactions on a single-event level, providing sensitivity to heterogeneities and rare events. Fundamentally, it allows following the movement of single molecules of two different protein species in live cells with a localization precision beyond the diffraction limit of light in real time. It hence provides information about the diffusion behavior of every protein as well as about their dimerization kinetics. Here, we describe all the necessary steps to obtain two-color tracking data of plasma membrane-associated proteins in live cells using SNAP-tag and HaloTag fusion constructs and total internal reflection fluorescence (TIRF) microscopy. Also, we outline the main steps needed for analyzing the recorded data.

Key words Fluorescence microscopy, Single-particle tracking, Protein-protein interactions, Signaling, Membrane, Dimerization, Diffusion, Labeling protocol, HaloTag, SNAP-tag

1 Introduction

Single-molecule tracking has emerged as a valuable tool for observing and quantifying protein dynamics in vitro and in vivo [1]. While one-color tracking is typically being used to follow protein motion (e.g., position and transport of virus particles [2]) or to probe the cellular environment (e.g., spatial confinements in the cell membrane [3]), two-color tracking additionally allows quantifying specific interactions between two proteins in real time [4]. In contrast to ensemble-based interaction assays (e.g., co-immunoprecipitation), two-color single-molecule tracking reports on every single interaction event and is therefore sensitive to transient and rare interactions. This chapter wants to guide you through the different steps that are necessary to obtain two-color single-molecule tracking data of membrane-associated proteins and also outlines the basic components of data analysis (Fig. 1).

Holger Erfle (ed.), *Super-Resolution Microscopy: Methods and Protocols*, Methods in Molecular Biology, vol. 1663,
DOI 10.1007/978-1-4939-7265-4_11, © Springer Science+Business Media LLC 2017

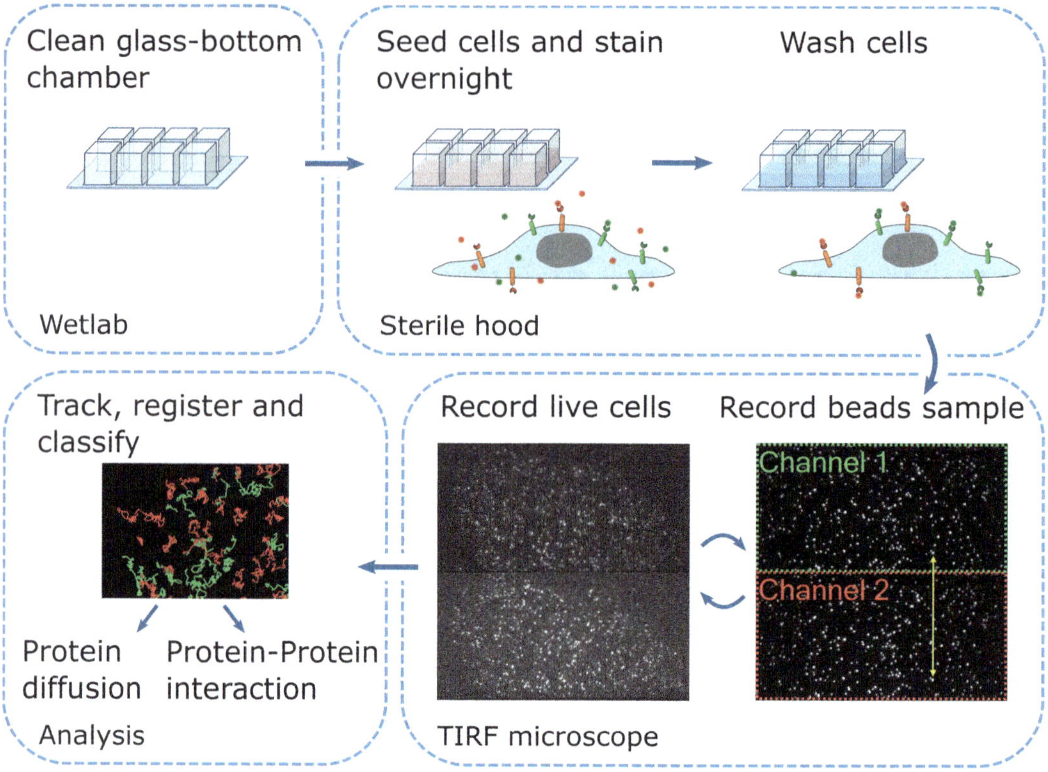

Fig. 1 Workflow to obtain and analyze two-color single-molecule data. Cells that express the proteins of interest as SNAP-tag and HaloTag fusion constructs are seeded in a clean glass-bottom chamber and stained with the respective dye substrates overnight. Next, unbound dye is washed off in three steps and the cells are imaged on a total internal fluorescence reflection (TIRF) microscope with each spectral channel assigned to half of the camera chip. Multi-color beads are imaged as calibration sample for later image registration. The analysis of the data includes tracking of single molecules in each channel, registration of both channels, and finally the detection and quantification of dimerization events

In order to record single-molecule trajectories, specific labeling of the proteins of interest within a living cell has to be established. Enzymatic protein tags like SNAP-tag [5] and HaloTag [6] have emerged as robust fluorescence labels combining some of the advantages of fluorescent proteins and organic dyes [7]. While they are fused to a protein of interest on DNA level, they can be used with a variety of dye substrates with different photophysical properties. Most organic dye substrates are considerably more photostable than fluorescent proteins, which allows for longer observation times. However, the additional staining step may create unspecific background signal as the dye substrates may bind to off-targets (e.g., the glass surface of the coverslip or the cell membrane). Both protein tags undergo a non-reversible, catalytic labeling reaction and there is a tradeoff between the degree of labeling (the fraction of protein tags that has actually bound a substrate) and the background signal (the amount of substrate that has bound to off-targets). While this tradeoff is different for the two protein tags (from our experience, the HaloTag reacts faster than the SNAP-tag) and for different dye substrates (some dye substrates tend to be more "sticky" or show distinct off-target effects [8]), we have generally experienced that a long incubation time with low substrate concentration favors a high degree of labeling while keeping background signal low. For tracking of intracellular proteins, the dye substrates have to be cell-permeable. In this protocol we use Tetramethylrhodamine (TMR) and Siliconrhodamine (SiR) [9] as cell-permeable and photostable dyes in the yellow and the far-red wavelength regime, respectively.

Imaging of single molecules requires a microscopy technique with strong background rejection. For targets within or near the plasma membrane of cells, objective-type total internal reflection fluorescence (TIRF) microscopy provides a robust and widely used option. The TIRF microscope should be equipped with a single-molecule sensitive camera that enables frame rates up to 50 Hz (e.g., an electron multiplying charge-coupled device (emCCD) or a scientific complementary metal-oxide-semiconductor (sCMOS) camera) in order to be able to perform robust single-molecule tracking and temporally resolve short-lived interactions. For simultaneous imaging of two spectral channels, emission light from the sample has to be spectrally split and each channel has to be imaged separately. This can be done by using two separate cameras or by allocating half of the camera chip to each channel. Since using one camera might be more affordable for many users, we will focus on the split-chip option here.

Analysis of the recorded movies can be divided into three main steps, namely (1) single-particle tracking, (2) image registration, and (3) classification of the dimerization state. Unfortunately, we do not know of any free software package that combines all three analysis steps in an automated manner. Implementation of such an analysis from scratch or based on pre-existing algorithms will therefore require some advanced knowledge in machine-leaning techniques and programming. A variety of open-source algorithms are available to perform the single-particle tracking step, e.g., the ImageJ plugin TrackMate (http://imagej.net/TrackMate) or the Matlab-based tracking software u-track (http://lccb.hms.harvard.edu/software.html). A quantitative comparison of different single-particle tracking routines can be found in the final publication of the Particle Tracking Challenge 2012 [10]. The first step in single-particle tracking is the detection of particles in diffraction-limited images. Most approaches include intensity-based shape detection (e.g., a two-dimensional Gaussian fitting with background comparison) to localize particles with sub-pixel precision below the diffraction limit of light. In a next step, particles of consecutive frames have to be linked to each other for generating particle trajectories. While there are different strategies in place to decide which particles correspond to each other in different frames, the accuracy of this linkage will generally be increased if particles move slowly and their density is low [10]. The single-particle tracking step will already give you information about diffusion properties of your proteins of interest and possibly allows detecting homo-dimerization events through merging and splitting of trajectories within one channel and the analysis of their relative intensity [11, 12].

In order to detect and quantify hetero-dimerization events, trajectories from both spectral channels need to be overlaid using a registration step that can be calibrated, e.g., with multicolor beads. Being visible in both the channels, the beads act as landmarks that are used to generate a transformation matrix describing the projection of one channel to the other. In most cases, this projection can be sufficiently described by a linear transformation that contains combinations of translation, scaling, rotation and shearing. The final part of the analysis is the detection of dimerization events. This classification is usually based on the distance between two particles and can be implemented by setting a fixed distance threshold that has to be met for a certain time period [11, 13] or by using a hidden Markov model (HMM) with the distance between two particles as the observation layer [14]. In order to estimate the expected distance distribution between particles in a dimer, it is useful to use constitutive dimers (either biologically or artificially) as control. The kinetics of a dimerization reaction can then be obtained from the transition probabilities between the different HMM states or by fitting the distribution of dimer durations to an exponential model (this gives the dissociation constant describing the stability of the dimer) [4].

2 Materials

2.1 (Optional) Dye Substrate Coupling

Given some basic knowledge and equipment for organic synthesis, the dye substrates for SNAP-tag and HaloTag can be coupled from the respective substrate building block (Benzylguanine amine for SNAP-tag and HaloTag Amine Ligand for HaloTag) and an organic dye of your choice as *N*-hydroxysuccinimid (NHS) ester. Alternatively, you can also buy the ready-to-use dye substrates from New England Biolabs (NEB) and Promega for SNAP-tag and HaloTag, respectively. Synthesis of the dye substrates from an organic dye NHS ester and the Benzylguanine (BG) amine or HaloTag amine substrate is a relatively simple, one-step reaction [9, 15]. Performing this coupling reaction yourself may be beneficial if you want to test organic dyes that are not commercially available or if you have already been working with organic dyes and have them as NHS esters.

1. Carboxytetramethylrhodamine *N*-hydroxysuccinimide ester (TMR NHS ester).
2. HaloTag Amine (O2) Ligand (Promega).
3. Siliconrhodamine *N*-hydroxysuccinimide ester (SiR NHS ester, Spirochrome).
4. O6-Benzylguanine amine (BG).
5. Dimethylformamid anhydrous (DMF).
6. *N*,*N*-Diisopropylethylamine (Dipea).
7. Basic high performance liquid chromatography (HPLC) setup, equipped with a standard reversed phase C18 column and preferably a fluorescence detector.
8. Rotary evaporator or vacuum concentrator.
9. Absorption spectrometer.

2.2 Chamber Cleaning and Cell Seeding

1. A stable cell line expressing one protein of interest as a SNAP-tag fusion construct and the other one as a HaloTag fusion construct at low expression levels (*see* **Note 1**).
2. Glass-bottom, 8-well cell chamber (e.g., Nunc Lab-Tek chambered coverglass) (*see* **Note 2**).
3. 0.1 M hydrofluoric acid in ultrapure water (*see* **Note 3**).
4. Dulbecco's phosphate buffered saline (DPBS; 1×): 0.2 g/L KCl, 0.2 g/L KH_2PO_4, 8.0 g/L NaCl, 1.15 g/L Na_2HPO_4 (anhydrous).
5. 0.25% Trypsin Ethylenediaminetetraacetic acid (EDTA): 400 mg/L KCL, 60 mg/L KH_2PO_4, 350 mg/L $NaHCO_3$, 8 g/L NaCl, 90 mg/L $Na_2HPO_4 \cdot 7H_2O$, 1 g/L D-Glucose, 380 mg/l EDTA 4Na $2H_2O$, 2.5 g/L Trypsin.
6. Normal growth medium: Dulbecco's Modified Eagle Medium (DMEM), high glucose, no phenol red, supplemented with 10% fetal bovine serum (FBS), 1 mM sodium pyruvate and 2 mM L-glutamine (*see* **Note 4**).
7. Multicolor beads, 0.2 µm (*see* **Note 5**).

2.3 Cell Staining and Washing

1. 100 nM HaloTag TMR ligand (10×): Dilute the HaloTag TMR ligand from the coupling reaction with normal growth medium to yield a 100 nM working solution. Alternatively: Dilute 1 µl of HaloTag TMR Ligand 5 mM stock solution (Promega) in 50 µl dimethylformamid (DMF) to obtain a 100 µM stock. Dilute 1 µl of this stock in 1 ml normal growth medium to obtain a 100 nM working solution (*see* **Note 6**).
2. 100 nM Siliconrhodamine-Benzylguanine (SiR-BG) (10×): Dilute the SiR-BG from the coupling reaction with normal growth medium to yield a 100 nM working solution.

Alternatively, dissolve one vial of SNAP-Cell 647-SiR (30 nmol, New England Biolabs) in 300 μl DMF to obtain a stock solution of 100 μM. Dilute 1 μl of this stock solution in 1 ml normal growth medium to obtain a 100 nM working solution.

3. Starvation medium: Dulbecco's Modified Eagle Medium (DMEM), high glucose, no phenol red, supplemented with 1 mg/ml BSA, 1 mM sodium pyruvate and 2 mM L-glutamine (*see* **Note 7**).

2.4 Imaging

1. A TIRF microscope, equipped with a 561 nm and a 640 nm laser line, a dichroic mirror to split the excitation and emission path (e.g., a dichroic mirror that transmits all wavelengths except for the laser lines), a 100× Oil high-NA TIRF objective, a double-notch filter to block remaining laser light in the emission path, a beam splitter with band-pass filters for 576 nm emission (TMR) and 661 nm emission (SiR), an emCCD or sCMOS camera and preferably a heating chamber with CO_2 supply (*see* **Note 8**). If no CO_2 supply can be provided, use CO_2-independent medium during imaging, e.g., Leibovitz's L-15 medium.

3 Methods

3.1 (Optional) Dye Coupling

1. Prepare a solution of 1 μmol SiR NHS ester in 50 μl DMF in a 1.5 ml tube. Transfer 5 μl and an additional 5 μl DMF in a new 1.5 ml tube for a HPLC test run. To the remaining 45 μl, add a threefold molar excess of BG amine (3 μmol) solved in 50 μl DMF, followed by 10 μl Dipea. Incubate both tubes for 3 h at 40 °C in a shaker.
2. Purify the SiR-BG reaction product via HPLC with the unreacted SiR NHS ester as a reference.
3. Dry the SiR-BG product fraction with a rotary evaporator or a vacuum concentrator.
4. Dissolve the dry SiR-BG in 100 μl DMF. Prepare a 1:100 dilution in water and measure the concentration with an absorption spectrometer (*see* **Note 9**). Adjust the concentration of the SiR-BG stock to around 1 mM with DMF.

Using TMR NHS ester as organic dye and the HaloTag amine ligand, this protocol can likewise be applied to couple the HaloTag TMR ligand. More generally, the SNAP-tag and HaloTag do not only work with fluorescent dye substrates. In principle, any residue can be coupled to the respective substrate to functionalize your protein of interest.

3.2 Chamber Cleaning, Cell Seeding, and Cell Staining (Day 1)

1. Add 200 μl of 0.1 M hydrofluoric acid to each well of the glass-bottom, 8-well cell chamber. Incubate for 2 min at room temperature (*see* **Note 10**). Remove the hydrofluoric acid and wash each chamber twice with 200 μl DPBS. Collect the used hydrofluoric acid in a 50 ml Falcon tube that contains 10 ml of a saturated calcium chloride solution. Repeat **step 1** two more times. For the last washing step, wash each chamber three times with 200 μl DPBS.

   All the following steps should be performed in a sterile laminar flow hood.
2. Remove the DPBS from the wells of the clean glass-bottom, 8-well chamber and seed 20,000–30,000 cells in a total volume of 160 μl normal growth medium in as many wells as you have conditions to test. As a negative staining control you should reserve one well to wild-type cells (i.e., cells that do not express the SNAP-tag and HaloTag fusion constructs).
3. Add 20 μl of 100 nM HaloTag TMR ligand and 20 μl of 100 nM SiR-BG to each well to yield a final concentration of 10 nM HaloTag TMR ligand and 10 nM SiR-BG (*see* **Note 11**). Mix carefully with a 200 μl pipette. As an autofluorescence control, you should leave one well with cells expressing the SNAP-tag and HaloTag fusion construct unstained.
4. Incubate the chamber overnight in the incubator (*see* **Note 12**).

3.3 Beads Sample for Image Registration

These steps can be performed in the wet lab.

1. Vortex the multicolor beads for 2 min, then sonicate them for 10 min. Prepare 800 μl of a 20 μM, a 10 μM, and a 5 μM dilution in DPBS.
2. Clean a glass-bottom, 8-well cell chamber as mentioned above. Add the differently diluted multicolor beads to different wells of the chamber. Seal the wells with Parafilm. This beads sample can be reused over months. Store it dark either in the fridge or at room temperature.

3.4 Cell Washing and Imaging (Day 2)

All the following steps should be performed in a sterile laminar flow cabinet.

1. Carefully replace the medium in each well with 250 μl starvation medium: Tilt the chamber toward yourself, place the pipette tip in the lower left corner and slowly remove the medium (*see* **Note 13**). Try to remove the medium as thoroughly as possible to maximize the washing effect. Discard the medium in a 50 ml waste Falcon tube, take a new tip and slowly fill the well again from the lower left corner with starvation medium. Incubate the chamber for 15 min in the incubator.

2. Repeat **step 1** two more times. After the last washing step, incubate the chamber for 40 min in the incubator instead of 15 min. You can use this 40 min incubation break to adjust the microscope and record the beads calibration movies (**step 3**). The two 15 min washing steps aim to dilute residual dye substrate within the well, while the longer 40 min incubation step allows unbound dye substrate molecule to diffuse out of the cell and be washed out in the very last washing step.
3. Replace the medium with 250 µl CO_2-independent medium or starvation medium if your microscope is equipped with a CO_2-controlling heating chamber. The cells are now ready to be imaged.

   These steps take place at the microscope.
4. Switch on the heating chamber and allow the microscope to warm up. Load the beads sample and select a well that shows a high density of single, isolated beads. If you use one camera, adjust the two channels in such way that the chip of the camera is horizontally split and each half of the chip sees the same field of view but in different spectral channels (*see* Fig. 1 bottom right, "Record beads sample"). If you use two cameras, allocate each channel to one camera and align the fields of view. Record a short movie (10 frames at 50 ms exposure time) from at least three different positions of the well. These movies will be used as a calibration sample for the registration step (*see* **Note 14**).
5. Load the chamber with the stained live cells and set the imaging setting. We recommend laser intensities in the low milliwatt regime (measured with a photometer during widefield excitation), 20 ms exposure time, and 1000 frames recording time (*see* **Note 15**).

   The unstained cells as well as the stained wild type cells should be as dark as possible. If you observe a significant number of moving particles in the wild-type control, repeat the experiment with a reduced concentration of dye substrate. You can perform a titration series as mentioned in **Note 9** to find the optimal dye concentration.

   If the wild-type cells show acceptably low unspecific background, move to the stained cells expressing the SNAP-tag and the HaloTag fusion constructs and record at least ten movies per condition (*see* **Note 16**).

The recorded data can then be analyzed as outlined in the last two paragraphs of the introduction.

4 Notes

1. Generating a stable cell line will save you time in the long-term because the reproducibility of the expression levels is high compared to a transient transfection of both constructs. A low expression level is important because high particle densities will result in broken or incorrectly assigned trajectories during single-particle tracking. In the final movie, you should be able to detect and follow individual particles by eye for a couple of frames. If needed, you can use fluorescence-activated cell sorting (FACS) or raise single cell clones to select a cell population with low and consistent expression levels.
2. Using a glass-bottom chamber is preferable because of the large difference between the refractive indexes of glass and aqueous medium, which facilitates the total internal reflection needed for TIRF microscopy.
3. Although being strongly diluted, 0.1 M hydrofluoric acid still needs to be handled as a hazardous substance and all necessary safety measures for working with hydrofluoric acid need to be met. This includes wearing personal safety gear like safety goggles and impervious gloves (neoprene or nitrile) as well as providing first-aid measures like calcium gluconate gel.
4. The choice of the best-suited growth medium may depend on the cell type that you use. Most cell types however grow well in DMEM. In any case, you should use phenol red-free medium to reduce background fluorescence.
5. You can use any other beads, as long as they sediment to the bottom of the chamber and fluoresce in all spectral channels that will be used for co-tracking. The beads should not be functionalized and not be smaller than 0.2 μm since a very small size as well as some functionalizations may prevent the sedimentation of the beads to the glass surface.
6. Store all dye substrate stocks at −20 °C and use up the working solution within a few days. This also applies to the SiR-BG dilutions.
7. Starvation medium is used for the later washing steps in order to additionally deplete the cells of growth factors and other FBS-originated signaling molecules that could interact with the proteins of interest.
8. Regulating the temperature to rather strict 37 °C is not only important for cell viability. Since temperature has a direct effect on diffusion and reaction kinetics of molecules, changes in temperature will influence the measured diffusion properties and dimerization rates of your proteins of interest.

9. In solvents of lower dielectric constants like DMF, SiR-BG and to a lower extent also TMR form a non-fluorescent, nonabsorbing spirolactone [9]. When diluted in water, both dyes are predominantly in their open form and their concentration can be measured.
10. If you incubate too long with hydrofluoric acid, the wells may get leaky as the glue between the glass slide and the plastic wells may be partly dissolved.
11. You can add the dye substrates right after seeding the cells. The cells do not have to be attached yet since you stain overnight anyway. As for the dye substrate concentration, the labeling and washing efficiency may be slightly different for different cell lines or different batches of SNAP-tag or HaloTag substrates. In order to individually optimize the substrate concentrations, you can perform a concentration titration series based on wild-type cells. For this, seed wild-type cells to all 8 wells of a cleaned glass-bottom, 8-well cell chamber and incubate each well overnight with a different concentration of the two dye substrates. Vary the concentrations in a range from 1 to 500 nM and include an unstained control. Although SNAP-tag and HaloTag are labeled independent of each other, they are eventually imaged with the same camera settings and should therefore have somewhat comparable intensities per cell. Therefore, we recommend broadly aligning the concentrations of both dye substrates within the titration series. Wash and image the cells as described. The optimal concentration for two-color tracking is the highest concentration of each substrate that still does not lead to significant unspecific staining.
12. In some cases, you may detect vesicles after overnight staining when imaging the cells. This may result from endocytosis of dye substrate or from endocytosis of your protein of interest as part of their natural turnover. You can reduce vesicle formation by decreasing the incubation time of the dye substrates. From own experience, we recommend incubating the cells with SNAP-tag substrate for at least 3 h, while the HaloTag can already be labelled after an incubation time of 15 min. If you reduce the incubation time, you should allow the cells to adhere overnight before you start the staining step.
13. Some cell lines do not adhere well on glass and are easily washed off. Therefore, try to reduce lateral flow as much as possible. One way to achieve this is to keep the pipette tip just below the surface of the medium and "follow" the medium as you remove it.
14. Single-molecule two-color tracking requires a registration precision in the order of one pixel. Depending on your microscope setup and the duration of the measurement, the alignment of

the channels may show small drifts over time (e.g., due to temperature changes or at defined time point if someone slams the door). Therefore, we advise recording a few beads movies every time you go to a new condition.

15. Depending on your setup, you may need to adjust these parameters to find a good tradeoff between signal-to-noise ratio and bleaching rate. Increasing the laser power will increase the signal-to-noise ratio of single molecules up to a point where the fluorophores are saturated. At the same time it will also lead to faster bleaching and therefore shorten the trajectories. If your protein of interest diffuses relatively slow and the interaction that you want to observe is long-lived, you have the option to sacrifice temporal resolution for an increased signal-to-noise ratio by increasing the exposure time of the camera.

16. The more movies you record, the better your statistics will be. However also keep in mind that the cells will most likely show reduces viability over time (this depends very much on how well your setup can regulate temperature and CO_2 level). So if you have many conditions to test, you should reduce the number of movies per experiment and increase the number of individual experiments instead.

References

1. Cognet L, Leduc C, Lounis B (2014) Advances in live-cell single-particle tracking and dynamic super-resolution imaging. Curr Opin Chem Biol 20:78–85. doi:10.1016/j.cbpa.2014.04.015
2. Brandenburg B, Zhuang X (2007) Virus trafficking – learning from single-virus tracking. Nat Rev Microbiol 5:197–208. doi:10.1038/nrmicro1615
3. Fujiwara T, Ritchie K, Murakoshi H et al (2002) Phospholipids undergo hop diffusion in compartmentalized cell membrane. J Cell Biol 157:1071–1082. doi:10.1083/jcb.200202050
4. Valley CC, Arndt-Jovin DJ, Karedla N et al (2015) Enhanced dimerization drives ligand-independent activity of mutant epidermal growth factor receptor in lung cancer. Mol Biol Cell 26:4087–4099
5. Juillerat A, Gronemeyer T, Keppler A et al (2003) Directed evolution of O 6-alkylguanine-DNA alkyltransferase for efficient labeling of fusion proteins with small molecules in vivo. Chem Biol 10:313–317
6. Los GV, Encell LP, McDougall MG et al (2008) HaloTag: a novel protein labeling technology for cell imaging and protein analysis. ACS Chem Biol 3:373–382. doi:10.1021/cb800025k
7. Jing C, Cornish VW (2011) Chemical tags for labeling proteins inside living cells. Acc Chem Res 44:784–792. doi:10.1021/ar200099f
8. Bosch PJ, Corrêa IR, Sonntag MH et al (2014) Evaluation of fluorophores to label SNAP-tag fused proteins for multicolor single-molecule tracking microscopy in live cells. Biophys J 107:803–814. doi:10.1016/j.bpj.2014.06.040
9. Lukinavičius G, Umezawa K, Olivier N et al (2013) A near-infrared fluorophore for live-cell super-resolution microscopy of cellular proteins. Nat Chem 5:132–139. doi:10.1038/nchem.1546
10. Chenouard N, Smal I, de Chaumont F et al (2014) Objective comparison of particle tracking methods. Nat Methods 11:281–289. doi:10.1038/nmeth.2808
11. Hern JA, Baig AH, Mashanov GI et al (2010) Formation and dissociation of M1 muscarinic receptor dimers seen by total internal reflection fluorescence imaging of single molecules. Proc Natl Acad Sci U S A 107:2693–2698. doi:10.1073/pnas.0907915107
12. Suzuki KGN, Kasai RS, Hirosawa KM et al (2012) Transient GPI-anchored protein

homodimers are units for raft organization and function. Nat Chem Biol 8:774–783. doi:10.1038/nchembio.1028

13. Wilmes S, Beutel O, Li Z et al (2015) Receptor dimerization dynamics as a regulatory valve for plasticity of type I interferon signaling. J Cell Biol 209:579–593. doi:10.1083/jcb.201412049

14. Low-Nam ST, Lidke KA, Cutler PJ et al (2011) ErbB1 dimerization is promoted by domain co-confinement and stabilized by ligand binding. Nat Struct Mol Biol 18:1244–1249. doi:10.1038/nsmb.2135

15. Hermanson GT (2013) Bioconjugate techniques. Academic Press, New York

Chapter 12

Fully Automated Targeted Confocal and Single-Molecule Localization Microscopy

Jan Philipp Eberle, Walter Muranyi, Holger Erfle, and Manuel Gunkel

Abstract

Single-molecule localization microscopy (SMLM) enables imaging of biological structures in the nanometre range. Long measurement times are the consequence of this kind of microscopy due to the need of acquiring thousands of images. We built a setup that automatically detects target structures using confocal microscopy and images them with SMLM. Utilizing the Konstanz Information Miner (KNIME), we were able to connect a confocal microscope with an SMLM unit for targeted screening. In this process, we developed KNIME plugins to communicate with the microscope components and combined them to a workflow. Thus, measuring biological nanometre-sized structures in a sufficient number to get statistical significance becomes feasible. For proof of principle HIV-1 assembly complexes in HeLa cells derived from transfection of replication deficient viral construct were imaged by a fully automated screen.

Key words Super-resolution microscopy, Single-molecule localization microscopy (SMLM), Konstanz Information Miner (KNIME), High content screening (HCS), Multiscale imaging, Online feedback image analysis

1 Introduction

In fluorescence microscopy versatile methods have been developed that provide quantitative measurements of biological samples. SMLM is one of them, allowing imaging structures with a lateral resolution of about 20 nm [1]. However, many structures have to be detected and analyzed to obtain statistical significant results. This demands time-consuming measurements for the experimenter, especially if the target structures are rare events. In order to solve this issue, we present a fully automated setup for targeted confocal and single-molecule localization microscopy.

Beneath statistical significance and sparing time, additional benefits come along with an automatized microscope that executes imaging routines. One of them is to prevent the sample from unnecessary photobleaching during manual stage positioning and imaging. Online image analysis of fast low resolution recordings

Holger Erfle (ed.), *Super-Resolution Microscopy: Methods and Protocols*, Methods in Molecular Biology, vol. 1663,
DOI 10.1007/978-1-4939-7265-4_12, © Springer Science+Business Media LLC 2017

allows selective imaging of target positions with slow super-resolution. In this way, required memory space as well as overall imaging time is reduced. Machine learning techniques for classification like random forests, enable objective decisions based on training data. Thus, experimenter biased imaging that could influence the results is minimized.

For implementing fully automated target imaging, two main tasks were tackled. Image acquisition was automatized by accessing all devices with a central control unit and online image analysis was applied for a decision-making process. Combining both tasks resulted in a feedback loop, enabling target identification and imaging while screening the sample.

There have already been several approaches for automatizing a microscope with a feedback loop for targeted screening. For example, CellProfiler was used for online image analysis in feedback confocal microscopy [2]. Another example is Micropilot, a software used to automate 3D multicolor time-lapse confocal microscopy or fluorescence recovery after photobleaching (FRAP) microscopy [3]. In contrast to these approaches, here, a microscopy routine is presented for imaging targets with super-resolution.

KNIME [4] was used for connecting confocal microscopy, providing overview images, with SMLM for target imaging. Since KNIME is a visual programming language, also non-programmers are able to use it. Its modular data pipelining concept allows adapting on different biological tasks and imaging routines. KNIME already provides image processing and classification methods from an active community that can be combined to image analysis pipelines for target identification. The connection to the confocal and SMLM unit had to be established and programmed from scratch. In this context, we created KNIME plugins to extend its functionality for controlling a confocal and an SMLM unit. These plugins can be integrated and used together with the image analysis pipeline within KNIME workflows to design experiments for targeted imaging.

The resulting application is target or event triggered SMLM imaging. Due to the modularity of KNIME, the experiment can be adapted to a variety of biological tasks. One example is fully automated SMLM imaging of cellular phenotypes, providing data of substructures for statistical analysis. Also possible is live cell imaging of events like cell division with resolution in the nanometre range.

For proof of principle experiments, we used HeLa cells that were transfected with HIV construct resulting in expression of human immunodeficiency virus type 1 (HIV-1) proteins. Target structures were assembly complexes of the structural Gag protein matrix (MA) together with envelope (Env) glycoproteins of HIV-1. Those structures have already been manually observed using SMLM [5]. In this way, Gag proteins have been found to build punctuate structures of about 130 nm in diameter and Env proteins to form clusters with radii of 50–150 nm.

In the following, we will describe our method for fully automated targeted confocal microscopy and SMLM. This method uses confocal microscopy, primarily to provide precise autofocus planes via low background images of target structures.

2 Materials

2.1 Biological

In general, all plate types are supported by the presented method as long as they are compatible with the automated stage of the used confocal Leica TCS SP5. Furthermore, the samples and their preparation should allow imaging with confocal microscopy and SMLM.

1. 8-well chambered cover glasses (LabTek, Thermo Fisher).
2. HeLa cells.
3. Growth medium: Dulbecco's modified Eagle's medium (DMEM; Invitrogen), supplemented with 10% fetal calf serum (FCS), 100 U/ml penicillin and 100 μg/ml streptomycin.
4. Plasmid pCHIV, expressing all HIV-1 proteins except for the negative regulatory factor (Nef), but is replication deficient due to deletion of the long terminal repeats (LTR) has been described previously [5, 6].
5. Transfection solution: FuGene HD transfection reagent (Roche Diagnostics).
6. Fixation solution: 4% paraformaldehyde (PFA).
7. Blocking solution: 2% bovine serum albumin (BSA).
8. Washing solution: phosphate-buffered saline (PBS).
9. Permeabilization reagent: 0.1% Triton X-100.
10. Staining of surface resident Env molecules using the human monoclonal anti-gp120 antibody 2G12 ([7], Polymun Scientific) as primary antibody and goat antihuman Alexa Fluor® 647 as secondary antibody (Invitrogen).
11. Staining the MA domain of HIV-1 using a mouse anti-MA APR342 (CFAR, UK) as primary antibody and goat anti-mouse Alexa Fluor® 532 (Invitrogen) as secondary antibody.
12. Switching buffer: 100 mM beta-mercaptoethylamine (MEA) in PBS, pH 8.0.

2.2 Microscope Setup

The microscope setup (Fig. 1a) consists of a commercial Leica TCS SP5 as confocal unit which was extended by a custom-built SMLM unit. Both units are controlled by two separate computers that communicate via KNIME (Fig. 1b and c) with each other.

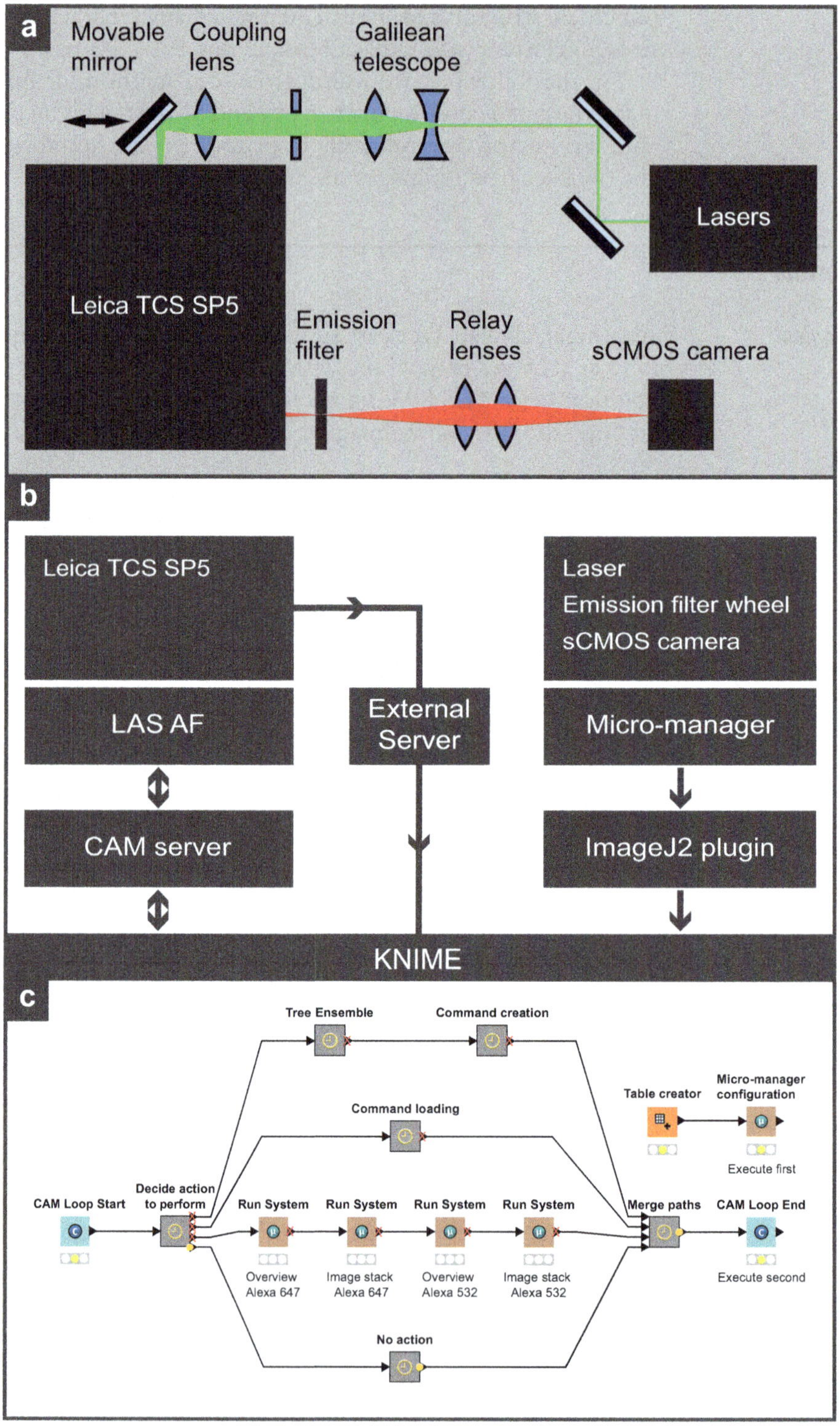

Fig. 1 Fully automated targeted confocal and SMLM setup. (**a**) Beam path including confocal (Leica TCS SP5) and self-built SMLM setup. (**b**) Device control of confocal and SMLM unit via KNIME. (**c**) KNIME workflow for confocal screening and SMLM imaging of target structures

2.2.1 Hardware Confocal Unit

The Leica TCS SP5 represents the point scanning confocal unit:

1. Laser: argon-ion laser (514 nm) and helium-neon laser (633 nm).
2. Objective: 63× objective lens (Leica HCX PL APO, NA 1.40).
3. Confocal detection unit: acousto-optical beam splitter as emission filter, photomultiplier tubes (PMTs) as detectors.
4. Live cell chamber.
5. Wide-field extension: used as connection to the SMLM setup.
6. External server for data storage.
7. Software: Leica Application Suit Advanced Fluorescence (LAS AF) version 2.7.1.9530 extended with the Matrix Screener.

2.2.2 Software Confocal Unit

1. *LAS AF*: The confocal Leica TCS SP5 is controlled by the LAS AF, a software provided by Leica Microsystems. Starting the LAS AF initializes the components of the confocal microscope and allows making general settings such as choosing the objective, lasers, and pixel depth of images. With the LAS AF, basic confocal imaging is applicable.
2. *Matrix Screener*: The Matrix Screener is an extension for the LAS AF and allows setting up screening experiments. This comprises stage calibration on the sample, description of imaging tasks, and planning the screen. For the latter one, the sample is sectioned in fields (Fig. 2a) which are scanned subsequently by starting the experiment. Imaging tasks, called jobs, can be executed for each field. For instance, autofocus jobs allow extracting the optimal focus plane calculated by gradient, contrast, or noise-based algorithms. Confocal jobs acquire single or multicolor confocal images. Camera jobs are used for switching between the detection beam paths of confocal and SMLM unit with a movable mirror. WaitForCAM jobs pause the confocal screen, until an external command triggers the scan to continue. Several jobs can be executed subsequently when they are connected to a pattern. Thus, it is possible to determine a plane where the target object is in focus with an autofocus job and image in that specific plane with a confocal job. Furthermore, the Matrix Screener includes a data exporter, which saves the acquired images to an external server directly after measuring.
3. *CAM Server*: The Matrix Screener contains the computer-aided microscopy (CAM) server, which allows controlling the confocal unit via external commands. By sending CAM commands to the CAM server, already defined jobs or patterns can be executed. CAM commands are strings containing certain keywords that are interpreted from the Matrix Screener. Finally, an experiment can be designed by applying CAM commands in

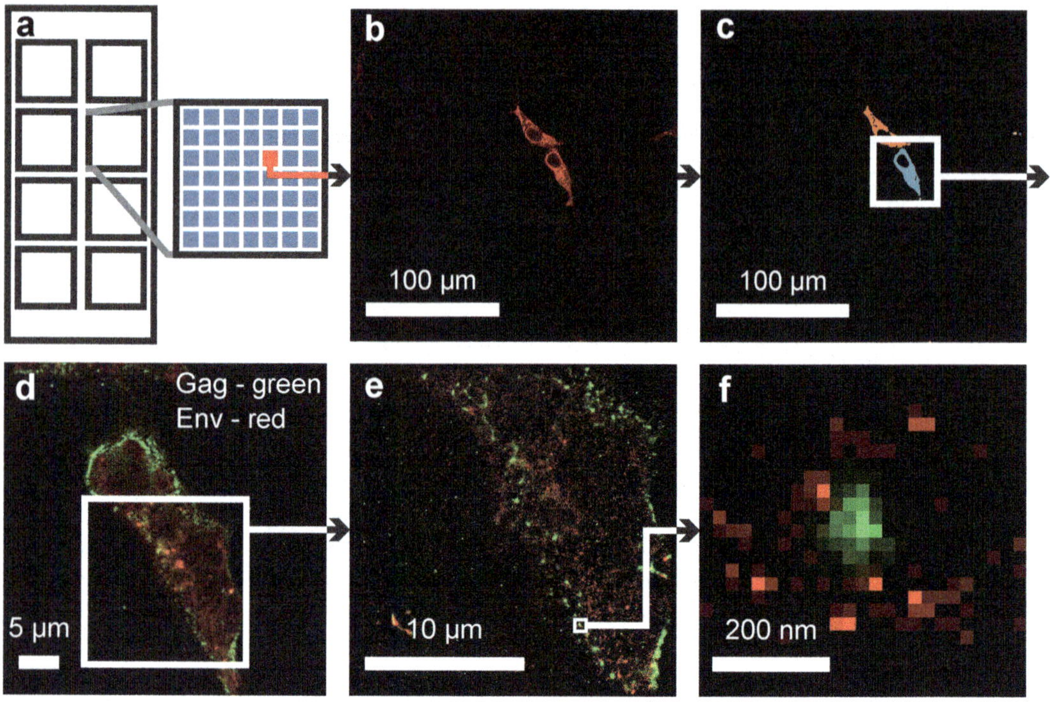

Fig. 2 Steps through automated targeted imaging. HIV-1 expressing HeLa cells with visualized Env-Alexa Fluor® 647 and Gag-Alexa Fluor® 532. (**a**) Sample rasterized in fields. (**b**) Overview image of Env-Alexa Fluor® 647, acquired with the confocal unit. (**c**) Segmented overview image. (**d**) Overlay of confocal images with higher sampling, showing Env-Alexa Fluor® 647 (*red*) and Gag-Alexa Fluor® 532 (*green*) labeled target structures. (**e**) Overlay of SMLM images from pixel-independent data analysis. (**f**) SMLM image from pixel-dependent data analysis at a target position of (**e**). SMLM images are shown as 2D histograms with 20 nm pixel size

a certain order. This can be achieved by interrupting the confocal scan with WaitForCAM jobs and executing subsequent patterns through CAM commands.

2.2.3 Hardware SMLM Unit

The SMLM unit is connected to the confocal Leica TCS SP5 via ports on the sides of the microscope (Fig. 1a). The illumination beam is coupled into the Leica microscope at the back port by a movable mirror. This enables switching from normal wide-field illumination to highly inclined and laminated optical sheet (HILO) microscopy which decreases background noise. The resulting fluorescence is directed through the right side port to a scientific complementary metal-oxide-semiconductor (sCMOS) camera. In the following, the elements are listed that assemble the SMLM setup:

1. Movable mirror: mirror with motorized actuator (Thorlabs Z825B).
2. Lasers in a laser box (Omicron LightHUB-6®):

   (a) Diode lasers from Omicron: LuxX® 647, LuxX® 488-200, LuxX® 405-60.
   (b) Single frequency continuous wave (CW) diode pumped lasers (DPL) from Cobolt: Cobolt Fandango™ with 515 nm and 150 mW, Cobolt Jive™ with 561 nm and 150 mW.
3. Galilean telescope to widen the Gaussian beam: Achromatic lenses with focal lengths of 80 and −20 mm result in a full width at half maximum of about 35 μm for the Gaussian beam after the objective.
4. Coupling lens: Achromatic lens with a focal length of 600 mm for coupling the laser beam into the microscope.
5. Filter wheel (Thorlabs FW102C) with emission filters (AHF Analysentechnik AG, Germany):
   (a) 647 LP Edge Basic longpass filter.
   (b) 650/45 ET bandpass filter.
   (c) 600/50 ET bandpass filter.
   (d) 520/40 ET bandpass filter.
6. Relay lenses for extending the beam path: Two achromatic lenses with focal lengths of 150 mm.
7. Hamamatsu Orca Flash 4.0 sCMOS camera with a resulting pixel size of 103.0 nm/pixel.
8. Software: KNIME, μManager.

2.2.4 Software SMLM Unit

Micro-Manager: μManager [8] is a free open-source software package for the device control of light-microscopes by computers (*see* **Note 1**) which allows setting up imaging routines as well. Here, μManager 1.4 is used to control laser, filter wheel and camera in an automated microscope via Java. Also the movable mirror is controlled with μManager utilizing Java to maintain the same angle for HILO microscopy.

2.3 Software and Tools for Automatization

1. *KNIME*: We used the free data analytics platform KNIME to connect image acquisition with image analysis. With its modular data pipelining concept, KNIME allows managing diverse tasks in a workflow. Each task in KNIME is represented by a module called node. Nodes can be connected with each other by their input and output ports. In doing so, data tables are transferred between nodes and a workflow can be established. For designing workflows with a clear structure, multiple nodes can be wrapped into a "Meta Node".

   The basic functionality of KNIME is further extended by external plugins and community contributions such as KNIME image processing (KNIP). In this way, image processing becomes accessible in addition to machine learning algorithms like random forests which are available as basic functionality. KNIME is an open-source platform based on Java that allows designing individual nodes for special tasks. An easy way in doing so is writing an ImageJ2 plugin that is then automatically translated into a KNIME node.

2. *Tool for integrating the SMLM Unit*: The SMLM unit was integrated in KNIME through a self-built ImageJ2 plugin that uses μManager to control the SMLM devices. Here, an already existing Imagej2 plugin for KNIME [9] was used and extended. The created plugin provides three KNIME nodes for managing the connection between KNIME and the SMLM devices. A configuration node (MMConfiguration) for initializing all devices and a stop system node (MMStopSystem) for de-initializing them after measuring were created. Furthermore, a run node (MMRunSystem) was implemented for executing the devices after applying settings on camera, filter-wheel and lasers. In detail, the camera subarray for image acquisition, the pixel exposure time, the number of frames acquired per image stack, the filter wheel state, and the power of two lasers can be set via this node. Beneath device settings also an image save name is generated in a predefined structure for consistent naming of files. Moreover, it can be chosen to save the images directly using Java advanced imaging (JAI) or to maintain them in KNIME.
3. *Tool for interacting with the Confocal Unit*: For the interaction between KIME and the confocal unit, a loop start node and a loop end node were created. They provide a connection to the CAM server using a Java socket. For this purpose, open-source code of KNIME loop nodes was used as a template for our own nodes. The CAMLoopStart node allows setting the IP address of the computer, on which the LAS AF software runs, the port number of the CAM server and the first CAM command which is sent to the server. By running this start node the connection to the CAM server is initialized and the experiment, defined in the Matrix Screener, is started. For each iteration the loop start node listens to the CAM server and returns its messages to the KNIME workflow. For example, a message can be a string containing the save path of an image that was acquired. The CAMLoopEnd node is able to send CAM commands to the CAM server for applying additional imaging tasks on a scan field.

3 Methods

3.1 Sample Preparation

1. HeLa cells were seeded at a density of 1×10^4 cells per well and grown at 37 °C and 5% CO_2 in DMEM growth medium.

2. Transfection was performed with 0.2 μg plasmid per well on the following day.
3. After 24 h, cells were fixed and blocked, followed by staining of the surface resident Env molecules using the primary antibody for 40 min. The cells were washed and blocked before incubation with the secondary antibody.
4. Cells were washed again, permeabilized and blocked for 10 min. Then, they were incubated with the primary antibody to detect the MA domain of HIV-1 for 40 min followed by the secondary antibody.
5. Prior to imaging, the medium was exchanged by a switching buffer that contained the primary thiol MEA essential for switching of some cyanine dyes and their derivatives such as certain Alexa fluorophores [10].

3.2 Microscopy

In the following, we describe how the system is prepared and how the connection between confocal and SMLM unit is managed to set up a fully automated microscopy procedure. The basic steps are:

1. Start the screen at a defined position and acquire a confocal image as overview.
2. Apply image analysis for target identification on that overview image and extract the target positions.
3. Image targets, utilizing confocal microscopy with higher sampling rate and perform SMLM subsequently.
4. Iterate this procedure over all scan positions.

3.2.1 Preparation of the System

1. Start the LAS AF and the Matrix Screener. This initializes the confocal unit and allows specifying the path for saving the acquired images on an external server.
2. Define patterns with autofocus jobs, imaging jobs, camera jobs, and WaitForCAM jobs. For instance, the following settings can be used:

   16 bit pixel depth, 1 airy unit pinhole size, and 600 Hz scan frequency per row. For Alexa Fluor® 647 a laser wavelength of 633 nm and an emission filter with a spectral range of 647–730 nm was used. For Alexa Fluor® 532 a laser wavelength of 514 nm and an emission filter with a spectral range of 532–620 nm was set. The defined imaging and autofocus jobs with individual parameters are listed in Table 1. JobAfGlobal was used for the autofocus map (*see* **step 5**) based on signals from Alexa Fluor® 647. The remaining jobs build the basis of three patterns:

   (a) *Overview pattern*: The aim of this pattern is to discover target structures, marked by Alexa Fluor® 647. Therefore, JobAf647 provides the focus plane for the confocal

Table 1
Job description for confocal imaging using the Matrix Screener

Job name	Job type	Image (pixel)	Pixel size (μm)	Focal intensity (kW/cm^2)	Axial steps	Step size (μm)	PMT gain (V)
JobAfGlobal	Autofocus	64 × 64	3.91	0.003	25	2	1071
JobAf647	Autofocus	64 × 64	3.91	0.065	20	2	1071
JobAfTar532	Autofocus	64 × 64	0.653	0.117	40	0.5	1071
JobAf532	Autofocus	64 × 64	3.91	0.003	40	0.5	1071
Job647	Imaging	2048 × 2048	0.120	12.5	1	–	850
JobTar647	Imaging	400 × 400	0.103	6.00	1	–	700
JobTar532	Imaging	400 × 400	0.103	0.850	1	–	700

Parameters for confocal image acquisition and finding the autofocus plane are summarized

overview image, acquired with Job647 (*see* **Note 2**). A WaitForCAM job pauses the confocal scan until the KNIME loop has analyzed the overview image for targets.

(b) *Confocal imaging pattern*: This pattern acquires confocal images of centred target structures with higher resolution of both markers Alexa Fluor® 647 and Alexa Fluor® 532. The autofocus plane is set by JobAfTar532. Then JobTar647 and JobTar532 yield the confocal images. A WaitForCAM job triggers KNIME to send the following switching pattern.

(c) *Switching pattern*: Starting this pattern, the target is shifted to the centre position of the SMLM laser profile. Subsequently, JobAf532 provides the focus plane of the target structures and a camera job (JobCam) switches the detection beam path to the SMLM unit. At last, a WaitForCAM job pauses the confocal scan for SMLM imaging.

3. Exchange PBS with switching buffer and place the sample on the stage.
4. Calibrate the stage on the sample with the Matrix Screener, activate the CAM server, and set up an experiment.
5. Use JobAfGlobal to create a focus map that contains z-offset values for each field for focussing structures coarsely.
6. Start KNIME on the second PC and use the workflow, which is described in the following, to run the experiment.

3.2.2 Workflow for Fully Automated Targeted Confocal and SMLM Microscopy

In order to implement the basic steps for a fully automated microscopy procedure, we utilized KNIME and created a workflow (Fig. 1c) that combines microscopy and image analysis in a clear modulated form. To this end, we used the self-built nodes for integrating the confocal and the SMLM unit in KNIME. In the following, we will give a description of how the workflow runs the microscope in a fully automated imaging routine:

1. First the SMLM unit is configured using MMConfiguration, then the workflow can be started by executing the CAM loop.
2. CAMLoopStart initializes the connection to the CAM server by starting the confocal screen which has been defined in the LAS AF. In this process, the overview pattern is executed to generate an overview image (Fig. 2b) and for saving it on an external server.
3. The CAM loop iterates until the CAM server sends an image path, directing to the overview image on the external server.
4. KNIME accesses the images on the external server by using this path. Subsequent image analysis (*see* **Note 3**) in KNIME decides if the overview image contains a target structure to perform SMLM imaging. If a structure is found (Fig. 2c), their centroids allow calculating a stage-shift. This shift is then used to centre the target in the laser profile of the SMLM setup and in the coordinate system of the confocal image.
5. Confocal image acquisition with higher sampling (Fig. 2d) of the centred target is applied by the confocal imaging pattern.
6. Receiving the confocal image path of the target from the server triggers KNIME to execute the switching pattern. This results in a stage-shift to centre the target in the SMLM setup and executes JobCAM which directs the fluorescence onto the sCMOS camera.
7. Hereupon, SMLM imaging via MMRunSystem is performed. In the node settings we chose a 200×200 pixel array and an exposure time of 50 ms for acquiring 7000 frames per image stack. At each target position two image stacks and two wide-field images were acquired for dual-color SMLM images. For exciting Alexa Fluor® 647 we used a laser wavelength of 647 nm and an intensity of 0.4 kW/cm^2 and for exciting Alexa Fluor® 532 we applied a laser wavelength of 515 nm and an intensity of 1.9 kW/cm^2.
8. At last, the images are saved and a text file stores all settings that were made for imaging with the MMRunSystem node.
9. The loop continues at the next scan position with **step 3**. The times for imaging one scan position are listed in **Note 4**.

10. The CAM loop stops either if the scan is finished and an ending-command is received via CAMLoopStart, or if this node waits for more than a certain number of iterations for a server message.

3.3 Data Analysis of the SMLM Raw Data

The SMLM raw data was saved as multi-page tiff files. Because we are using an sCMOS camera, each pixel has an individual gain value. This should be considered in a pixel-dependent SMLM raw data analysis such as the MATLAB script provided by Huang et al. [11]. As a first attempt to evaluate the raw data, we employed the ImageJ plugin ThunderSTORM [12], although it only uses averaged pixel-gain and offset values (Fig. 2e). In doing so, we kept the default settings which allowed reliable spot detection. Here, ThunderSTORM provided lateral localization uncertainties of (15 ± 8) nm for Alexa Fluor® 647 and (19 ± 12) nm for Alexa Fluor® 532. For regions with interesting structures of Gag and Env, pixel-dependent SMLM data analysis with the mentioned script was executed. With this method, uncertainties of (16 ± 5) nm for Alexa Fluor® 647 and (13 ± 7) nm for Alexa Fluor® 532 were determined by a Cramér-Rao lower bound that is implemented in the script. In this way, structures were found (Fig. 2f) similar to those reported by Muranyi et al. [5]. A cluster analysis using the density-based spatial clustering of applications with noise (DBSCAN) algorithm [13] allowed determining the radius of Gag protein (green) accumulations to (56 ± 5) nm. The obtained value lies in the two sigma range of a radius with 65 nm which has been reported for those punctuate structures.

4 Notes

1. It depends on the device and your operating system if there is a driver, µManager can utilize. This needs to be considered in the decision which hardware to use in the microscopy setup. A list of supported devices can be found on [14].
2. The right focal plane is found if an autofocus routine provided by the Leica microscope fits to the target structures. For that, several autofocus routines are available in the Matrix Screener.
3. The image analysis starts with image preprocessing by a Gaussian filter and background subtraction. Then an image segmentation (Fig. 2c) using Otsu's method and a connected component analysis are performed. Subsequently, a trained tree ensemble classifies the obtained segments. The used tree ensemble learner generates similar models to the ones of a random forest learner, but its implementation in KNIME enables exporting the trained models. That is why it was favored over the random forest implementation. For the tree

ensemble the default settings in KNIME were kept such as the information gain ratio as split criterion and a number of 100 models. Depending on the target structures, intensity features like minimum, maximum and mean intensities of the segments are sufficient to predict targets reliably.

4. Here, the times for imaging one field with a single laser wavelength containing a target structure are listed. For finding the autofocus plane and acquiring an overview image, about 8 s are needed. Subsequent image processing and confocal imaging with higher sampling rate requires about 24 s. SMLM imaging dominates the measurement time with about 5 min and 50 s for 7000 frames and an exposure time of 50 ms. Saving an image stack with a size of $200 \times 200 \times 7000$ pixels in a 16 bit format takes at last about 1 min using JAI. For instance, a dual-color measurement that scanned 153 positions in confocal mode as well as identified and imaged 18 targets via SMLM ran for about 7 h and 50 min.

Acknowledgment

The authors thank Michael Berthold and the KNME/KNIP developers for their support with the creation of the KNIME nodes. This work was funded within the project CancerTelSys (grant number 01ZX1302) in the e:Med program, within the project RNA-Code (grant number 031A298) in the e:Bio program and within the project HD-HuB (grant number 031A537C), all of the German Federal Ministry of Education and Research (BMBF). The Advanced Biological Screening Facility is also supported by the CellNetworks-Cluster of Excellence (grant number EXC81).

References

1. Rust MJ, Bates M, Zhuang X (2006) Stochastic optical reconstruction microscopy (STORM) provides sub-diffraction-limit image resolution. Nat Methods 3(10):793–795
2. Tischer C, Hilsenstein V, Hanson K, Pepperkok R (2014) Chapter 26 – Adaptive fluorescence microscopy by online feedback image analysis. In: Jennifer CW, Torsten W (eds) Methods in cell biology, vol 123. Academic Press, New York, pp 489–503
3. Conrad C, Wunsche A, Tan TH, Bulkescher J, Sieckmann F, Verissimo F, Edelstein A, Walter T, Liebel U, Pepperkok R, Ellenberg J (2011) Micropilot: automation of fluorescence microscopy-based imaging for systems biology. Nat Methods 8(3):246–249
4. Berthold MR, Cebron N, Dill F, Gabriel TR, Kötter T, Meinl T, Ohl P, Sieb C, Thiel K, Wiswedel B (2008) KNIME: the Konstanz Information Miner. In: Preisach C, Burkhardt H, Schmidt-Thieme L, Decker R (eds) Data analysis, machine learning and applications: proceedings of the 31st annual conference of the Gesellschaft für Klassifikation e.V., Albert-Ludwigs-Universität Freiburg, March 7–9, 2007. Springer, Berlin, Heidelberg, pp 319–326
5. Muranyi W, Malkusch S, Müller B, Heilemann M, Kräusslich H-G (2013) Super-resolution microscopy reveals specific recruitment of HIV-1 envelope proteins to viral assembly sites dependent on the envelope C-terminal tail. PLoS Pathog 9(2):1–13

6. Lampe M, Briggs JAG, Endress T, Glass B, Riegelsberger S, Kräusslich H-G, Lamb DC, Bräuchle C, Müller B (2007) Double-labelled HIV-1 particles for study of virus–cell interaction. Virology 360(1):92–104
7. Trkola A, Purtscher M, Muster T, Ballaun C, Buchacher A, Sullivan N, Srinivasan K, Sodroski J, Moore JP, Katinger H (1996) Human monoclonal antibody 2G12 defines a distinctive neutralization epitope on the gp120 glycoprotein of human immunodeficiency virus type 1. J Virol 70(2):1100–1108
8. Edelstein AD, Tsuchida MA, Amodaj N, Pinkard H, Vale RD, Stuurman N (2014) Advanced methods of microscope control using μManager software. J Biol Methods 1 (2):10
9. Schindelin J, Dietz C, Gunkel M (2015) Micro-manager integration for KNIME image processing. https://github.com/knime-ip/knip-micromanager. Accessed 20 Dec 2016
10. Dempsey GT, Bates M, Kowtoniuk WE, Liu DR, Tsien RY, Zhuang X (2009) Photoswitching mechanism of cyanine dyes. J Am Chem Soc 131(51):18192–18193
11. Huang F, Hartwich TMP, Rivera-Molina FE, Lin Y, Duim WC, Long JJ, Uchil PD, Myers JR, Baird MA, Mothes W, Davidson MW, Toomre D, Bewersdorf J (2013) Video-rate nanoscopy using sCMOS camera-specific single-molecule localization algorithms. Nat Methods 10(7):653–658
12. Ovesný M, Křížek P, Borkovec J, Švindrych Z, Hagen GM (2014) ThunderSTORM: a comprehensive ImageJ plug-in for PALM and STORM data analysis and super-resolution imaging. Bioinformatics 30(16):2389–2390
13. Ester M, Kriegel H-P, Sander J, Xu X (1996) A density-based algorithm for discovering clusters in large spatial databases with noise. In: KDD-96. AAAI Press, Portland, OR, pp 226–231
14. Micro-Manager (2011) Device support. https://micro-manager.org/wiki/Device_Support. Accessed 20 Dec 2016

Chapter 13

Brain Slice Staining and Preparation for Three-Dimensional Super-Resolution Microscopy

Christopher L. German, Manasa V. Gudheti, Annette E. Fleckenstein, and Erik M. Jorgensen

Abstract

Localization microscopy techniques—such as photoactivation localization microscopy (PALM), fluorescent PALM (FPALM), ground state depletion (GSD), and stochastic optical reconstruction microscopy (STORM)—provide the highest precision for single-molecule localization currently available. However, localization microscopy has been largely limited to cell cultures due to the difficulties that arise in imaging thicker tissue sections. Sample fixation and antibody staining, background fluorescence, fluorophore photoinstability, light scattering in thick sections, and sample movement create significant challenges for imaging intact tissue. We have developed a sample preparation and image acquisition protocol to address these challenges in rat brain slices. The sample preparation combined multiple fixation steps, saponin permeabilization, and tissue clarification. Together, these preserve intracellular structures, promote antibody penetration, reduce background fluorescence and light scattering, and allow acquisition of images deep in a 30 μm thick slice. Image acquisition challenges were resolved by overlaying samples with a permeable agarose pad and custom-built stainless-steel imaging adapter, and sealing the imaging chamber. This approach kept slices flat, immobile, bathed in imaging buffer, and prevented buffer oxidation during imaging. Using this protocol, we consistently obtained single-molecule localizations of synaptic vesicle and active zone proteins in three dimensions within individual synaptic terminals of the striatum in rat brain slices. These techniques may be easily adapted to the preparation and imaging of other tissues, substantially broadening the application of super-resolution imaging.

Key words Super-resolution microscopy, Single-molecule localization, Brain slice, Tissue imaging

1 Introduction

Single-molecule localization techniques, including PALM, FPALM, STORM, and GSD, currently provide the highest precision for protein localization in micrographs, with 20 nm precision in the *X* and *Y* axes and 50 nm precision in the *Z* axis [1]. Despite the successful application of super-resolution imaging to thin, sparse samples such as primary cell cultures, the widespread adoption of localization microscopy to thicker samples, such as tissues or organs, is limited by substantial obstacles presented during sample

Holger Erfle (ed.), *Super-Resolution Microscopy: Methods and Protocols*, Methods in Molecular Biology, vol. 1663, DOI 10.1007/978-1-4939-7265-4_13,

preparation and imaging. As a result, only a few studies have been published using tissue slices for single-molecule super-resolution imaging [2–4] (*see* **Note 1**).

Imaging brain slices using single-molecule localization microscopy poses five fundamental challenges that this protocol addresses. First, fine synaptic terminal structures must be preserved, but thick sections must be sufficiently permeabilized to allow antibody penetrance deep into the tissue. This can be accomplished through cardiac perfusion and extensive fixation with paraformaldehyde followed by saponin permeabilization during staining. Second, prolonged fixation with aldehydes and thick tissues create background autofluorescence that must be reduced to ensure precise single-molecule localization and allow imaging at depth, which can be achieved through sodium borohydride incubation prior to antibody staining and tissue clarification prior to imaging. Third, light scattering must be reduced to allow deep imaging (~5–10 μm) within a tissue slice and maximize the precision of particle localization, which can be achieved by clarifying slices with Scale U2 buffer prior to imaging accomplished. Fourth, the tissue must be immersed within a reducing buffer to allow fluorophore blinking and maintenance of the dark state [5], but remain completely immobile on the coverslip while imaging. Fifth, oxygen promotes dye photobleaching [6], so the sample chamber must be sealed for the duration of the imaging session and/or the imaging buffer must be replaced regularly. To solve these two problems, this protocol utilizes an agarose pad and stainless-steel adapter to prevent sample movement during imaging, yet allow imaging buffer to access the tissue and be changed repeatedly. To prevent imaging buffer oxygenation during imaging, the chamber is sealed with a parafilm-lined culture plate lid. Using the methods outlined within this protocol, three-dimensional super-resolution images can be obtained deep within tissue slices. The effectiveness of this protocol in imaging thick brain slices is demonstrated herein through staining and imaging of the vesicular monoamine transporter-2 (VMAT2), a marker of dopaminergic synaptic vesicles, and Bassoon, a structural protein of presynaptic active zones, within individual synaptic terminals of the striatum from 30 μm thick rat brain slices. The sample preparation, staining, and imaging steps of this protocol can be adapted to thick tissue slices of other complex organ systems.

2 Materials

2.1 Equipment

1. Perfusion pump.
2. Rodent and small animal guillotine.
3. Freezing stage microtome.

4. Single-molecule localization super-resolution microscope (Vutara 350; Bruker Nanosurfaces, Salt Lake City, UT, USA):
   (a) Irradiation lasers at 405, 488, 561, and 647 nm (15 kW/cm^2).
   (b) Water-immersion objective (60×, 1.2NA).
   (c) EM-CCD camera.

2.2 Brain Collection and Slicing

1. Male Sprague-Dawley rats: 300–400 g in weight (*see* **Note 2**).
2. Pentobarbital sodium salt: 250 mg/kg in 0.9% sterile saline solution. Prepare in small batches, vortex into solution, and use immediately.
3. Handheld wire cutters.
4. Straight operating scissors; 4.5–5.5 in.
5. Small dressing forceps.
6. Razor blades.
7. Phosphate-buffered saline (PBS): 137 mM NaCl, 2.7 mM KCl, 10.1 mM Na_2HPO_4, and 1.8 mM KH_2PO_4 in ddH_2O, pH 7.5. Store at 4 °C.
8. Four percent paraformaldehyde (4% PFA): 4% (wt/vol) paraformaldehyde in PBS, pH 7.5, store at 4 °C.
9. Thirty percent sucrose (wt/vol) in 4% PFA, store at 4 °C.
10. Dry ice.
11. Shandon cyromatrix freezing medium.
12. 90% ethanol or methanol.
13. Paint brush.

2.3 Slice Staining

1. Blocking buffer: 5% (vol/vol) normal donkey serum (species from which secondary antibody is derived), 1% bovine serum albumin (wt/vol), and 0.1% saponin (wt/vol) in PBS, pH to 7.5.
2. PBS with 0.1% saponin (PBS-saponin).
3. Primary antibodies
   (a) Rabbit anti-vesicular monoamine transporter-2 polyclonal antibody (VMAT2; made by Bethyl Laboratories, Montgomery, TX, USA; *see* **Note 3**).
   (b) Guinea pig anti-Bassoon polyclonal antibody (Synaptic Systems, Göttingen, Germany).
4. Secondary antibodies
   (a) Goat anti-rabbit Alexa Fluor 647 IgG (Invitrogen, ThermoFisher Scientific, Pittsburgh, PA, USA).
   (b) Donkey anti-guinea pig Cy3B IgG (conjugated in house, *see* **Note 4**; IgG antibody from Jackson Immunoresearch,

West Grove, PA, USA; Cy3B NHS-ester from GE Healthcare Life Sciences, Pittsburgh, PA, USA).

5. Parafilm M.
6. Scale U2 buffer: 4 M urea, 30% glycerol (vol/vol), 0.1% triton X-100 (vol/vol), ddH_2O.

2.4 Super-Resolution Imaging

1. Agarose stock: 2% in ddH_2O, microwave until the agarose goes into solution (45 s to 2 min), stir briefly, and aliquot into vials or test tubes. Store at 4 °C until use.
2. 10 cm petri dish.
3. 6 cm petri dish cover.
4. 25 mm #1 circular coverslips (0.12 mm thick).
5. Imaging buffer: 20 mM cysteamine, 1% 2-mercaptoethanol (vol/vol), 168.8 active units glucose oxidase per ml of buffer, 1404 active units catalase per ml of buffer, 10 mM NaCl, 10% (wt/vol) glucose, 50 mM Tris–HCl, ddH_2O, pH 8.0.
6. Stainless-steel imaging chamber (Life Sciences A-7816).
7. Stainless-steel imaging insert (machined in-house; *see* Fig. 1d for schematics): the bottom of the insert was painted with Rust-Oleum Painter's Touch Ultra Cover Semi-gloss Black to prevent light backscattering during laser excitation.

3 Method

3.1 Brain Collection and Slicing

1. Deeply anesthetize a rat with an intraperitoneal 250 mg/kg sodium pentobarbital injection.
2. Once the animal no longer responds to both a tail/foot pinch and an audible startle (usually within 5 min of injection), clean the chest fur and skin with ethanol, open the chest cavity to expose the heart. Cut a small incision into the right ventricle of the heart and perfuse 100 ml of ice-cold 4% PFA at a rate of 25 ml/min into the left ventricle of the heart (*see* **Note 5**).
3. Decapitate the rat by guillotine following perfusion. To dissect the brain, open the skull by placing the wire cutter shears in the optic foramen and fracturing the nasal bridge of the frontal bone. Next, place the wire snip shears in the auditory canal and the foramen magnum and fracture the bones within the masseter region. Starting in the foramen magnum, press the bottom shear of the operating scissors against the parietal bone of the skull and slowly cut along the sagittal fissure until the snipped region of nasal bridge, shortly past bregma, is reached. Gently pry away the parietal and frontal bones to expose the brain. Scoop the brain out the skull with the forceps and place it into ice-cold 4% PFA.

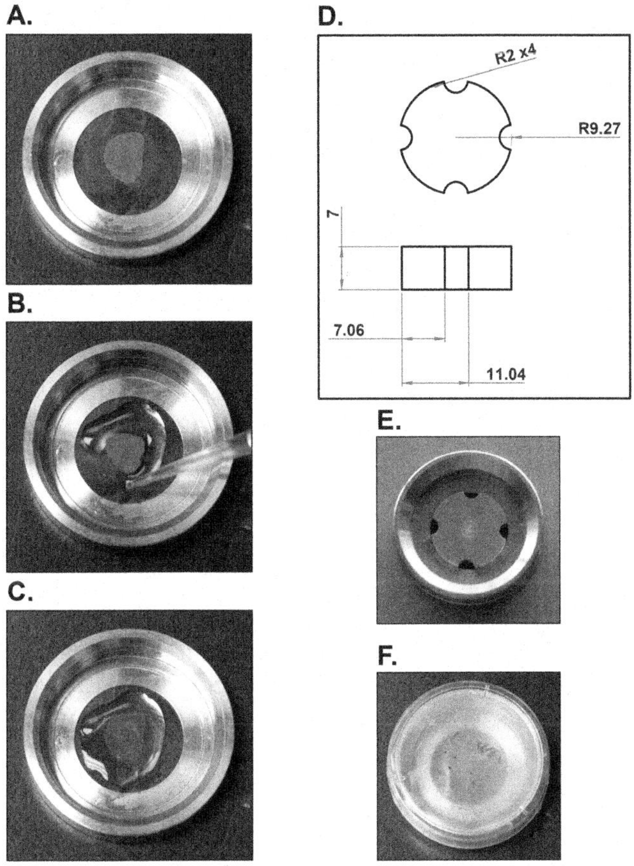

Fig. 1 Brain slice mounting for super-resolution imaging. (**a**) A brain slice is floated, mounted on circular glass coverslips and then allowed to completely dry. (**b**, **c**) Liquid 2% agarose is pipetted along the edges of the brain slice and then across the entire surface to form an immobile pad. (**d**) Schematics of the stainless-steel imaging chamber insert with all dimensions in mm and radii with a curvature measurement of 2 mm. (**e**) The stainless-steel insert is placed in the imaging chamber along with imaging buffer and (**f**) sealed with a parafilm-lined lid of a 35 mm culture plate for imaging

4. Incubate brains in 4% PFA for at least 24 h at 4 °C. Brains can be stored for prolonged periods (months to years) at 4 °C provided they are kept submerged 4% PFA.
5. Cryoprotect the brain by transferring it to a 50 ml conical tube containing 30% sucrose in PFA and incubating until the brain sinks to the bottom of the tube (approximately 72 h).
6. Using a razor blade, remove the cerebellum and olfactory bulbs prior to slicing the brain.
7. Cool the microtome stage with dry ice placed into the bath containing 90% ethanol and place a sufficient amount of Shandon cryomatrix freezing medium on the stage to embed at least 3 mm of the caudal brain region.

8. Place the brain, rostral end up, into the freezing medium just as it begins to freeze. Quickly apply a few mm of medium to the base of the brain to prevent separation from the stage during slicing.
9. Flash freeze the brain by covering it with crushed dry ice for 1.5 min.
10. Cut 30 μm coronal slices through the striatum (or brain region of interest), remove them from the microtome blade with the paintbrush, and place them into 24-well plate wells containing PBS. Store slices at 4 °C until use.

3.2 Slice Staining

1. Incubate sections with 0.1% $NaBH_4$ in PBS while gently rotating for 15 min at room temperature. *See* **Note 6**.
2. Wash sections once for 10 min and then twice for 5 min with PBS while gently rotating at room temperature.
3. Incubate sections in blocking buffer for at least 2 h while gently rotating at room temperature. 250 μl of buffer is sufficient per each well in a 24-well plate.
4. Remove blocking solution and incubate with primary antibody mixture in PBS-saponin (200 μl total volume; *see* **Note 7**) while gently rotating overnight at 4 °C. To prevent antibody evaporation during this long incubation stage, wells were sealed with a sheet of parafilm placed between the lid and plate.
5. Wash sections once for 10 min then twice for 5 min with PBS-saponin while gently rotating at room temperature.
6. Incubate sections in secondary antibody mixture (200 μl; PBS-saponin) for 1.5 h in the dark (or covered) while rotating gently at room temperature.
7. Wash sections twice for 5 min with PBS-saponin and three times for 5 min with PBS while rotating gently at room temperature.
8. Post-fix sections in 4% PFA for 1 h while rotating gently at room temperature (*see* **Note 8**).
9. Wash sections three times for 5 min with PBS while rotating gently at room temperature.
10. Incubate samples in Scale U2 buffer overnight (12 h minimum) and store covered at 4 °C until imaging (*see* **Note 9**).

3.3 Super-Resolution Imaging

1. Remove 2% agarose stock from 4 °C storage, liquefy in a microwave for 45 s, and incubate at 55 °C until use.
2. Submerge a coverslip and brain slice in a PBS-filled 10 cm petri dish and use the paint brush to gently float the brain slice onto the coverslip.

3. While still wet, gently move the brain slice to the center of the coverslip and flatten. Allow the brain slice to dry completely on the coverslip before proceeding (Fig. 1a; *see* **Note 10**).
4. Place the coverslip into the imaging chamber and rapidly pipette liquid (55 °C; approximately 150 μl) 2% agarose onto the slice, starting at the edges and working in a spiral pattern (Fig. 1b and c). This step must be performed rapidly or the agarose will solidify within the pipette.
5. Allow the agarose to cool to room temperature for 5 min. Since rapid temperature reduction deforms the agarose pad, failure to perform this step leads to agarose dissociation from the coverslip when ice-cold imaging buffer is applied.
6. Place the stainless-steel insert (Fig. 1d; *see* **Note 11**) over the sample (Fig. 1e, fill the chamber with imaging buffer (approximately 750 μl), and seal the chamber with a parafilm-lined 30 mm culture plate lid (Fig. 1f; *see* **Note 12**).
7. Image the brain slice, replacing imaging buffer every 1 h. *See* Fig. 2 for a representative image of VMAT2 and Bassoon within striatal terminals, and **Note 13** for further discussion. To store slices for future imaging sessions, replace the imaging buffer with PBS and incubate at 4 °C.

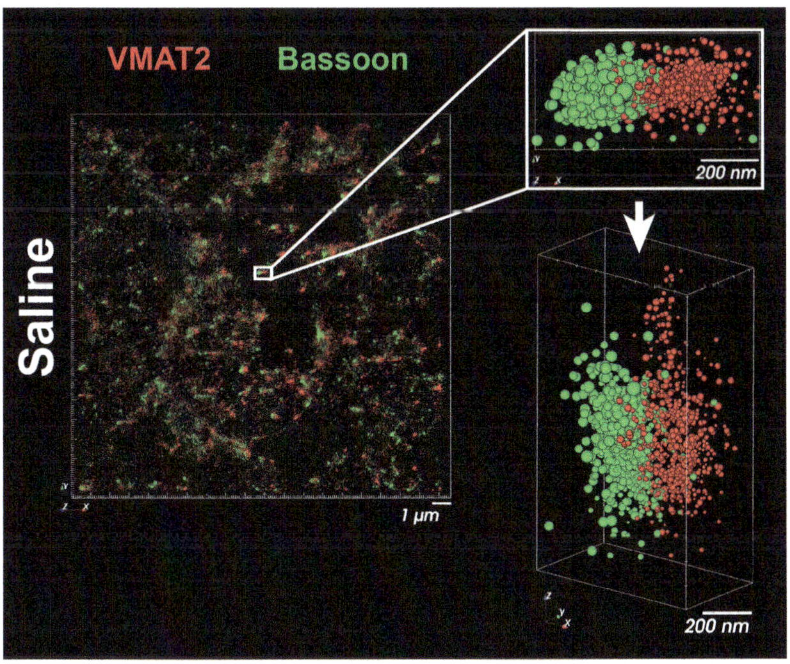

Fig. 2 The distribution of VMAT2 (*red*) and Bassoon (*green*) within individual presynaptic terminals of the striatum

4 Notes

1. Other approaches have imaged thinner tissue sections by sandwiching them between the slide and coverslip with a small amount of imaging buffer and sealing the edges of the coverslip with nailpolish [3]. We found this approach problematic for at least two reasons. First, prolonged imaging sessions (greater than an hour) required frequent replacement of the imaging buffer to ensure fluorophore stability and blinking, which was not possible when the coverslip was sealed. Second, the coverslip alone did not provide sufficient weight to flatten the slice along the surface of the slide, which led to slight tissue folding, flotation, and movement of the tissue. The tissue folding and flotation led to *z*-axis inconsistency, while slight tissue movement led to dramatic reduction in the accuracy of particle localization in all axes.
2. Rats were housed with food and water ad libitum in a temperature-controlled environment and all experiments were approved by the University of Utah Institutional Animal Care and Use Committee.
3. The VMAT2 rabbit polycolonal antibody was produced by Bethyl Labs and raised against a 20-amino acid c-terminal sequence (CTQNNIQSYPIGEDEESESD) of human VMAT2. Proper immunoreactivity of this antibody against VMAT2 was validated through western blotting, immunoprecipitation, and tandem mass spectrometry identification of immunoreactive bands (data not shown).
4. Cy3b donkey anti-guinea pig antibody was created by cross-linking Cy3b NHS-ester to donkey anti-guinea pig antibody which was then purified over NAP-5 columns per manufacturer's direction.
5. The integrity of fine central nervous system structures, such as pre- and post-synaptic terminals, was accomplished through thorough fixation, cryoprotection, and gentle detergent use (saponin) during staining. 4% PFA perfusion and the prolonged, 24 h incubation that follows ensure structures deep within the brain are well fixed, and sucrose cryoprotection is a critical step that prevents ice crystal formation and membrane lysis during flash freezing.
6. Free aldehyde fluorescence can be quite pronounced in super-resolution images because of the prolonged fixation and thickness of sections; thus, autofluorescence was quenched with $NaBH_4$ treatment prior to staining.
7. Antibody penetration into thick tissue samples requires tissue permeabilization using a detergent. Saponin, unlike Triton X-

100, is a mild and reversible nonionic detergent that improves staining efficiency but does not damage the structure of organelles and vesicles in presynaptic terminals [7], and was included in all blocking, staining, and wash steps.

8. The post-fixation step is critical for preventing tissue destruction and loss of signal when clarifying slices with the Scale U2 steps that follow.
9. Tissue clarification using Scale U2 buffer dramatically reduced both light scattering and tissue autofluorescence, and was necessary for imaging deeper than a few microns from the slice surface. *See* [8] for further discussion of Scale buffer formulation and use.
10. Allowing the brain slice to dry to the coverslip is important for two reasons. First, it completely flattens the tissue to the surface of the coverslip, which aids in *z*-plane imaging. Second, it prevents tissue folding when the warm agarose is subsequently added.
11. Keeping brain slices flat and immobile on the coverslip while immersed in buffer during image acquisition is a significant challenge for localization microscopy of tissues. To solve these problems, slices were overlayed with a 2% agarose pad dense enough to immobilize the tissue but porous enough to allow buffer penetrance. Since the agarose pad can occasionally detach and/or move during imaging, a stainless-steel insert was designed and fabricated to sit on the top of the agarose pad, provide sufficient weight to prevent movement, and still allow imaging buffer to access the tissue slice.
12. Throughout the course of imaging, the glucose oxidase and catalase necessary for fluorophore stability converts oxygen and glucose into gluconic acid [6]. The progressive acidification of the sample impairs fluorophore blinking, necessitating the replacement of imaging buffer every 45 min. Covering the imaging chamber with Parafilm M and sealing it with the petri dish cover reduces oxygen exposure during imaging, prolonging imaging time before the buffer must be changed.
13. VMAT2 is clustered adjacent to Bassoon throughout the striatum in pairs approximately 500 nm in width, consistent with the scale of dopaminergic terminals identified by electron microscopy [9]. Prior to these data, the capacity to resolve the distribution of active zone components within individual terminals of complex tissues, such as brain slices, has largely been limited to immuno-electron microscopy.

Acknowledgments

We thank Sam Feinman for his assistance in creating the imaging adapter blueprints in SolidWorks. E.M. Jorgensen is a Howard Hughes Medical Institute (HHMI) investigator. This work was funded by the NIH grants R01NS034307 (E.M.J. and M.V.G.), DA019447 (A.E.F.), DA13367 (C.L.G. and A.E.F.), and DA11389 (C.L.G. and A.E.F.).

Conflicts of Interest: M.V. Gudheti is an employee of Bruker Nano Surfaces.

References

1. Maglione M, Sigrist SJ (2013) Seeing the forest tree by tree: super-resolution light microscopy meets the neurosciences. Nat Neurosci 16 (7):790–797
2. Baddeley D et al (2011) 4D super-resolution microscopy with conventional fluorophores and single wavelength excitation in optically thick cells and tissues. PLoS One 6(5):e20645
3. Dani A et al (2010) Superresolution imaging of chemical synapses in the brain. Neuron 68 (5):843–856
4. Nanguneri S et al (2012) Three-dimensional, tomographic super-resolution fluorescence imaging of serially sectioned thick samples. PLoS One 7(5):e38098
5. Endesfelder U et al (2011) Chemically induced photoswitching of fluorescent probes—a general concept for super-resolution microscopy. Molecules 16(4):3106–3118
6. Shi X, Lim J, Ha T (2010) Acidification of the oxygen scavenging system in single-molecule fluorescence studies: in situ sensing with a ratiometric dual-emission probe. Anal Chem 82 (14):6132–6138
7. Goldenthal KL et al (1985) Postfixation detergent treatment for immunofluorescence suppresses localization of some integral membrane proteins. J Histochem Cytochem 33 (8):813–820
8. Hama H et al (2011) Scale: a chemical approach for fluorescence imaging and reconstruction of transparent mouse brain. Nat Neurosci 14 (11):1481–1488
9. Nirenberg MJ et al (1997) Vesicular monoamine transporter-2: immunogold localization in striatal axons and terminals. Synapse 26 (2):194–198

Chapter 14

Correlative In-Resin Super-Resolution Fluorescence and Electron Microscopy of Cultured Cells

Errin Johnson and Rainer Kaufmann

Abstract

Correlative super-resolution light and electron microscopy (super-resolution CLEM) is a powerful and emerging tool in biological research. The practical realization of these two very different microscopy techniques with their individual requirements remains a challenging task. There is a broad range of approaches to choose from, each with their own advantages and limitations. Here, we present a detailed protocol for in-resin super-resolution CLEM of high-pressure frozen and freeze substituted cultured cells. The protocol makes use of a strategy to preserve the fluorescence and photo-switching capabilities of standard fluorescent proteins, such as GFP and YFP, to enable single-molecule localization microscopy (SMLM) in-resin sections followed by transmission electron microscopy (TEM) imaging. This results in a fivefold improvement in resolution in the fluorescence image and a more precise correlation of the distribution of fluorescently labeled molecules with EM ultrastructure compared with conventional CLEM.

Key words Correlative light and electron microscopy, Super-resolution microscopy, Transmission electron microscopy, Single-molecule localization microscopy, SMLM, GFP, Tannic acid, High pressure freezing, In-resin fluorescence

1 Introduction

Correlative light and electron microscopy (CLEM) encompasses a range of different approaches which enable fluorescently labeled proteins/compounds of interest to be placed in their ultrastructural context (*see*, e.g., Müller-Reichert and Verkade [1]). The use of super-resolution fluorescence microscopy techniques in CLEM bridges the resolution gap between conventional light and electron microscopy (EM) [2–6], resulting in a more precise correlation between specific fluorescently labeled molecules and the surrounding ultrastructure. As both super-resolution microscopy and EM offer a plethora of implementation options, often with different requirements regarding sample preparation for each technique, the combination of possibilities is numerous and the practical implementation remains somewhat challenging [7–9].

Holger Erfle (ed.), *Super-Resolution Microscopy: Methods and Protocols*, Methods in Molecular Biology, vol. 1663, DOI 10.1007/978-1-4939-7265-4_14,

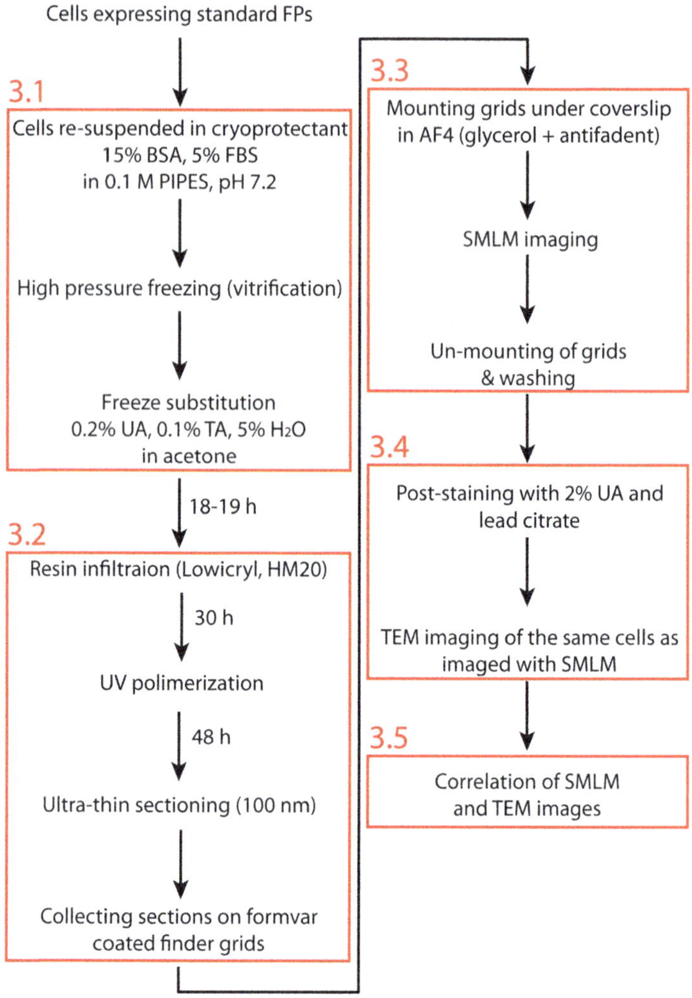

Fig. 1 Overview of protocol for in-resin super-resolution CLEM. Highlighted in *red* are the corresponding paragraphs in Subheading 3. *UA* uranyl acetate, *TA* tannic acid

We have developed a method that preserves the fluorescence and photo-switching capabilities of standard fluorescence proteins throughout the process of high-pressure freezing (HPF), freeze substitution (FS), and resin embedding [10], such that sections can be imaged with single-molecule localization microscopy (SMLM) followed by transmission electron microscopy (TEM). An overview of the protocol is given in Fig. 1.

The advantages of this procedure are: (1) noninvasive fluorescence labeling of the living cell, (2) superior structural preservation of the sample compared to chemical fixation, and (3) minimal structural changes of the sample between fluorescence and electron microscopy imaging. Being independent of special fluorophores

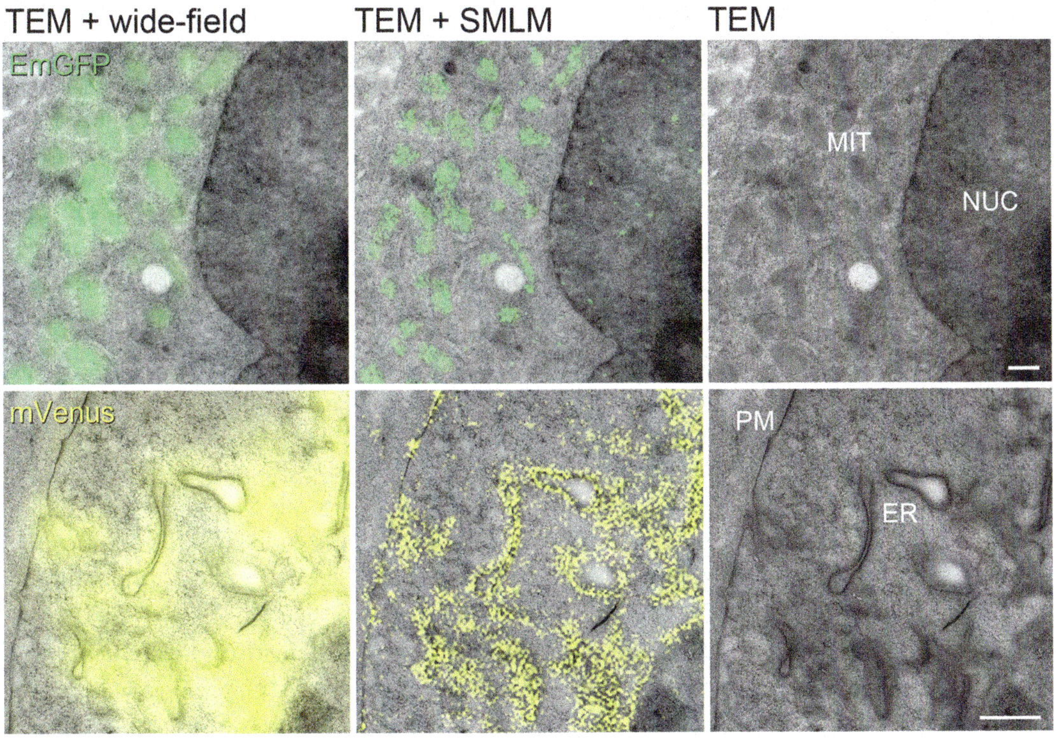

Fig. 2 Example images of in-resin super-resolution CLEM. *Top row*: EmGFP labelled mitochondria (CellLight Mictochondria-GFP, Bacman 2.0) in HeLa cells. *Bottom row*: HEK293T cells transfected with EphA2/A4 receptor proteins fused to mVenus (from Johnson et al. (2015) [10]). In the *left column* overlays of conventional wide-field fluorescence images with the corresponding TEM images are shown; in the *middle column* overlays of SMLM and TEM images; and in the *right column* the TEM images alone. Scale bars are 0.5 μm. *MIT* mitochondria, *NUC* nucleus, *PM* plasma membrane, *ER* endoplasmic reticulum

[6], the method enables a wide range of applications and broad choice of fluorescent proteins for labelling (Fig. 2, for more examples *see* [10]).

2 Materials

2.1 High-Pressure Freezing and Freeze Substitution

1. Culture cells expressing fluorescent fusion proteins (*see* **Notes 1** and **2**) grown in 6-well cell culture plates.
2. High Pressure Freezer (HPF, *see* **Note 3**).
3. Liquid nitrogen and appropriate personal protective equipment.
4. Trypsin (TryplE Express, Gibco).
5. 0.2 M PIPES buffer stock solution, pH 7.2: On a magnetic stirrer, add 302.37 g PIPES to 300 ml ultrapure water in a 1 L beaker containing a magnetic flea. Add 10 M NaOH until the pH reaches 7.2; the solution should go from cloudy to clear.

Make up to 500 ml final volume with ultrapure water. Store at 4 °C.

6. Cryoprotectant: 15% (w/v) Albumin bovine serum (BSA), Fraction V, ≥98% (Sigma) + 5% (v/v) Fetal Bovine Serum (FBS, Sigma) in 0.1 M PIPES buffer. In a 50 ml centrifuge tube, add 2.25 g BSA + 750 μl FBS and make up to 15 ml with 0.1 M PIPES. Vortex until the BSA is fully dissolved. Alternatively, 20% BSA in 0.1 M PIPES pH 7.2 may be used, prepared according to McDonald et al. [11] (*see* **Note 4**).
7. Acetone, 99.6% ACS.
8. 10% (w/v) Tannic Acid (TA) in acetone stock: Add 0.025 g low molecular weight TA (Electron Microscopy Sciences; *see* **Note 5**) to 0.25 ml acetone and vortex briefly to mix (*see* **Note 6**).
9. 5% Uranyl Acetate (UA) in methanol stock: Carefully weigh out 0.5 g UA (Agar Scientific; *see* **Note 7**) and add to a 15 ml centrifuge tube with 10 ml methanol. Cap and shake vigorously to dissolve. Wrap tube in foil and store at −20 °C until required (*see* **Note 8**).
10. 1.5 or 2 ml cryo-tubes for freeze substitution (FS), one for each different cell line you are using.
11. Automated Freeze Substitution (AFS) unit with UV light attachment (*see* **Note 9**). Fill with liquid nitrogen and set the temperature to −130 °C.
12. FS medium in acetone: 0.1% TA (v/v) + 0.2% uranyl acetate (in methanol) (v/v) + 5% water (v/v). To each cryo-tube add 10 μl 10% TA stock, 40 μl 5% UA stock, 50 μl water, and 900 μl acetone (*see* **Notes 10** and **11**). Mix well. Make this up fresh each time and transfer immediately to the precooled AFS unit (*see* **Note 12**).

2.2 Resin Infiltration, Embedding, and Sectioning

1. Lowicryl HM20 resin (*see* **Note 13**). In a fumehood, weigh 5.96 g Crosslinker D, 34.04 g Monomer E and 0.2 g Initiator C into a 50 ml plastic centrifuge tube. Vigorously shake the mixture to dissolve the initiator (~30 s) and store at −20 °C.
2. Acetone, 99.6% ACS.
3. Pure ethanol.
4. Glass bottles with screw caps, 15 ml.
5. Aluminium dishes, 43 mm (Agar Scientific).
6. Reagent bath (Leica) and flow through rings (Leica).
7. Aluminum foil.
8. Razor blade.
9. Liquid nitrogen and appropriate personal protective equipment.

10. Perfect loop for transferring sections to grids (TAAB Laboratories).
11. Formvar coated Finder grids. Quickly dip a clean glass slide in and out of a solution of 0.6% Formvar in chloroform or 1,2-dichloroethane (w/v), then allow the slide to dry. Score the edges and base of one side of the slide, breath on it briefly to loosen the film, and submerge it at a shallow angle in a bath of ultrapure water. The film should float off on the surface of the water (*see* **Note 14**) and be uniform pale grey in color (*see* **Note 15**). Carefully place 200 mesh copper F1 alphanumeric finder grids (Agar Scientific) onto the film, dull side down. Gently roll a small square of parafilm over the top of the grids and formvar, then transfer the parafilm to the bench (grid side up) to dry.
12. Diamond knife and ultramicrotome (*see* **Note 16**).

2.3 SMLM Imaging

1. Glass slides (conventional or what is compatible with SMLM system).
2. High precision coverslips no. 1.5 (170 μm thickness, 5 μm tolerance).
3. Mounting medium (Citifluor, AF4, *see* **Note 17**).
4. Nail polish.
5. Tweezers for handling grid (type 5 tweezer, 45° angled) and coverslip (flat end tweezer).
6. SMLM microscope with possibility for low magnification mapping of the grid.
7. Ultrapure water.

2.4 Post Staining and TEM Imaging

1. 2% Uranyl acetate aqueous stock: Carefully weigh out 1 g UA (Agar Scientific; *see* **Notes 7** and **8**) and add to 50 ml ultrapure water in a brown glass bottle containing a magnetic flea. Mix on a magnetic stirrer for ~30 min and then let it sit overnight before use. Store at room temperature or 4 °C (*see* **Note 18**).
2. Reynolds lead citrate: Prepare according to Reynolds, 1963 [12]. Briefly, add 1.33 g lead nitrate $(PbNO_3)_2$ and 1.76 g sodium citrate $(Na_3C_6H_5O_7{\cdot}2H_2O)$ to a 50 ml plastic tube containing 30 ml degassed water (*see* **Note 19**). Shake vigorously for 1 min and then shake intermittently for 30 min. The solution will appear milky. Add 8 ml of freshly prepared 1 N sodium hydroxide (*see* **Note 20**). Invert slowly; the solution should become clear. Add 12 ml degassed water and invert slowly to mix. Wrap tube in foil to protect it from light (or carefully transfer to a brown bottle) and store at 4 °C (*see* **Note 21**).
3. Ultrapure water.
4. Sodium hydroxide pellets.

5. Parafilm, 4 in. × 250 ft.
6. Fine forceps (Antimagnetic no. 7).
7. TEM equipped with a digital camera.

2.5 Correlation of SMLM and TEM Images

1. Software for correlating images (e.g., MATLAB, Mathworks) using fiducial markers or characteristic features as control points (*see* **Note 22**).

3 Methods

3.1 High-Pressure Freezing and Freeze Substitution

1. Before starting, image the cell culture plate containing the transfected cells on an inverted epifluorescence microscope to verify that transfection efficiency is sufficient (>50%), that fluorescence intensity is as expected, and that the cells are looking healthy (*see* **Note 23**). Transfer the plate back to the 37 °C incubator.
2. Place the trypsin in the 37 °C incubator, along with one 15 ml tube of 0.1 M PIPES and the tube containing the cryoprotectant (15% BSA + 5% FBS in 0.1 M PIPES buffer).
3. Top up the hydraulic fluid on the HPF, switch it on and fill with liquid nitrogen (*see* **Note 24**).
4. Program the AFS unit as follows: warm from −130 to −90 °C at a rate of 20 °C/h, hold at −90 °C for 6 h, warm to −45 °C at a rate of 5 °C/h, hold at −45 °C (protocol adapted from Hawes et al. [13]), and save the program for use in **step 9**. Ensure that the standby temperature is set to −130 °C.
5. Trypsinize the cells (*see* **Notes 25–27**). Remove the culture media from each well (*see* **Note 28**) and rinse cells with 1 ml pre-warmed 0.1 M PIPES buffer. Remove the buffer and add 0.5 ml pre-warmed trypsin to each well and swirl it around to ensure that the cells are covered. Incubate at 37 °C for 4 min, then gently tap the plate to release the cells. Add 1 ml of pre-warmed cryoprotectant solution and resuspend the cells gently. Transfer the cell suspensions to a 15 ml centrifuge tube and spin in a bench top centrifuge for 2 min at 1000 rpm. Proceed without delay to the HPF.
6. Remove the supernatant and load approx. 0.5 μl of the resulting cell slurry to the membrane carrier (*see* **Note 29**), immediately load into the HPF and freeze the sample. Repeat this step twice to produce three replicate frozen carriers per sample.
7. Repeat **steps 5** and **6** for the remaining samples.
8. Under liquid nitrogen, transfer the carriers to the cap of a 50 ml centrifuge tube (*see* **Note 30**) and then use tweezers to carefully carry the cap over to the AFS unit, which has been

precooled to −130 °C and contains the cryo-tubes with FS solution (*see* Materials 2.1). Transfer the carriers to the appropriate tube and carefully cap it (*see* **Note 31**). Repeat for all remaining samples.

9. Start the FS cycle programmed in **step 4**. When the program reaches −45 °C, wait 1–2 h before proceeding with the next step (*see* **Note 32**). In the meantime, place a glass vial containing 10 ml acetone in the AFS chamber.

3.2 Resin Infiltration, Embedding, and Sectioning

1. Remove the FS solution, add ~1 ml precooled acetone to each tube and incubate for 15 min (*see* **Note 33**). In the meantime, place two glass vials containing 10 ml ethanol in the AFS chamber, along with the aluminum dish.
2. Remove the acetone, add ~1 ml precooled ethanol to each tube, and incubate for 15 min. In the meantime, assemble the reagent bath and flow through ring (*see* **Note 34**) or a similar system and place this/these in the AFS chamber.
3. Add just enough precooled ethanol to the reagent bath (through the center flow ring) to cover the base of the flow ring tubes where the carriers will sit. Carefully tip the carriers from the first cryotube into the aluminum dish. Using precooled tweezers, transfer each carrier (sample side up) to a separate position in the reagent bath, starting with 0 and working clockwise (*see* **Notes 35** and **36**). Repeat for all the samples.
4. Fill the bath(s) with precooled ethanol and incubate for 15 min.
5. Infiltrate with 25%, 50%, and 75% HM20 Lowicryl resin-ethanol solution (precooled to −45 °C) for 2–3 h each. Leave overnight in precooled 100% resin, then perform a further three to four changes of precooled 100% resin over ~8 h (*see* **Note 37**).
6. Program the AFS unit for UV resin polymerisation as follows: −45 °C for 24 h, warm to 0 °C at a rate of 3.7 °C/h, then hold at 0 °C for 12 h.
7. Attach the UV head to the AFS chamber and start the polymerization program to polymerize the samples with 360 nm UV light for a total of 48 h (*see* **Note 38**).
8. Remove samples from the AFS and leave in the fume hood at room temperature for 1–2 days (*see* **Note 39**).
9. Using a razor blade, carefully cut away the reagent bath and tubes from the blocks.
10. Fill a small ice box with liquid nitrogen. Using tweezers, briefly dip the very top of the block (containing the HPF carrier) into the liquid nitrogen (*see* **Note 40**). Immediately scrape away the resin covering the carrier and run the tweezers around the edge

of the carrier, then use the tweezers to pop off the carrier to expose the cell pellet.

11. Section the blocks on the ultramicrotome using a diamond knife. Collect 100–150 nm sections onto 0.6% formvar coated 200 mesh Finder grids (*see* Materials 2.2) using a perfect loop.
12. Sections are now ready for SMLM imaging. This should be done on the same day as sectioning, as fluorescence and blinking ability in sections reduces significantly after 24 h (*see* **Note 41**).

3.3 SMLM Imaging

1. Mount grids between a slide and glass coverslip with 20 μl of a glycerol-based antifade mounting medium (e.g., Citifluor AF4, *see* **Note 42**).
2. Seal/secure the coverslip with nail polish.
3. Use a low magnification objective lens to identify the areas on the grid containing the sections and possible areas of interest.
4. Before starting the SMLM data acquisition take overview images with transmitted light and fluorescence to capture the grid square including a reference (e.g., finder grid reference) to allow finding the same area later in the TEM.
5. Acquire SMLM data set with appropriate parameters for the particular fluorophores used (*see* **Note 43**).
6. Perform single-molecule localization procedure and SMLM image reconstruction (*see* **Note 44**).
7. After SMLM data acquisition remove the nail polish (*see* **Note 45**).
8. Carefully lift the coverslip and pass the grid over three drops of water to remove the mounting medium (*see* **Note 46**).

3.4 Post Staining and TEM Imaging

Sections are post-stained with salts of uranium and lead to increase contrast (*see* **Note 47**) and then imaged on the TEM. We use a Tecnai 12 TEM (FEI) with a Lab6 filament, operated at 120 kV, equipped with a bottom mount OneView CMOS digital camera (Gatan).

1. Heat ~50 ml ultrapure water to boiling point in a microwave. Set aside to cool slightly.
2. Place two clean lengths of parafilm onto the bench (*see* **Note 48**).
3. Apply ~30 μl droplets of 2% aqueous uranyl acetate to one of the parafilm sheets and float the grids (one per droplet) section side down on the droplets for 10 min (*see* **Note 49**). Use a box or foil covered dish to protect the stain from light during this time. In the meantime, place the NaOH pellets in a ring wide enough for two 30 μl droplets on the second parafilm sheet, apply the lead droplets within this ring, and place a cover over them.

4. When the uranyl staining is complete, apply 5× 30 μl droplets of warm water (**step 1**, *see* **Note 50**) per grid in a line on the first parafilm sheet. Transfer each grid directly to its first water droplet (section side down) and incubate for ~30 s.
5. Gently blot the grid perpendicular to a wedge of filter paper and transfer it to the second droplet for ~1 min. Repeat for the remaining droplets.
6. Immediately transfer the grid to lead stain and incubate covered for 10 min (*see* **Note 51**). Wash as for **steps 4** and **5**, except that the first water droplet should be placed inside the circle of NaOH pellets.
7. Blot the grid and transfer (section side up) onto a piece of filter paper to dry before transferring to a grid box.
8. Image grids on the TEM. Use the transmitted light and fluorescence images obtained in Methods 3.3, **step 4**, to find back the cell(s) imaged with SMLM in Methods 3.3, **steps 5** and **6**.

3.5 Correlation of SMLM and TEM Images

1. Identify and mark the control points (i.e., fiducial markers or other structures which can be clearly identified in SMLM and TEM) in the SMLM and TEM images.
2. Derive coordinate transformation between SMLM and TEM images from the selected control points (*see* **Note 52**).
3. Apply coordinate transformation to SMLM image and overlay both images (*see* **Note 53**).

4 Notes

1. GFP, mVenus, and mRuby2 work well with this technique. The latter was the best of the red fluorescent proteins we have tried, e.g., mCherry produced a lower in-resin SMLM image quality.
2. We used HEK239T and HeLa cells cultured and transfected as described [10] or with CellLight® Mitochondria-GFP, BacMam 2.0 (ThermoFisher Scientific) according to the manufacturer's instructions. Cells were processed 20–28 h post-transfection.
3. We use the Leica EM PACT2 HPF with Leica Membrane carriers, 1.5 mm × 0.1 mm.
4. We tried various concentrations of BSA, FBS, and goat serum in a range of different buffers including 0.1 M PIPES, 0.1 M PBS, and 0.1 M TBS, but found that 0.1 M PIPES with 15–20% BSA and 5% FBS gave the best results.
5. We highly recommend using low molecular weight TA $(C_{14}H_{10}O_9)_n$ (MW 1000–1500) from Electron Microscopy Sciences. It penetrates better into cells and extracts less protein

(Simionescu and Simionescu [14]) than the high molecular weight tannic TA ($C_{76}H_{52}O_{46}$; MW 1701). The latter is more readily available, so ensure that you double check the MW before using.

6. Make the TA stock immediately before use.
7. UA is a radioactive heavy metal and should be handled with caution in accordance with your local safety procedures.
8. UA is photo-reactive and care should be taken to protect it from light in order to avoid the formation of precipitates. The 5% UA methanol stock solution can be stored at −20 °C for several months.
9. We use the Leica AFS2 unit with LED UV light attachment.
10. The FS solution will turn a reddish brown when the UA and TA are combined and some precipitate may form, this is normal and will not affect the FS process.
11. We also tried Platinum Blue and Lanthanum stains at 1:20 dilutions in acetone as a substitute for UA, but found that the contrast at the TEM was not as good as it is with UA.
12. Cap the tubes as loosely as possible to make it easier to open them in the AFS chamber at −130 °C.
13. We initially used the Lowicryl HM20 monostep resin (Agar Scientific), which is ready to use and is stored at 4 °C, but found that polymerization was more consistent when we mixed the resin components ourselves (Lowicryl HM20 kit, Agar Scientific) in small volumes as required. Unused resin was stored at −20 °C for up to 2 months.
14. If the film does not release from the slide, the most likely cause is hygroscopic oil contamination on the glass slide. Open a fresh box of slides, wipe the new slide with lint free tissue, and try again.
15. The color of the formvar film is an indication of its thickness. If it is multi-colored or patterned, it should not be used. Film thickness will increase the longer the slide is left in the formvar solution, so it should be dipped quickly in and out. The slide can be rested at the neck of the formvar bottle for a few moments to allow the chloroform vapors to thin the film. Dry the slide in a draught free area (i.e., not in the fumehood) to ensure that the film dries evenly.
16. We use a Diatome 45° knife with a 3 mm blade and a Leica UC7 Ultramicrotome.
17. With the AF4 antifade mounting medium we achieved better results for SMLM imaging as compared with PBS or glycerol alone. It is important to choose a medium that does not

polymerize and that can be washed of the grid before post staining for TEM imaging.

18. Filter 1 ml UA stock solution using a 0.22 μm membrane syringe filter and centrifuge briefly before use to minimize the amount of precipitates in the solution.
19. Degas water by boiling it for a few minutes in the microwave. Cool before use.
20. Prepare a fresh 1 N sodium hydroxide solution using degassed water: add 1 g pellets to 25 ml water in a 50 ml tube and invert gently to dissolve.
21. Keep an eye on the stain; if a precipitate develops over time or the solution becomes cloudy, discard it and make a new solution as above.
22. Various registration tools are available in a broad range of image processing environments (e.g., *eC-CLEM* (Icy); *Name Landmarks and Register* (ImageJ)). We use the *Control Point Selection Tool* of MATLAB (Mathworks).
23. If something has gone wrong with the cell culture (e.g., >30% dead/dying cells) or transfection (e.g., lower than expected proportion of cells expressing the fluorescent protein fusion), abort the experiment and set up new cultures.
24. When handling liquid nitrogen, wear insulated gloves and eye protection, and follow your local safety rules for liquid nitrogen use.
25. Our protocol uses trypsin to dislodge adherent cells from the culture dish so that they can be pelleted for loading into the HFP membrane carriers. However, for samples where trypsinization would disrupt the localization of the fluorescent fusion protein, cells may be grown on 1.2 mm sapphire discs, which can be transferred directly to the HPF carrier and frozen [15]. We have found these difficult to work with and if you have the Leica EM HPM100, Leica EM ICE, or equivalent, then the 3 or 6 mm sapphire discs would be a better option, as would aclar discs [16], which is also the cheaper option. Alternatively, cells can instead be grown on transwell membranes (Millipore 24 well millicell Hanging Cell Culture Insert, 0.4 μm PET), which can then be carefully cut to fit into the HPF carrier and frozen [17]. In this case we recommend the use of 2% low melting point agarose in cell culture medium containing 20% FBS or BSA as the cryoprotectant.
26. Unlike HelLa cells, HEK293T cells are weakly adherent and can be dislodged from the culture plate by gentle tapping and pipetting, then spun in a benchtop centrifuge at 500 rpm for 1 min. Cells can then be resuspended in 100 μl pre-warmed 20% BSA in 0.1 M PIPES buffer and pelleted for 30 s at

~10,000 × *g*. However, this method resulted in tight cell clumps that were poorly infiltrated with resin in many places, a problem that is mitigated by trypsinizing the HEK293T cells instead.

27. Trypsinize cells from 3 wells of the 6-well culture plate and pool them, keeping 3 wells as a backup in case anything goes wrong.
28. When performing solution changes, leave a thin layer covering the cells so that they do not dry out.
29. Visually inspect the carrier to ensure that it is not under or over-filled, both of which result in poor freezing quality (*see* also Vanhecke et al. [18]).
30. The cap and tweezers must be dry and precooled in liquid nitrogen prior to coming into contact with the carriers, to prevent the sample warming and subsequent formation of crystalline ice.
31. Take care not to breathe on the samples or warm them in any way. We use tweezers to loosely cap the cryo-tubes once the carriers are in. We typically have three carriers per tube.
32. All the subsequent steps are performed in the AFS at −45 °C.
33. We originally performed all three wash steps and the resin infiltration steps with precooled acetone, but found that resin polymerization was often variable since acetone can influence polymerization of Lowicryl resin by acting as a radical scavenger [19]. Performing the first wash with acetone and then the subsequent washes and infiltration steps with ethanol produced more evenly polymerized blocks and better quality sections.
34. The flow through ring holds ten HPF carriers. The base of the reagent bath is numbered 0–9. Mark the top rim of the bath to indicate position 0. If you have more than ten carriers and therefore need a second or third bath, label the outside of the reagent baths with a diamond scribe so you know which is which. Do not use permanent marker, as this could be washed off by the solvent or resin.
35. Ensure that the carrier is flush at the bottom of the tube. Membrane carriers have a small notch, which should be aligned with the gap at the base of the flow through tube.
36. If the disc of cells dislodges from the carrier and floats out, gently prod it back into place using tweezers before transferring to the reagent bath.
37. On the last change of resin, ensure that the tubes in the flow through ring are completely filled with resin but that the area surrounding the central flow through tube does not contain

excess resin. This will make it easier to cut the polymerized blocks out of the tubes.

38. Before starting the program, place a small square of aluminum foil over the reagent bath. This should be removed after 24 h. This indirect exposure to UV light for the first 24 h of polymerization helps to ensure a more even polymerization of the resin [19].
39. At the end of the polymerization program, the resin may be quite pink. In our hands, this is normal and the pink color will fade over the next few days.
40. It is important that only the very tip of the block comes into contact with the liquid nitrogen, since submerging more of the block often leads to fractures in the resin.
41. However, fluorescence/blinking ability is quite stable in the blocks themselves. We store our blocks at room temperature in the dark. Blocks can be re-sectioned and imaged months later with no appreciable drop in fluorescence level or blinking ability.
42. Four small droplets around the grid will prevent drifting of the grid toward the rim when lowering the coverslip.
43. For standard FPs (e.g., GFP, mVenus, mRuby2) we achieved good results with laser intensities of 2–10 kW/cm^2 and an integration time of the camera of 50 ms for one frame.
44. Choose an algorithm that is suitable for the data (i.e., background level, density of molecules in "on" state (on/off duty cycle), etc.) to minimize artifacts due to effects less common in conventional SMLM imaging, such as increased background fluorescence in the resin or the formvar coating. We are using *fastSPDM* [20], a maximum-likelihood-based algorithm with a sliding window for background subtraction.
45. If the nail polish is difficult to remove, putting the slide in warm PBS or applying some of it around the edges of the coverslip can help to loosen the nail polish.
46. Thorough washing is important to remove the antifade from sections, as glycerol residue can pick up the post-stain and leave unsightly sports across the section.
47. We use longer post-staining times than we would for a standard TEM prep to increase contrast as much as possible.
48. A few squirts of water underneath the parafilm will help keep it in place on the bench. We use a permanent marker to draw rows on the parafilm so grids do not get mixed up during staining and washing.
49. We do not recommend staining more than ten grids at a time, as in our experience this increases the risk of staining artifacts.

50. The water should be warm (~30–40 °C), which reduces the formation of stain artifacts. However, if it is too hot, the stain will be stripped from the section. Run cold water over the bottle for a minute or two to cool it down.
51. Take care not to breath on or near the lead droplets during the staining procedure. Exposure to CO_2 causes the formation of lead carbonate stain precipitates and there is nothing more frustrating than lead precipitates on a cell with excellent fluorescence.
52. If no significant shrinkage has occurred during TEM data acquisition, a linear conformal transformation is sufficient. Otherwise, an affine transformation should be applied to improve the correlation accuracy.
53. For the visualization choose a dynamic range for the TEM image to fill the lower half (e.g., 0–127 for an 8 bit image) to allow both images to use a maximum and equal range. Adapt accordingly for cases where one imaging modality requires more attention/emphasis.

Acknowledgments

We are grateful to Elena Seiradake and Marek Drozdz for culture and transfection of cells, and to Ilan Davis, Ian Dobbie, Kay Grünewald, Christoph Hagen, E. Yvonne Jones, Richard Parton, and Jordan Raff for their continuous support and constructive discussions. This work was supported by Wellcome Trust Senior Research Fellowships (090895/Z/09/Z; 096144/Z/11/Z), the Wellcome Trust core award to the Wellcome Trust Centre for Human Genetics (090532/Z/09/Z), a Cancer Research UK programme grant (A10976) and the Micron Strategic Award from the Wellcome Trust (091911).

References

1. Müller-Reichert T, Verkade P (eds) (2014) Correlative light and electron microscopy II. Academic Press, Amsterdam
2. Betzig E, Patterson GH, Sougrat R, Lindwasser OW, Olenych S, Bonifacino JS, Davidson MW, Lippincott-Schwartz J, Hess HF (2006) Imaging intracellular fluorescent proteins at nanometer resolution. Science 313 (5793):1642–1645
3. Watanabe S, Punge A, Hollopeter G, Willig KI, Hobson RJ, Davis MW, Hell SW, Jorgensen EM (2011) Protein localization in electron micrographs using fluorescence nanoscopy. Nat Methods 8(1):80–84
4. Kopek BG, Shtengel G, Xu CS, Clayton DA, Hess HF (2012) Correlative 3D superresolution fluorescence and electron microscopy reveal the relationship of mitochondrial nucleoids to membranes. Proc Natl Acad Sci U S A 109(16):6136–6141
5. Löschberger A, Franke C, Krohne G, van de Linde S, Sauer M (2014) Correlative super-resolution fluorescence and electron microscopy of the nuclear pore complex with molecular resolution. J Cell Sci 127(20):4351–4355
6. Segala MGP, Sun MG, Shtengel G, Viswanathan S, Baird MA, Macklin JJ, Patel R, Allen JR, Howe ES, Piszczek G (2015)

Fixation-resistant photoactivatable fluorescent proteins for correlative light and electron microscopy. Nat Methods 12(3):215

7. Lidke DS, Lidke KA (2012) Advances in high-resolution imaging–techniques for three-dimensional imaging of cellular structures. J Cell Sci 125(11):2571–2580
8. de Boer P, Hoogenboom JP, Giepmans BN (2015) Correlated light and electron microscopy: ultrastructure lights up! Nat Methods 12 (6):503–513
9. Wolff G, Hagen C, Grünewald K, Kaufmann R (2016) Towards correlative super-resolution fluorescence and electron cryo-microscopy. Biol Cell. doi:10.1111/boc.201600008
10. Johnson E, Seiradake E, Jones EY, Davis I, Grünewald K, Kaufmann R (2015) Correlative in-resin super-resolution and electron microscopy using standard fluorescent proteins. Sci Rep 5. doi:10.1038/srep09583
11. McDonald K, Schwarz H, Müller-Reichert T, Webb R, Buser C, Morphew M (2010) "Tips and tricks" for high-pressure freezing of model systems. Methods Cell Biol 96:671–693
12. Reynolds ES (1963) The use of lead citrate at high pH as an electron-opaque stain in electron microscopy. J Cell Biol 17(1):208–212
13. Hawes P, Netherton C, Mueller M, Wileman T, Monaghan P (2007) Rapid freeze-substitution preserves membranes in high-pressure frozen tissue culture cells. J Microsc 226(2):182–189
14. Simionescu N, Simionescu M (1976) Galloylglucoses of low molecular weight as mordant in electron microscopy. I. Procedure, and evidence for mordanting effect. J Cell Biol 70 (3):608–621
15. Reipert S, Fischer I, Wiche G (2004) High-pressure freezing of epithelial cells on sapphire coverslips. J Microsc 213(1):81–85
16. Jimenez N, Humbel B, Van Donselaar E, Verkleij A, Burger K (2006) Aclar discs: a versatile substrate for routine high-pressure freezing of mammalian cell monolayers. J Microsc 221 (3):216–223
17. Morphew M, McIntosh J (2003) The use of filter membranes for high-pressure freezing of cell monolayers. J Microsc 212(1):21–25
18. Vanhecke D, Zuber B, Brugger S, Studer D (2012) Safe high-pressure freezing of infectious micro-organisms. J Microsc 246 (2):124–128
19. Schwarz H, Humbel BM (2014) Correlative light and electron microscopy using immunolabeled sections. In: Electron microscopy: methods and protocols. Springer, New York, pp 559–592
20. Grüll F, Kirchgessner M, Kaufmann R, Hausmann M, Kebschull U (2011) Accelerating image analysis for localization microscopy with FPGAs. In: 2011 21st international conference on field programmable logic and applications, IEEE, p 1–5

Chapter 15

Synthesis of Janelia Fluor HaloTag and SNAP-Tag Ligands and Their Use in Cellular Imaging Experiments

Jonathan B. Grimm, Timothy A. Brown, Brian P. English, Timothée Lionnet, and Luke D. Lavis

Abstract

The development of genetically encoded self-labeling protein tags such as the HaloTag and SNAP-tag has expanded the utility of chemical dyes in microscopy. Intracellular labeling using these systems requires small, cell-permeable dyes with high brightness and photostability. We recently discovered a general method to improve the properties of classic fluorophores by replacing *N*,*N*-dimethylamino groups with four-membered azetidine rings to create the "Janelia Fluor" dyes. Here, we describe the synthesis of the HaloTag and SNAP-tag ligands of Janelia Fluor 549 and Janelia Fluor 646 as well as standard labeling protocols for use in ensemble and single-molecule cellular imaging.

Key words Fluorophore, HaloTag, SNAP-tag, Fluorescence microscopy, Organic chemistry, Single-molecule imaging

1 Introduction

The fulcrum of any fluorescence microscopy experiment is the strategy used to attach a fluorophore to the biomolecular target. Genetically encoded fluorophores, such as green fluorescent protein (GFP), allow the tagging of proteins with genetic specificity. Nevertheless, the use of fluorescent proteins is hampered by poor photostability and limited options for the far-red region of the spectrum [1, 2]. The self-labeling tag concept pioneered by Johnsson involves fusing the protein of interest with an enzyme variant that reacts specifically and irreversibly with a small molecule "ligand" motif that is attached to a chemical fluorophore [3, 4]. This allows the tagging of fusion proteins with synthetic dyes inside living cells, thus combining the superior brightness of chemical fluorophores with the genetic specificity of fluorescent proteins.

The Janelia Fluor (JF) dyes are derivatives of the classic fluorophore tetramethylrhodamine (TMR) where the *N*,*N*-dimethyamino groups are replaced with azetidine substituents [5]. This

Holger Erfle (ed.), *Super-Resolution Microscopy: Methods and Protocols*, Methods in Molecular Biology, vol. 1663, DOI 10.1007/978-1-4939-7265-4_15, © Springer Science+Business Media LLC 2017

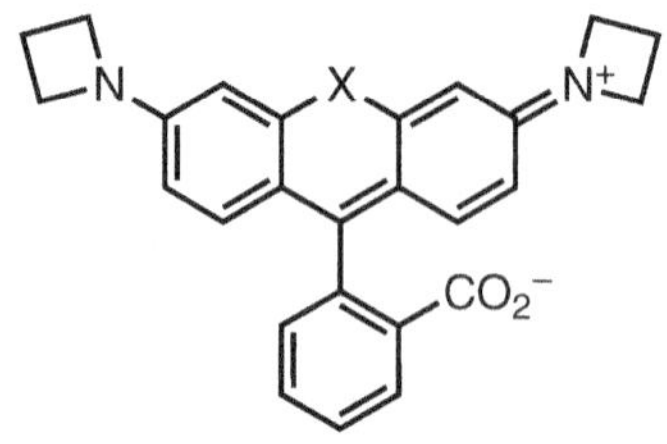

dye[a]	X	λ_{max} (nm)	ε ($M^{-1}cm^{-1}$)	λ_{em} (nm)	Φ
1, JF_{549}	O	549	101,000	571	0.88
2, JF_{646}	$Si(CH_3)_2$	646	152,000[b]	664	0.54

[a]All properties taken in 10 mM HEPES, pH 7.3 unless noted
[b]Measurement taken in EtOH containing 0.1% v/v TFA

Fig. 1 Spectral properties of Janelia Fluor 549 (**1**) and Janelia Fluor 646 (**2**)

novel auxochrome unit presumably reduces nonradiative decay by twisted internal charge transfer (TICT), resulting in higher quantum yields and better photostability. JF_{549} (**1**, Fig. 1) is the direct analog of TMR with a similar absorption maximum (λ_{max}) and fluorescence emission maximum (λ_{em}) but a higher extinction coefficient (ε) and quantum yield (Φ). JF_{646} (**2**) is an analog of JF_{549} that contains a dimethylsilicon moiety in place of the xanthene oxygen in JF_{549}, which elicits 100 nm bathochromic shifts in absorption and emission maxima. Figure 1 gives the structures of the two JF dyes and lists key spectral properties.

The small size, excellent brightness, and cell permeability of the Janelia Fluor dyes make them useful for cellular imaging when combined with the SNAP-tag or HaloTag system, particularly for single-molecule experiments [6–8]. Below we describe the synthesis and purification of the SNAP-tag [3] and HaloTag [9] ligands of Janelia Fluor 549 and Janelia Fluor 646 (Fig. 2) and standard protocols for cellular labeling in live and fixed cells for standard and single-molecule imaging.

2 Materials

All the chemical reactions should be sealed with septa through which a nitrogen or argon atmosphere is introduced and stirred using a magnetic stir plate.

2.1 Janelia Fluor Dye SNAP-tag or HaloTag Synthesis

1. 5 or 10 mL round-bottom flask equipped with rubber septum and stir bar.
2. Carboxy-JF dyes [5]: 6-carboxy-Janelia Fluor 549 or 6-carboxy-Janelia Fluor 646.

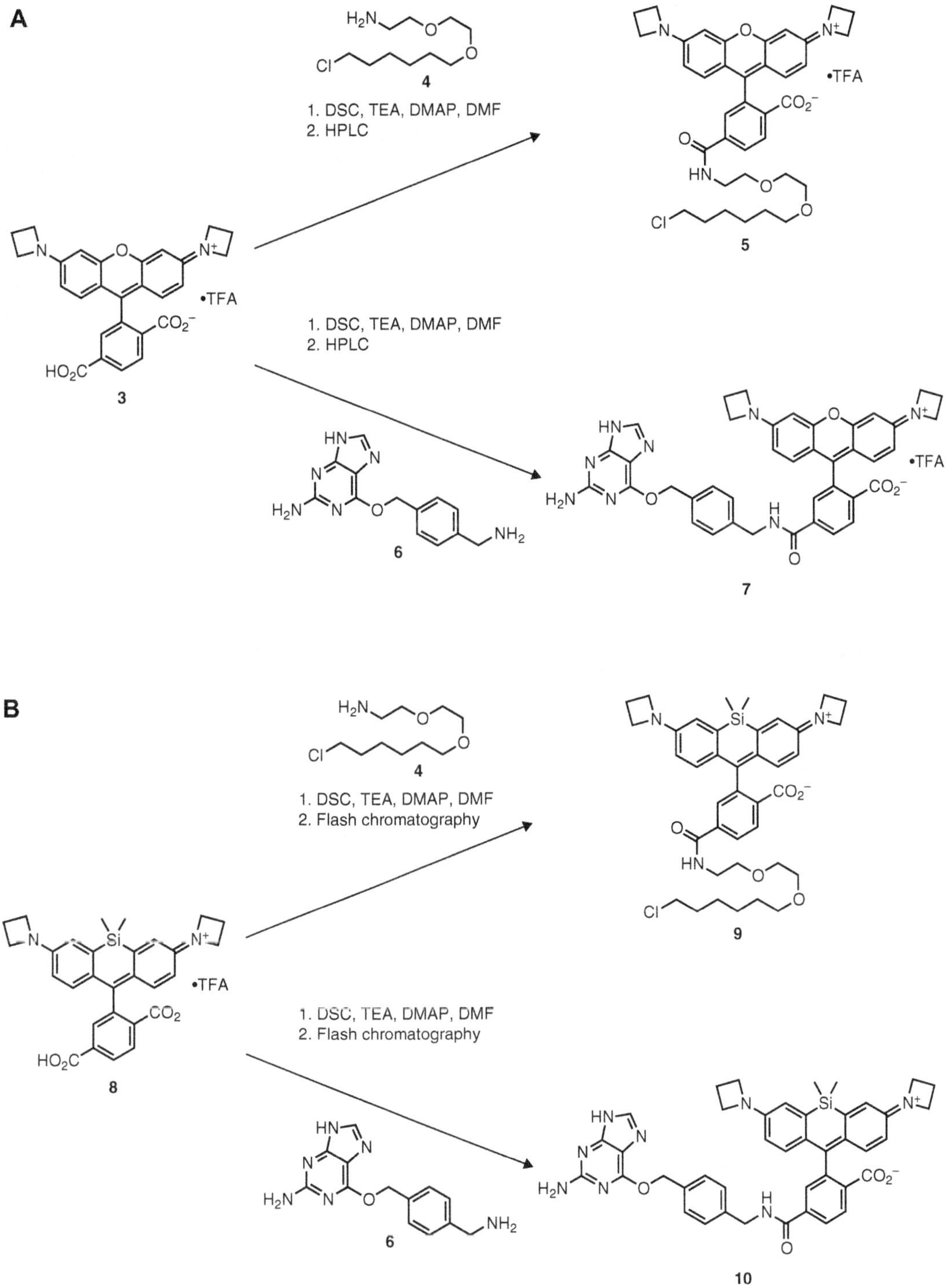

Fig. 2 (**a**) Synthesis of Janelia Fluor 549 HaloTag ligand (**5**) and Janelia Fluor 549 SNAP-tag ligand (**7**). (**b**) Synthesis of Janelia Fluor 646 HaloTag ligand (**9**) and Janelia Fluor 646 SNAP-tag ligand (**10**)

3. *N*,*N*′-Disuccinimidyl carbonate (DSC).
4. Triethylamine (Et_3N).
5. *N*,*N*-dimethylaminopyridine (DMAP).
6. Anhydrous *N*,*N*-dimethylformamide (DMF; e.g., Aldrich Sure/Seal).
7. HaloTag Amine (O2) Ligand (Promega) or BG-NH_2 (New England BioLabs).
8. Glass or aluminum thin-layer chromatography (TLC) plates precoated with silica gel 60.
9. Saturated $NaHCO_3$ solution.
10. Magnesium Sulfate ($MgSO_4$).
11. Celite.
12. Toluene.
13. Dichloromethane (CH_2Cl_2).
14. Ethyl acetate (EtOAc).
15. Methanol (CH_3OH).
16. Acetic acid (AcOH).
17. Acetonitrile (CH_3CN).
18. Trifluoroacetic acid (TFA).
19. Rotary evaporator.
20. Silica gel and glass column or flash chromatography system.
21. Preparative high pressure liquid chromatography (HPLC) system.
22. Lyophilizer.

2.2 Preparation of Aliquots and DMSO Stocks

1. High quality (i.e., HPLC grade) methanol (CH_3OH) or acetonitrile (CH_3CN).
2. Solvent Safe pipette tips (Molecular BioProducts).
3. 0.5 mL screw top polypropylene tubes for storage of DMSO stocks.
4. Vacuum concentrator (i.e., SpeedVac) system.
5. Ampule of cell-culture grade anhydrous DMSO.

2.3 Cellular Labeling Using HaloTag or SNAP-tag Ligands

1. MatTek dishes.
2. Cell lines expressing HaloTag or SNAP-tag fusion either through transient or stable transfection.
3. Appropriate completed media for cell type.
4. Phosphate-buffered saline (PBS, pH 7.2).

5. Paraformaldehyde (PFA) fixative: 2–4% paraformaldehyde, 100 mM phosphate buffer, pH 7.2.

2.4 Live Cell Single-Molecule Imaging Using HaloTag or SNAP-tag Ligands

1. 25 mm diameter coverslips (Electron Microscopy Sciences).
2. Ceramic coverslip holder (i.e., cover glass staining rack; Thomas Scientific).
3. 125 mL polypropylene jar.
4. 100% ethanol.
5. 1 M KOH: 14.03 g KOH, 250 mL water.
6. Sonicator.
7. Cell lines expressing HaloTag or SNAP-tag fusion either through transient or stable transfection.
8. 6-well plates.
9. Attofluor metal cell chamber for microscopy (Life Technologies).
10. Appropriate complete media for cell type.

2.5 Fixed-Cell STORM Imaging

1. Paraformaldehyde (PFA) fixative: 4% PFA in phosphate-buffered saline (PBS, pH 7.5).
2. 2 mL of freshly prepared glucose oxidase/catalase/MEA buffer: PBS (pH 7.5), 50 mM mercaptoethylamine, 10% (w/v) glucose, 0.5 mg/mL glucose oxidase, and 28,400 U/mL catalase.
3. Nail Polish.

3 Methods

3.1 Preparation of Janelia Fluor 549 HaloTag Ligand

1. Combine 10 mg (17.6 μmol) of 6-carboxy-JF_{549} (**3**) with 9.9 mg of DSC (38.7 μmol, 2.2 eq) in 1 mL of DMF in the small round-bottom flask. Add 14.7 μL of Et_3N (106 μmol, 6 eq) and 0.2 mg of DMAP (1.76 μmol, 0.1 eq) and stir at room temperature for 1 h while shielded from light.
2. Add a solution of 9.8 mg of HaloTag Amine (O2) Ligand **4** (44.0 μmol, 2.5 eq) in 100 μL DMF and stir the reaction for an additional 4 h at room temperature (*see* **Note 1**).
3. Test reaction progress by thin-layer chromatography (TLC) using $CH_2Cl_2/CH_3OH/AcOH$ (e.g., 9/1/0.1 v/v/v) as the eluent.
4. Dilute reaction mixture with saturated $NaHCO_3$ and extract twice with CH_2Cl_2.
5. Combine the organic extracts and dry over $MgSO_4$. Filter the solution into a flask, add 1 g of Celite, and concentrate the solution (rotary evaporator) to deposit the material onto Celite.

6. Load dry Celite mixture onto a silica gel column and perform chromatography using a linear gradient of 0–10% MeOH in CH_2Cl_2 with a constant 1% v/v AcOH additive.
7. Pool and concentrate product-containing fractions and perform a second purification by reverse phase preparative HPLC using a linear gradient of 10–95% MeCN in H_2O with a constant 0.1% v/v trifluoroacetic acid (TFA) additive (*see* **Note 2**).
8. Remove organic solvent (rotary evaporator) and then lyophilize the remaining solution to yield **5** (dark red solid). Typical yield is 8.5 mg (62%).

3.2 Preparation of Janelia Fluor 549 SNAP-tag Ligand

1. Combine 10 mg (17.6 μmol) of 6-carboxy-JF_{549} (**3**) with 9.9 mg of DSC (38.7 μmol, 2.2 eq) in 1 mL of DMF in a round-bottom flask. Add 14.7 μL of Et_3N (106 μmol, 6 eq) and 0.2 mg of DMAP (1.76 μmol, 0.1 eq) and stir at room temperature for 1 h while shielded from light.
2. Add a solution of 11.9 mg of BG-NH_2 **6** (44.0 μmol, 2.5 eq) in 100 μL DMF and stir the reaction for an additional 2 h at room temperature (*see* **Note 1**).
3. Test reaction progress by thin-layer chromatography (TLC) using $CH_2Cl_2/CH_3OH/AcOH$ (e.g., 9/1/0.1 v/v/v) as the eluent.
4. Concentrate the reaction mixture and purify by reverse phase preparative HPLC using a linear gradient of 10–95% MeCN in H_2O with a constant 0.1% v/v trifluoroacetic acid (TFA) additive (*see* **Note 2**).
5. Remove organic solvent (rotary evaporator) and then lyophilize the remaining solution to yield 7 (dark red solid). Typical yield is 11.5 mg (80%).

3.3 Preparation of Janelia Fluor 646 HaloTag Ligand

1. Combine 30 mg (49.1 μmol) of 6-carboxy-JF_{646} (**8**) with 28 mg of DSC (108 μmol, 2.2 eq) in 2 mL of DMF in the round-bottom flask. Add 41 μL of Et_3N (295 μmol, 6 eq) and 0.6 mg of DMAP (4.91 μmol, 0.1 eq) and stir at room temperature for 1 h while shielded from light.
2. Add a solution of 27 mg of HaloTag Amine (O2) Ligand **4** (44.0 μmol, 2.5 eq) in 250 μL DMF and stir the reaction for an additional 2 h at room temperature (*see* **Note 1**).
3. Test reaction progress by thin-layer chromatography (TLC) using toluene/EtOAc (e.g., 1/1 v/v) as the eluent.
4. Dilute reaction mixture with saturated $NaHCO_3$ and extract twice with EtOAc.
5. Combine the organic extracts and dry over $MgSO_4$. Filter the solution into a flask, and concentrate (rotary evaporator).

6. Load onto a silica gel column and perform chromatography using a linear gradient of 10–100% EtOAc in toluene (*see* **Note 3**).
7. Pool and concentrate product-containing fractions to yield **9** as the inner salt (blue foam). Typical yield is 25 mg (73%).

3.4 Preparation of Janelia Fluor 646 SNAP-tag Ligand

1. Combine 25 mg (40.9 μmol) of 6-carboxy-JF_{646} (**8**) with 23 mg of DSC (108 μmol, 2.2 eq) in 2 mL of DMF in the round-bottom flask. Add 34 μL of Et_3N (246 μmol, 6 eq) and 0.5 mg of DMAP (4.09 μmol, 0.1 eq) and stir at room temperature for 1 h while shielded from light.
2. Add 28 mg of BG-NH_2 **6** (102 μmol, 2.5 eq) and stir the reaction for an additional 2 h at room temperature (*see* **Note 1**).
3. Test reaction progress by thin-layer chromatography (TLC) using EtOAc/CH_3OH (e.g., 9/1 v/v) as the eluent.
4. Dilute reaction mixture with saturated $NaHCO_3$ and extract twice with CH_2Cl_2.
5. Combine the organic extracts and dry over $MgSO_4$. Filter the solution into a flask, and concentrate (rotary evaporator).
6. Load onto a silica gel column and perform chromatography using a linear gradient of 0–10% CH_3OH in EtOAc (*see* **Note 3**).
7. Pool and concentrate product-containing fractions to yield **10** as the inner salt (blue solid). Typical yield is 24.7 mg (80%).

3.5 Preparation of Ligand Aliquots and Ligand Stock Solutions in DMSO

1. Prepare a 1 mM stock solution of the purified HaloTag or SNAP-tag ligand in either methanol (CH_3OH; for JF_{549}-based ligands) or acetonitrile (CH_3CN; for JF_{646}-based ligands).
2. Aliquot material to 25 nmol (25 μL) or 100 nmol (100 μL) per vial using a pipette and Solvent Safe tips.
3. Remove solvent using vacuum concentrator (i.e., SpeedVac) system, avoiding active heating. These vials of dry material can be stored at −20 °C and are stable for at least 12 months.
4. For DMSO stock solutions, prepare a 1–5 mM stock in DMSO using a freshly opened ampule of DMSO.
5. These stock solutions can also be stored at −20 °C but can be aliquoted further to avoid repetitive freeze–thaw cycles (*see* **Note 4**).

3.6 Live-Mammalian Cell Labeling

1. Mammalian cells expressing HaloTag or SNAP-tag fusions are plated in 35 mm glass-bottom MatTek dishes and allowed to grow to appropriate density (e.g., 3–4 × 10^4 cells/cm^2) for imaging.
2. Dilute ligand DMSO stock solution into a completed cell culture medium at 37 °C. For HaloTag labeling the standard

final dye solution is at 10–100 nM. For SNAP-tag ligands the standard final dye solution is at 1–10 μM (*see* **Note 5**).

3. Add 2 mL of dye–medium for each MatTek plate to be stained and incubate for 30 min for HaloTag ligands or 60 min for SNAP-tag ligands (37 °C, 5% CO_2 incubator).
4. Remove the medium and wash cells with 2 mL of 37 °C PBS (*see* **Note 6**).
5. Replace PBS with 2 mL of completed growth medium at 37 °C; allow 15 min at 37 °C prior to imaging.

3.7 Fixed-Cell Labeling

1. Mammalian cells expressing HaloTag or SNAP-tag fusions are plated in 35 mm MatTek dishes and allowed to grow to appropriate density (e.g., 3–4 × 10^4 cells/cm^2) for imaging.
2. Cells are then fixed by replacing media with 2–4% PFA and incubating for 30 min.
3. Wash cells with PBS and incubate for 30 min with a labeling solution comprised of the ligand DMSO stock solution diluted into PBS. For HaloTag labeling the standard final dye solution is at 10–100 nM. For SNAP-tag ligands the standard final dye solution is at 1–10 μM. Allow the staining to proceed for 30 min.
4. Alternatively, *for the HaloTag ligands only*, the DMSO ligand stock solution can be diluted directly into 2–4% PFA fixative. The cell culture medium is replaced with this dye fix solution and incubated with this dye-fix solution for 30 min.
5. Wash fixed and labeled cells with PBS and image (*see* **Note 7**).

3.8 Live-Cell Single-Molecule Imaging

1. Small impurities are common on the surface of glass coverslips and generate unwanted fluorescent spots that interfere with single-molecule experiments. To remove these impurities first fit the 25 mm diameter coverslips in the ceramic holder and place the assembly into the polypropylene jar.
2. Fill the jar with 100% ethanol, sonicate for 15 min, and rinse with MilliQ water.
3. Fill the jar with 1 M KOH, sonicate for 15 min, and rinse with MilliQ water.
4. Repeat **steps 2** and **3** (sonication cycle of ethanol followed by 1 M KOH) three times. The cleaned slides can be stored for weeks at 4 °C in 100% ethanol before use. Prior to use, each coverslip is rinsed with Millipore water and allowed to dry for 5 min in a cell culture hood under sterilizing UV light.
5. Seed cells on these pre-cleaned coverslips deposited in standard 6-well culture plates and grow to appropriate density for imaging.

6. Label and wash cells with HaloTag or SNAP-tag ligands as described in Subheading 3.6 using lower ligand concentrations of 0.5–50 nM.
7. Mount coverslips into the metal cell chambers in preparation for imaging. Variations on this imaging procedure have been described in detail [5].

3.9 Fixed-Cell Direct Stochastic Optical Reconstruction Microscopy (dSTORM) Imaging

1. Grow and label cells on pre-cleaned 25 mm diameter coverslips placed in 6-well culture plates as described in Subheading 3.8.
2. Fix cells with a solution of 4% paraformaldehyde in PBS buffer (pH 7.5).
3. Mount coverslips into metal cell chambers.
4. Place 1 mL of freshly prepared glucose oxidase/catalase/MEA buffer into the metal cell chamber (*see* **Note 8**). Before use, deoxygenate by slowly bubbling nitrogen through the buffer solution.
5. Cap the chamber immediately by placing a second 25 mm diameter coverslip on the top of the chamber and sealing air-tight with nail polish.

4 Notes

1. The scale of the reaction can be adjusted depending on the amount of dye and ligand amine available.
2. The use of trifluoroacetic acid (TFA) in the HPLC eluent yields the dye as the TFA salt.
3. JF_{646} preferentially adopts a colorless lactone form, especially in organic solvents. This allows for purification and isolation of the JF_{646} ligands by silica gel chromatography with less polar eluents and permits the JF_{646} reactions to be run on a larger scale.
4. Si-rhodamine-based dyes such as Janelia Fluor 646 are particularly susceptible to freeze–thaw degradation.
5. The SNAP-tag ligands are less cell-permeable and react slower than the HaloTag ligands. Thus, the SNAP-tag system requires higher concentrations of ligand or longer labeling times.
6. The JF_{646} ligands are modestly chromogenic and fluorogenic (~20-fold increase upon binding to their cognate protein tag), which facilitates washing. In some experiments the washing steps can be optional depending on the amount of ligand used and the nonspecific staining profile of the cell type. SNAP-tag ligands have higher nonspecific binding and typically require additional washings with PBS or completed media.

7. Alternatively, the fluorescent ligand can be added up to 24 h after fixation in 2–4% paraformaldehyde, although we observe up to 50% loss of staining intensity in post-fixation labeling.
8. The metal cell chamber contains exactly 1.0 mL and allows the complete filling of the volume between two cover slips to exclude air bubbles.

References

1. Kremers G-J, Gilbert SG, Cranfill PJ, Davidson MW, Piston DW (2011) Fluorescent proteins at a glance. J Cell Sci 124:157–160
2. Cranfill PJ, Sell BR, Baird MA, Allen JR, Lavagnino Z, de Gruiter HM, Kremers G-J, Davidson MW, Ustione A, Piston DW (2016) Quantitative assessment of fluorescent proteins. Nat Methods. doi:10.1038/nmeth.3891
3. Keppler A, Gendreizig S, Gronemeyer T, Pick H, Vogel H, Johnsson K (2002) A general method for the covalent labeling of fusion proteins with small molecules in vivo. Nat Biotechnol 21:86–89
4. Xue L, Karpenko IA, Hiblot J, Johnsson K (2015) Imaging and manipulating proteins in live cells through covalent labeling. Nat Chem Biol 11:917–923
5. Grimm JB, English BP, Chen J, Slaughter JP, Zhang Z, Revyakin A, Patel R, Macklin JJ, Normanno D, Singer RH, Lionnet T, Lavis LD (2015) A general method to improve fluorophores for live-cell and single-molecule microscopy. Nat Methods 12:244–250
6. Liu Z, Legant WR, Chen BC, Li L, Grimm JB, Lavis LD, Betzig E, Tjian R (2014) 3D imaging of Sox2 enhancer clusters in embryonic stem cells. Elife 3:e04236
7. Knight SC, Xie L, Deng W, Guglielmi B, Witkowsky LB, Bosanac L, Zhang ET, El Beheiry M, Masson J-B, Dahan M (2015) Dynamics of CRISPR-Cas9 genome interrogation in living cells. Science 350:823–826
8. Swinstead EE, Miranda TB, Paakinaho V, Baek S, Goldstein I, Hawkins M, Karpova TS, Ball D, Mazza D, Lavis LD, Grimm JB, Morisaki T, Grontved L, Presman DM, Hager GL (2016) Steroid receptors reprogram FoxA1 occupancy through dynamic chromatin transitions. Cell 165:593–605
9. Los GV, Encell LP, McDougall MG, Hartzell DD, Karassina N, Zimprich C, Wood MG, Learish R, Ohana RF, Urh M (2008) HaloTag: a novel protein labeling technology for cell imaging and protein analysis. ACS Chem Biol 3:373–382

Chapter 16

Measuring Nanometer Distances Between Fluorescent Labels Step-by-Step

Susanna Maria Früh and Ingmar Schoen

Abstract

Super-resolution fluorescence microscopy methods are increasingly applied to study the structure of biological molecules within their natural context or at biomaterial interfaces. We here provide a protocol for Single-molecule High-Resolution Imaging with Photobleaching (SHRImP) that can be used to obtain information about the conformation of large proteins or other macromolecules at the single-molecule level. This procedure requires site-specific protein labeling with fluorescent dyes, immobilization and sample preparation, optimization of imaging buffer composition and microscope settings, and acquisition of short time-lapse movies that capture the stepwise bleaching behavior of individual molecules. We then describe a method for reliably determining the relative positions of labels from bleaching movies using the free image processing package Fiji (ImageJ) with the help of auxiliary macros that are provided as Supplementary Material. The presented approach allows for measuring intramolecular distance distributions in the range of a few to hundreds of nanometers and can be applied to a wide variety of biological systems.

Key words Fluorescence, Single-molecule, Localization, Protein conformation, TIRF, Site-specific labeling

1 Introduction

Proteins, DNA strands, synthetic polymers, and other macromolecules are nanometer-sized objects. Interactions with their environment can lead to changes in their conformation or their association, which often determines their function. In the simplest case, a distance between two sites on the same macromolecule can inform us about its conformation (this can be the end-to-end distance of a polymer, or two sites on a protein that are close in one and distant in a second conformation), whereas distances between different molecules can be representative for their association/clustering state. Only few experimental techniques are available that can

Electronic Supplementary Material: The online version of this chapter (doi: 10.1007/978-1-4939-7265-4_16) contains supplementary material, which is available to authorized users.

Holger Erfle (ed.), *Super-Resolution Microscopy: Methods and Protocols*, Methods in Molecular Biology, vol. 1663,
DOI 10.1007/978-1-4939-7265-4_16, © Springer Science+Business Media LLC 2017

study such distances for macromolecules of different types under physiological conditions. We here give a step-by-step protocol for one particularly suitable technique called *S*ingle-molecule *H*igh-*R*esolution *Im*aging with *P*hotobleaching (SHRImP) [1].

Distance measurements by SHRImP are based on the determination of the positions of two or more fluorescent labels on the molecule of interest. To this end, a time-lapse movie is acquired in which these labels are sequentially bleached. From the images before and after each individual bleaching event, the bleached fluorophore can be localized with sub-pixel precision, even when the two or more fluorophores are too close together to be resolved in the conventional image. In contrast to other techniques, the range of distances that can be measured is not limited. This is an advantage in comparison to Förster resonance energy transfer (FRET) which is restricted to distances ±4 nm around the Förster radius (2–10 nm) [2], or to photoactivated localization microscopy (PALM) or stochastic optical reconstruction microscopy (STORM) that use stochastic dye blinking and can only resolve labels that are spaced at least >40 or >15 nm apart, respectively [3]. Compared to localization-based distance measurements using fluorophores that can be distinguished by color [4–7], the labeling procedure is often simpler with same dyes, and no registration of color channels is needed. Naturally, also SHRImP has its limitations: It is practically restricted to few (2–5) fluorescent labels within a diffraction limited spot. The more labels, the more imprecise becomes their localization due to the increasing fluorescent background [8], and the more difficult it gets to distinguish individual bleaching steps. Furthermore, SHRImP typically only measures projected distances in the *xy*-plane; however, encoding the *z*-position into the shape of the point spread function (e.g., by astigmatism [9]) in principle allows for measuring distances and orientation in three dimensions. In summary, SHRImP is best suited to measure distances in the range 0–300 nm between two labels in the focal plane at a low density of labeled molecules.

The capabilities of SHRImP have been demonstrated in a wide range of applications, from synthetic constructs to proteins, from in vitro to cell culture. Artificial molecular rulers made from double-stranded DNA with designed fluorophore distances between 6–30 nm have been used to validate the technique [1, 10–13]. SHRImP in turn was used to validate the positioning precision of an AFM-based single-molecule cut'n'paste technique [14]. The structure of a synthetic dendrimer [15], as well as the conformation of a linear flexible polymer [16] and of the multimeric protein fibronectin [17] at surfaces have been determined by SHRImP and were shown to depend on surface chemistry. SHRImP distance measurements between the two heads of myosin VI bound to actin filaments in vitro were consistent with a hand-over-hand type of walking motion [18]. Ca^{2+}-dependent conformational changes of E-cadherin dimers in vitro, resulting in average EC5-EC5 distances between 18–32 nm, were resolved with a measurement precision better than 1 nm [7]. Recently, the in vitro assembly of intrinsically disordered domains of a bacterial protein into rod-like fibrils was verified, among other techniques, by SHRImP [19]. In cell culture, SHRImP was used to measure the separation between EGF receptors on cell membranes [20]. Moreover, SHRImP was used to measure the extension of end-labeled fibronectin dimers in extracellular fibronectin fibrils, and the observed distances between 100 and 150 nm were confirmed by dSTORM measurements [21].

Despite these versatile usages of SHRImP and the large available repertoire of site-specific labeling strategies, SHRImP is not yet a widespread technique, probably because it demands dedicated microscopy equipment and custom postprocessing strategies. This protocol will thus provide a guideline for producing good quality photobleaching data and performing data analysis to end up with robust distance measurements. SHRImP can be well performed on novel STORM or PALM microscopes that are becoming increasingly available. The sample preparation varies according to the different applications, thus we here explain the procedure for measuring protein conformation of immobilized molecules in vitro and provide comments on other samples/probes. As no dedicated software package for SHRImP analysis is available, a simple implementation using the open-source image analysis platform Fiji [22] is described that will enable non-specialists to get acquainted with SHRImP. For taking full advantage of SHRImP capabilities and more convenient processing, however, customized programs will be required (*see* **Note 1**). Finally, the counting of bleaching steps can also yield information about the stoichiometry of complexes [23–26], but while the experimental procedures are very similar to SHRImP, this application has its own challenges [27] and thus is beyond the scope of this article.

2 Materials

All buffers or solutions were prepared using Millipore water.

2.1 Protein Labeling

1. The protein of interest has to be labeled site-specifically with two (to five) fluorophores (Fig. 1a). In general, organic dyes are preferred over fluorescent proteins because they are smaller and brighter, and red dyes are preferred over blue dyes due to higher photostability and less (auto)fluorescent sample background. We here recommend Alexa Fluor 647; for other suitable dyes *see* **Note 2**.
2. Possible pathways for protein labeling are reviewed by Stephanopoulos et al. [28].

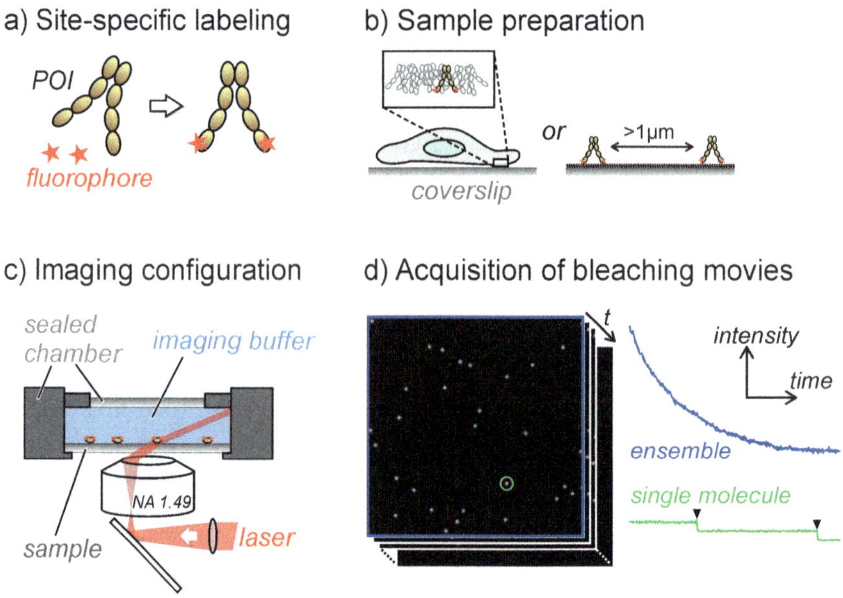

Fig. 1 Experimental procedure for SHRImP. (**a**) The protein of interest (POI) is labeled with fluorophores at specific sites using appropriate bioconjugation or genetic strategies. (**b**) The labeled molecules are incorporated into their native context (*left*) or immobilized on the coverslip (*right*) at dilute concentrations. (**c**) The sample is mounted in a sealed chamber with imaging buffer and observed on a fluorescence microscope under oblique illumination or in TIRF. (**d**) Bleaching movies are analyzed with respect to the bleaching kinetics of the ensemble (*blue*) which informs about fluorophore survival probabilities, and of single molecules (*green*) which informs about signal-to-noise ratio and the number of fluorophores in the diffraction-limited spot

2.2 Coverslips and Sample Holder

1. Round coverslips with 18 mm diameter (thickness 1.5, *Hecht-Assistant*).
2. Clean glass beakers and sonicator for cleaning.
3. Chemicals for cleaning: 20 mL acetone; 20 mL methanol; 20 mL ethanol; H_2O; 10 mL H_2O_2 (30%); 10 mL H_2SO_4 (95%).
4. Sonicate 100 coverslips in a glass beaker subsequently in 20 mL acetone, methanol and ethanol, then wash with H_2O for 20 min each. Freshly prepare 20 mL piranha solution (H_2O_2/H_2SO_4 ratio 1:1) by placing a beaker with H_2O_2 on ice in a chemical hood and dropwise addition of H_2SO_4. Sonicate coverslips for 20 min in piranha solution and store them as is. Immediately before usage, grab individual coverslips by a forceps, rinse extensively with H_2O, and blow dry.
5. A suitable chamber for coverslip mounting is Ludin Chamber type 1 (*Life Imaging Services*).

2.3 Imaging Buffer Components

1. Oxygen scavenger (OS): 10 mM Tris pH 8.2, 5 mg/mL glucose oxidase type VII (*Sigma-Aldrich*, #G2133), 1 mg/mL

catalase from bovine liver (*Sigma-Aldrich*, #C30), 0.1 mM tris (2-carboxyethyl)phosphine (TCEP), 12.5% glycerol. For 10 mL of a 10× stock, mix 100 μL of 1 M Tris pH 8.2 with 7 mL H_2O in a 15 mL Falcon tube. Add 50 mg glucose oxidase, 555 μL catalase (18 mg/mL stock concentration, adjust volume for other stock concentrations), and add 1 μL of 1 M TCEP. Mix gently and bring volume to 8.75 mL with H_2O. Finally add 1.25 mL glycerol, mix gently by inverting the Falcon tube several times, make 50 μL aliquots, and store at −20 °C.

2. 40% (w/v) D-(+)-glucose stock solution: dissolve at 37 °C, filter sterile, aliquot and store at −20 °C.
3. 100 mM stock solutions of ascorbic acid (AA) and methyl viologen (MV): Weigh a few mg into 1.5 mL reaction tubes and label with the appropriate amount of H_2O to be added for a 100 mM stock solution. Make fresh AA and MV solutions before every experiment and store on ice until usage. Do not use the AA solution for more than 3 h.
4. Imaging buffer: 200 mM Tris pH 8.2, 0.5 mg/mL glucose oxidase, 0.1 mg/mL catalase, 4% w/v glucose, 2 mM AA, 2 mM MV, 0.02 mM TCEP, and 2.5% glycerol. For preparing 500 μL imaging buffer, take 50 μL 10× OS and sequentially add 280 μL H_2O, 100 μL of 1 M Tris pH 8.2, 50 μL glucose stock, 10 μL AA stock, 10 μL MV stock. Mix by gently pipetting up and down. Store on ice.

2.4 Microscope

1. The principle requirements for SHRImP imaging are a microscope with high illumination intensity, a shutter, a stable stage, a high NA objective with at least 60× magnification, good optical filters, and a camera with single-photon sensitivity. Typically, inverted epifluorescence or TIRF microscopes that are equipped with modulated ~100 mW (640 or 647 nm) lasers, an active focus stabilization, a TIRF objective, and an EM-CCD camera will work (Fig. 1c). These requirements are essentially the same as for PALM or STORM imaging. While many studies have used custom-built microscopes in the past, several microscope vendors now offer suitable instruments, such as NSTORM (*Nikon*), SR GSD (*Leica*), ELYRA (Zeiss), IX83 with cellTIRF (*Olympus*).
2. Any suitable software for controlling the microscope and for data acquisition can be used. A good open-source alternative is MicroManager (http://www.micro-manager.org).

2.5 Analysis Software

Visual inspection and basic analysis of SHRImP movies can be done in the open-source image processing package Fiji [22] which is based on ImageJ (*NIH*). For a detailed description, please *see* the tutorial in the Supplementary information. For more advanced analysis solutions, *see* **Note 1**.

1. Download the Fiji installer for your operating system from http://fiji.sc and follow the installation guidelines.
2. Activate the GDSC SMLM plugins: Open Fiji, select "Help > Update...", click "Manage update sites," check "GDSC-SMLM," click "close," click "Apply changes," wait until successfully finished, then restart Fiji.
3. Download the file "SHRImP_Macros.ijm" from the Supplementary Material accompanying this article online and save it to disk.
4. For further analysis and histogram fitting, we recommend MATLAB (*MathWorks*), Mathematica (*Wolfram Research*), or other scientific analysis programs (e.g., *Origin*).

3 Methods

The following procedure describes distance measurements by SHRImP to study the conformation of proteins labeled with Alexa Fluor 647.

3.1 Getting Ready

1. Prepare your sample on the cleaned coverslips using your standard procedure (Fig. 1b; *see* **Note 3**). Adjust the dilution of the labeled protein of interest so that single molecules can be observed. Do not exceed a density of 40–80 molecules within a sample area of 50 μm × 50 μm. For determining an adequate dilution, *see* **Note 4**.
2. Freshly prepare imaging buffer from (thawed) stock solutions immediately before usage (*see* Subheading 2.3, **item 4**). This imaging buffer was optimized for SHRImP with Alexa Fluor 647. For other dyes, *see* **Note 2**.
3. Set up the microscope. Select the appropriate objective (*see* **Note 5**) and appropriate lasers and filters for the imaged fluorophore. Set the image format for acquisition and saving to the highest possible bit depth (14 or 16 bit grayscale). Use frame-transfer mode to minimize lag times between acquired images.
4. Start Fiji. Install the SHRImP macros: click "Plugins > Macros > Install...", select the file "SHRImP_Macros.ijm" from its stored location and confirm by clicking "Open."

3.2 Determine Appropriate Acquisition Settings

1. Mount sample in the coverslip holder (Fig. 1c). Exchange buffer to freshly prepared imaging buffer. Seal the chamber (*see* **Note 6**) and place it on the microscope stage.
2. Focus on the features of interest using the lowest possible laser intensity (0.01–0.1 mW). To maintain the focus during xy

translation and during the acquisition of movies, activate the focus lock.

3. Make sure that the stage position does not drift (*see* **Note 7**).
4. Stop the exposure and move to a sample region which has not been exposed before. Choose appropriate camera and laser settings for the detection of single fluorophores. If these are not known, tentatively set the camera exposure to 50 ms, the EM gain to half of its maximum value, and the laser power to ~30%.
5. Acquire a movie of 1000 frames (corresponding to a measurement duration of 30–90 s) and save it to disk.
6. Quantify the bleaching kinetics (Fig. 1d): Open the recorded movie in Fiji as an image stack, then run "Plugins > Macros > Bleaching kinetics." The percentage of spots in which all fluorophores are still ON after five frames should be greater than 0.9; if this number is too low, reduce the laser intensity and repeat **steps 4–6**. Then adjust the number of recorded frames such that at least 99% of all spots are completely bleached at the end of the movie. If fluorophores are extraordinarily stable, increase the laser intensity and/or increase the frame exposure time.
7. Adjust the EM gain setting to avoid any overexposure. Aim at using ~50% of the full dynamic range.
8. Assess the quality of individual bleaching traces (Fig. 1d). Open the movie in Fiji. Click "Image > Stacks > Plot *Z*-axis Profile" and activate the "Live" button. Make a small rectangular or round selection (e.g., 9 × 9 pixels) around a single fluorescent spot in the movie image. Inspect the stepwise bleaching behavior of this or other single spots by moving this ROI from one spot to another. The intensity should decrease in well-discernable steps separated by stable plateaus, and the number of steps per spot should match the expected number of fluorophores on your molecule of interest. Refer to the troubleshooting comments in case of intermittent dye blinking (*see* **Note 8**), poor signal-to-noise ratio (*see* **Note 9**), or too many steps (*see* **Note 4**).
9. Note down and save the optimized SHRImP acquisition settings for later usage with similar samples.

3.3 Acquire SHRImP Movies

1. Place mounted sample with imaging buffer on the microscope stage (*see* Subheading 3.4, **step 1**).
2. Adjust the focus and lock it. To avoid any light expose of the labeled molecules before acquiring SHRImP movies, either do the focusing at a different location in your sample (i.e., for surface-immobilized molecules) and then move to a new

stage position, and/or use an appropriate co-staining to identify the structures of interest (i.e., in cell culture) in a separate color channel (e.g., at 488 nm).

3. Switch to the optimized SHRImP settings (*see* Subheading 3.4, **step 9**).
4. Simultaneously start the acquisition of the SHRImP movie and exposure. Save the movie to a disk.
5. Repeat **steps 2–4** for several regions in your sample to obtain enough raw data for sufficient statistics.
6. Repeat **steps 1–5** for different samples to investigate the impact of sample-to-sample variation or to test the influence of parameters that are relevant for your molecule of interest.

3.4 Analyze SHRImP Movies

1. Start Fiji and install SHRImP macros (*see* Subheading 3.1, **step 4**).
2. Open one SHRImP movie as an image stack.
3. Detect spots (Fig. 2a). Run "Plugins > Macros > Preprocess Movie." Choose appropriate process parameters, confirm by "OK" and wait until processing is finished and the ROI manager has been automatically populated. Visually inspect the detection of spots by toggling "Show all" in the ROI manager on and off. If too few (many) spots are found, repeat the preprocessing with a lower (higher) threshold value. Adjust the "radius" parameter (in integer steps) such that an ROI fully contains a single fluorescent spot. A good choice for the radius is about 1.5 times the spot's full-width-at-half-maximum.
4. Click "Image > Stacks > Plot *Z*-axis Profile" and activate the "Live" button.
5. Click on the window with the original movie and then select an item in the ROI manager. Critically inspect the stepwise bleaching behavior in the *z*-axis profile window and in the original movie stack in a magnified view. Only select spots for analysis that are clearly separated from their neighbors, show distinct plateaus/steps in the intensity-time trace, and have the expected number of steps/dyes. *Do not* analyze bleaching traces that show intermittent blinking (*see* **Note 8**), which have very poor signal-to-noise ratio (*see* **Note 9**), which have a plateau shorter than ~4 frames, or with bleaching steps that differ in height by more than a factor of ~1.5. For detailed instructions, please *see* also the tutorial in the Supplementary Information.
6. Start the SHRImP analysis for the selected spot by clicking "Plugins > Macros > SHRImP Analysis of this ROI". In the dialog, enter the number of clearly discernably steps and confirm by "OK."

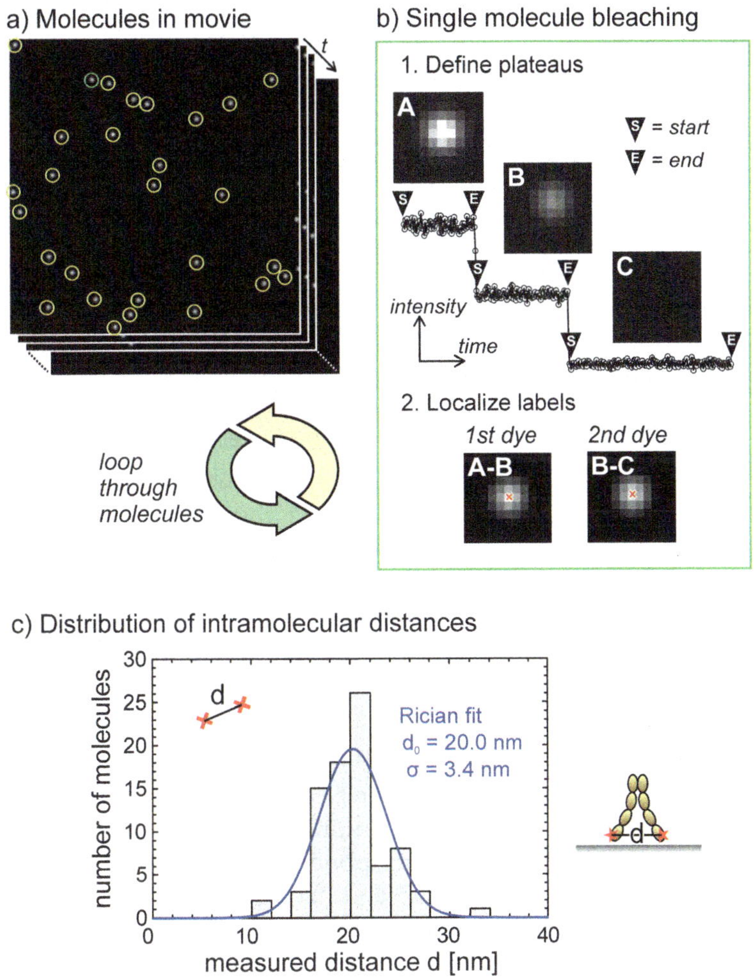

Fig. 2 SHRImP data analysis pipeline. (**a**) Fluorescent spots are detected in the recorded bleaching movie (*yellow* ROIs). (**b**) The step-by-step bleaching of a single spot (*green* ROI in **a**) is used to define plateaus in the intensity-time trace. From these, summed images are extracted that allow the localization of the individual fluorescent labels from subtracted images. This procedure is repeated for all spots in the movie that show a suitable bleaching behavior. (**c**) The distribution of pooled inter-label distances from many single molecules reveals information about the molecular conformation. The underlying mean distance is obtained by fitting the measured distances by a Rician distribution function

7. Define plateaus (Fig. 2b): Select the start frame of the first plateau by using the frame slider at the bottom of the movie window and confirm by "OK." Select the last frame of the first plateau in the same way. This is the last frame in which the intensity of the spot has still the same value as in the ones before. Use the curser in the intensity time trace and the active ROI in the movie for orientation. Proceed in the same way for

defining the start and end positions of the other plateau(s) and the background. Thereafter, wait until the automatic processing has completed.

8. The SHRImP analysis automatically adds the average plateau images (including information about spot ID and start/end frames) and subtracted images (from subsequent plateaus) to the "images4fit" stack. The fitting results are added to the "Fit Results" window. The "Log" window displays messages about the fitting success.
9. Repeat **steps 5–7** for all suitable ROIs in the ROI manager.
10. When you are finished with this movie, click "Plugins > Macros > Save Results & Clean up". Follow the advice in the dialog boxes and then wait until processing has finished.
11. The results of your analysis are saved at the location of your movie under the same name with extensions "*_ROIs.zip," "*_imgs4fits.tif," and "*_Fits.xls." These files enable you to restore the current state and to re-analyze your data at a later time point.
12. Repeat **steps 2–10** for other movies.

3.5 Analyze Distances

1. Open a saved "*_Fits.xls" file in Excel or another spreadsheet program.
2. The first column ("Source") contains information about the spot number and the fitted subtracted image (see the corresponding slice info in the "*_imgs4fit.tif" stack). The "X" and "Y" columns contain the fitted fluorophore positions in units of pixels. The "Error" column contains the R^2 value which informs about the goodness of the fit.
3. Only further analyze spots for which the R^2 value of each fluorophore is >0.95. Calculate the Euclidean distance between pairs of fluorophores within this single spot by $d = \sqrt{(X_i - X_j)^2 + (Y_i - Y_j)^2}$.
4. Repeat **step 3** for all fitted spots in this movie.
5. Rescale all distances by multiplying with the actual pixel size (*see* **Note 5**).
6. Repeat **steps 1–5** for different movies. Pool all distances from samples that correspond to the same condition.
7. Create a histogram (Fig. 2c).
8. When you expect a single population of distances, fit the data by a Rician probability distribution function: $p_{2D}(d) = \frac{d}{\sigma^2} \exp\left(-\frac{d_0^2 + d^2}{2\sigma^2}\right) I_0\left(\frac{d_0 d}{\sigma^2}\right)$ [8, 10] with the noncentrality parameter d_0 and the scale parameter σ. If several populations are expected, fit the histogram by the sum of several Rician distributions. The

fitted parameters give you the mean distance d_0 and the width σ (including measurement uncertainty) of the distance distribution (Fig. 2c).

9. Use this analysis and the fitted parameters and uncertainties to compare samples from different conditions.
10. Make sure that you do not mis- or over-interpret results due to limited measurement precision or biased analysis (*see* **Note 10**).

4 Notes

1. *Optimal analysis software.* The here described ImageJ-based SHRImP analysis localizes the fluorophores from subtracted images of average plateau images using Least Square Estimation. This procedure is very robust and yields valid results, however, it does not take full advantage of the information in SHRImP movies. An optimal analysis needs to carry out the fitting of all fluorophores in parallel on summed plateau images by Maximum Likelihood Estimation (MLE) [8]. This requires a conversion of camera counts into the number photons (*see* e.g. supplement of ref. 23) and custom-written software (*see* ref. 8 for details). The benefits are the best possible localization precision, reliable error estimates, and an easier automation of the analysis workflow. Such an analysis is essential for precision measurements of, i.e., small distances (below ~20 nm). We strongly recommend that custom-written analysis software is validated with simulated data (*see* **Note 10**).
2. *Fluorophore-imaging buffer combinations.* Besides Alexa Fluor 647, we have also used Atto 647N or Atto 655 with 640 nm excitation, or Cy3B with 532 nm excitation [17] with good success for SHRImP. Depending on the dye, other imaging buffer compositions to maximize brightness and reduce blinking and bleaching might be better suited than AA/MV. The here described Tris buffer incl. the oxygen scavenging system can serve as a starting point to test other additives such as Trolox (6-hydroxy-2,5,7,8-tetramethylchroman-2-carboxylic acid), β-mercaptoethanol, nitrobenzoic acid, or others. For guidelines, *see* the corresponding reviews [29, 30].
3. *Coverslips.* The described cleaning procedure minimizes background fluorescence and allows for versatile further usages of the coverslips (*see* Fig. 1b), i.e., for modifying the surface chemistry by silanization [17], for cell culture (after adhesive protein coatings) [21], or for PEG-biotin/neutravidin coatings for subsequent biotin-based probe immobilization [2]. Care has to be taken that surface coatings and sample preparation do not introduce additional background fluorescence. In

especially for cellular work, make sure that phenol red-containing culture medium is efficiently removed by washing with PBS before fixation or measurements.

4. *Molecule density.* For finding an appropriate density of surface-adsorbed proteins, make a serial dilution series around 1 ng/mL of your protein of interest. Incubate the solution on coverslips for 10 min and wash intensively. Take several snapshot images of your sample to estimate the density of single molecules. For adjusting the density of labeled proteins in cell culture, the concentration of exogenously labeled proteins can be adjusted in a similar way, whereas endogenous expression of labeled proteins at sufficiently low levels may demand low copy number stringent plasmids.
5. *Calibration of image pixel size.* The size of an image pixel is determined by the physical size of a camera pixel (typically 8–16 μm) divided by the magnification (typically 63–150×). The optimal image pixel size for the localization of single molecules is roughly 90–140 nm [32]. Use an appropriate objective magnification to reach this range. For a direct experimental calibration, use a μm-sized object of known size, e.g., a microscope calibration ruler or an EM grid (commercially available from several vendors).
6. *Photostability.* Irreversible bleaching is a consequence of chemical reactions between the fluorophore's triplet state with dissolved oxygen. To delay bleaching, triplet state quenchers are used as imaging buffer ingredients (*see* **Note 2**). Critical is also the enzymatic reduction of the dissolved oxygen concentration which can be substantially improved by sealing off the imaging buffer from surrounding air (Fig. 1c). Possible means are using the closed configuration of the Ludin chamber, sandwiching with an additional coverslip and sealing with nail polish, or coverage of the surface by paraffin oil. For testing other possible oxygen scavengers, *see* also ref. 31.
7. *Drift.* The SHRImP analysis requires that the stage drift in the *xy* plane during acquisition of a movie is substantially smaller than the to-be-measured distances. Even though typical acquisition times are short (30–90 s), stage drift can be a problem for the measurement of distances in the nanometer range. Measure the stage drift over 10 min using surface-immobilized fluorescent beads and estimate if the expected drift during SHRImP is within tolerable bounds, i.e., smaller than 10% of the to-be-measured distance. Stage drift especially occurs directly after stage movements and can be minimized by waiting for ~1 min before starting the acquisition at a new position.
8. *Fluorophore blinking.* Blinking of dyes hampers the SHRImP analysis because it prevents the unique assignment of specific

steps to specific dyes. On-off blinking of fluorophores can originate from non-fluorescent dark states [29, 30], from contact quenching or stacking of the dye by/with its environment/surface, or from binding/unbinding events. Regarding dark states, try to optimize the concentration and type of reducing/oxidizing agents in the buffer (*see* **Note 2**). Regarding surface interactions and unstable immobilization, consider changing the surface chemistry (e.g., by a dichlorodimethylsilane coating) and/or the immobilization method (physisorption, affinity-based, chemisorption).

9. *Poor signal-to-noise ratio.* The signal-to-noise ratio is determined by the fluorophore brightness and background fluorescence. The brightness of single fluorophores depends on the illumination intensity and a sufficient suppression of dark states by appropriate buffer agents (*see* **Notes 2** and **8**). For the minimization of the background, first try to find the origin (non-blocked excitation light, coverslip, buffer components, sample autofluorescence). If you detect excitation light, use better optical filters. If the coverslip fluoresces, optimize the cleaning and coating procedure (*see* **Note 3**). If it originates from the buffer solution, try filtering through 0.2 μm filters to remove impurities. Dissolved AA develops autofluorescence in the green (500–550 nm) over time and should be avoided when using green fluorophores. You might also want to try inclined (HiLo) or TIRF illumination to restrict the excitation to a region close to the coverslip (if this is compatible with your sample). If it originates from the sample itself, you might need to use a different spectral window (more in the red).

10. *Positive controls.* We recommend the usage of an experimental control with a defined distance between fluorophores to assess the validity and precision of SHRImP measurements on your setup. Nanometer rulers from double-stranded DNA (*see* refs. 1, 11, 17) cover distances up to ~15 nm, whereas DNA-origami standards can be customized in terms of the number of dyes and distances [12] (available from www.gattaquant.com). For validation of the SHRImP analysis, a simulated movie with known ground truth (Supplementary Information online) can be used.

References

1. Gordon MP, Ha T, Selvin PR (2004) Single-molecule high-resolution imaging with photobleaching. Proc Natl Acad Sci U S A 101(17): 6462–6465. doi:10.1073/pnas.0401638101
2. Roy R, Hohng S, Ha T (2008) A practical guide to single-molecule FRET. Nat Methods 5(6):507–516. doi:10.1038/Nmeth.1208
3. Allen JR, Ross ST, Davidson MW (2013) Single molecule localization microscopy for superresolution. J Opt 15(9). doi:10.1088/2040-8978/15/9/094001
4. Churchman LS, Okten Z, Rock RS, Dawson JF, Spudich JA (2005) Single molecule high-resolution colocalization of Cy3 and Cy5

attached to macromolecules measures intramolecular distances through time. Proc Natl Acad Sci U S A 102(5):1419–1423. doi:10.1073/pnas.0409487102

5. Margadant F, Chew LL, Hu X, Yu H, Bate N, Zhang X, Sheetz M (2011) Mechanotransduction in vivo by repeated talin stretch-relaxation events depends upon vinculin. PLoS Biol 9(12). doi:10.1371/journal.pbio.1001223
6. Michalet X, Lacoste TD, Weiss S (2001) Ultrahigh-resolution colocalization of spectrally separable point-like fluorescent probes. Methods 25(1):87–102. doi:10.1006/meth.2001.1218
7. Pertsinidis A, Zhang YX, Chu S (2010) Subnanometre single-molecule localization, registration and distance measurements. Nature 466(7306):647–U611. doi:10.1038/Nature09163
8. Schoen I (2014) Localization precision in stepwise photobleaching experiments. Biophys J 107(9):2122–2129. doi:10.1016/j.bpj.2014.09.035
9. Huang B, Jones SA, Brandenburg B, Zhuang XW (2008) Whole-cell 3D STORM reveals interactions between cellular structures with nanometer-scale resolution. Nat Methods 5 (12):1047–1052. doi:10.1038/nmeth.1274
10. Churchman LS, Flyvbjerg H, Spudich JA (2006) A non-Gaussian distribution quantifies distances measured with fluorescence localization techniques. Biophys J 90(2):668–671. doi:10.1529/biophysj.105.065599
11. Qu XH, Wu D, Mets L, Scherer NF (2004) Nanometer-localized multiple single-molecule fluorescence microscopy. Proc Natl Acad Sci U S A 101(31):11298–11303. doi:10.1073/pnas.0402155101
12. Schmied JJ, Gietl A, Holzmeister P, Forthmann C, Steinhauer C, Dammeyer T, Tinnefeld P (2012) Fluorescence and super-resolution standards based on DNA origami. Nat Methods 9(12):1133–1134. doi:10.1038/nmeth.2254
13. Steinhauer C, Jungmann R, Sobey TL, Simmel FC, Tinnefeld P (2009) DNA origami as a nanoscopic ruler for super-resolution microscopy. Angew Chem Int Ed 48 (47):8870–8873. doi:10.1002/anie.200903308
14. Kufer SK, Strackharn M, Stahl SW, Gumpp H, Puchner EM, Gaub HE (2009) Optically monitoring the mechanical assembly of single molecules. Nat Nanotechnol 4(1):45–49. doi:10.1038/Nnano.2008.333
15. Muls B, Uji-i H, Melnikov S, Moussa A, Verheijen W, Soumillion JP, Josemon J, Mullen K, Hofkens J (2005) Direct measurement of the end-to-end distance of individual polyfluorene polymer chains. ChemPhysChem 6 (11):2286–2294. doi:10.1002/cphc.200500235
16. Aoki H, Mori K, Takahashi T, Ito S (2013) Quantitative analysis of end-to-end distance of single polymer chain in ultra-thin film by super-resolution fluorescence imaging. Chem Phys 419:54–58. doi:10.1016/j.chemphys.2012.12.026
17. Klotzsch E, Schoen I, Ries J, Renn A, Sandoghdar V, Vogel V (2014) Conformational distribution of surface-adsorbed fibronectin molecules explored by single molecule localization microscopy. Biomater Sci 2:883–892. doi:10.1039/c3bm60262a
18. Balci H, Ha T, Sweeney HL, Selvin PR (2005) Interhead distance measurements in myosin VI via SHRImP support a simplified hand-over-hand model. Biophys J 89(1):413–417. doi:10.1529/biophysj.105.060608
19. Gruszka DT, Whelan F, Farrance OE, Fung HKH, Paci E, Jeffries CM, Svergun DI, Baldock C, Baumann CG, Brockwell DJ, Potts JR, Clarke J (2015) Cooperative folding of intrinsically disordered domains drives assembly of a strong elongated protein. Nat Commun 6:-7271. doi:10.1038/Ncomms8271
20. Needham SR, Hirsch M, Rolfe DJ, Clarke DT, Zanetti-Domingues LC, Wareham R, Martin-Fernandez ML (2013) Measuring EGFR separations on cells with similar to 10 nm resolution via fluorophore localization imaging with photobleaching. PLoS One 8(5). doi:10.1371/journal.pone.0062331
21. Früh SM, Schoen I, Ries J, Vogel V (2015) Molecular architecture of native fibronectin fibrils. Nat Commun 6:7275. doi:10.1038/Ncomms8275
22. Schindelin J, Arganda-Carreras I, Frise E, Kaynig V, Longair M, Pietzsch T, Preibisch S, Rueden C, Saalfeld S, Schmid B, Tinevez JY, White DJ, Hartenstein V, Eliceiri K, Tomancak P, Cardona A (2012) Fiji: an open-source platform for biological-image analysis. Nat Methods 9(7):676–682. doi:10.1038/Nmeth.2019
23. Früh SM, Spycher PR, Mitsi M, Burkhardt MA, Vogel V, Schoen I (2014) Functional modification of fibronectin by N-terminal FXIIIa-mediated transamidation. Chembiochem 15(10):1481–1486. doi:10.1002/cbic.201402099
24. Leake MC, Chandler JH, Wadhams GH, Bai F, Berry RM, Armitage JP (2006) Stoichiometry and turnover in single, functioning membrane protein complexes. Nature 443 (7109):355–358. doi:10.1038/nature05135

25. Plant LD, Xiong DZ, Dai H, Goldstein SAN (2014) Individual I-Ks channels at the surface of mammalian cells contain two KCNE1 accessory subunits. Proc Natl Acad Sci U S A 111(14): E1438–E1446. doi:10.1073/pnas.1323548111
26. Ulbrich MH, Isacoff EY (2007) Subunit counting in membrane-bound proteins. Nat Methods 4(4):319–321. doi:10.1038/Nmeth1024
27. Hines KE (2013) Inferring subunit stoichiometry from single molecule photobleaching. J Gen Physiol 141(6):737–746. doi:10.1085/jgp.201310988
28. Stephanopoulos N, Francis MB (2011) Choosing an effective protein bioconjugation strategy. Nat Chem Biol 7(12):876–884. doi:10.1038/Nchembio.720
29. van de Linde S, Sauer M (2014) How to switch a fluorophore: from undesired blinking to controlled photoswitching. Chem Soc Rev 43 (4):1076–1087. doi:10.1039/c3cs60195a
30. Zheng QS, Juette MF, Jockusch S, Wasserman MR, Zhou Z, Altman RB, Blanchard SC (2014) Ultra-stable organic fluorophores for single-molecule research. Chem Soc Rev 43 (4):1044–1056. doi:10.1039/c3cs60237k
31. Shi XH, Lim J, Ha T (2010) Acidification of the oxygen scavenging system in single-molecule fluorescence studies: in situ sensing with a ratiometric dual-emission probe. Anal Chem 82(14):6132–6138. doi:10.1021/ac1008749
32. Thompson RE, Larson DR, Webb WW (2002) Precise nanometer localization analysis for individual fluorescent probes. Biophys J 82 (5):2775–2783. doi:10.1016/S0006-3495(02)75618-X

Chapter 17

Correlative Single-Molecule Localization Microscopy and Confocal Microscopy

Christian Soeller, Yufeng Hou, Isuru D. Jayasinghe, David Baddeley, and David Crossman

Abstract

Single-molecule localization microscopy allows the ability to image fluorescence labeled molecular targets at nanoscale resolution. However, for many biological questions the ability to provide tissue and cellular context in addition to these high resolution data is eminently informative. Here, we describe a procedure to achieve this aim by correlatively imaging human cardiac tissue first at the nanoscale with direct stochastic optical reconstruction microscopy (dSTORM) and then at the diffraction limit with conventional confocal microscopy.

Key words Super-resolution microscopy, Confocal imaging, Fluorescence, Multiscale data, Heart, Antibodies, Correlative imaging, Human cardiac tissue

1 Introduction

Single-molecule localization microscopy (SMLM) provides high contrast nanoscale imaging of fluorescently labeled molecular targets [1–4]. This technique is rapidly increasing in popularity with instruments becoming available in many institutional imaging units allowing the non-specialist to incorporate this method in their experimental workflow. A highly desirable feature of this method is that many of the fluorophores used for localization microscopy can be imaged with diffraction-limited fluorescence techniques such as confocal microscopy. This feature provides the ability to place nanoscale subcellular data into an overall cellular and tissue level context. This is particularly helpful for the many biological questions that require integrating information across spatial scales. Below we describe a protocol on how to prepare, label, and image human cardiac tissue correlatively combining nano-scale imaging using direct stochastic optical reconstruction microscopy (dSTORM) with cellular and tissue level imaging by confocal

Holger Erfle (ed.), *Super-Resolution Microscopy: Methods and Protocols*, Methods in Molecular Biology, vol. 1663, DOI 10.1007/978-1-4939-7265-4_17,

microscopy. We assume that separate instruments are used for dSTORM super-resolution and confocal imaging, respectively. The procedure can be simplified if both modalities are available on the same microscope stand as has been recently described [5], but many facilities without such specialized instruments may have to resort to the approach described below which requires localizing the same tissue regions on different microscopes. In our experience, this can be readily achieved and we provide some guidance how to streamline this process.

In our research we generally focus on the biophysics of calcium release that controls contraction in the heart [6–8]. In the protocol below we therefore describe how to label human cardiac tissue for several macro-molecular membrane complexes that control calcium release and muscle contraction. This includes the protein junctophilin [9] (JPH), a protein critical in the formation of cardiac junctions (i.e. the calcium signaling sites where the plasma membrane closely opposes the sarcoplasmic reticulum, SR), the SR calcium release channel or ryanodine receptor [10] (RyR) and the cell membrane marker wheat germ agglutinin (WGA) [11]. Further application aspects for these targets are detailed in a recent publication [12] . Certainly, the approach can be applied to other molecular targets and tissue types with some adjustments to optimize the immuno-labeling protocols for the specific targets and tissues.

2 Materials

2.1 Materials for Cardiac Tissue Fixation and Sectioning

1. Fixative solution: 2% Paraformaldehyde (PFA) in phosphate-buffered saline is freshly prepared by diluting 8% PFA stock (this solution is stored under inert gas in sealed ampoules, and is available from Electron Microscopy Sciences). For example, a 10 ml vial is diluted to 40 ml with 4 ml 10× phosphate-buffered saline (PBS) and 26 ml of ultrapure water.
2. Sucrose solutions: 10, 20, and 30% (W/V) in PBS.
3. Liquid nitrogen and methylbutane.

2.2 Materials for Labeling Protocol

1. Triton X-100.
2. PBS.
3. Sodium azide (NaN3 added to prevent bacterial growth in antibody solutions).
4. Image-iT FX signal Enhancer (ThermoFischer Scientific).
5. Bovine Serum Albumin (BSA).
6. Primary antibodies: mouse monoclonal IgG1 anti-RyR2 (1:100 Cat# MA3–916; ThermoFischer Scientific) and a rabbit

polyclonal IgG anti-JPH2 (1:100, Cat# 40-5300; Thermo-Fischer Scientific).

7. Secondary antibodies: goat anti rabbit Alexa Fluor 750, goat anti mouse Alexa Fluor 680 and Alexa Fluor 488 (1:100 and 1:400, respectively).
8. Wheat germ agglutinin (WGA) Alexa Fluor 594 (1:50).

2.3 Switching Buffer

1. 100% Glycerol.
2. 10× PBS.
3. β-Mercaptoethylamine.

2.4 Bead Calibration Slide

1. Fluorescent microspheres 0.2 μm in diameter either Fluo-Spheres® or TetraSpeck™ (available from Thermo-Fisher Scientific). *See* **Note 1**.
2. Methanol or ethanol for dilution of beads.
3. Mounting medium: 90% glycerol and, 10% PBS.

3 Methods

3.1 Preparation of Cardiac Tissue

1. Small blocks of cardiac tissue are fixed in 2% (w/v) paraformaldehyde in PBS at 4 °C for 60 min (*see* **Note 2**).
2. Tissue blocks are then cryo-protected by sequential immersion in 10, 20, and 30% sucrose (w/v) dissolved in PBS. The first immersion is for at least 1 h with subsequent immersions continued until the tissue sinks to the bottom of the sampling tube indicating penetration of the sucrose.
3. Tissue blocks are frozen in liquid nitrogen cooled isopentane and stored in a −80 °C freezer prior to cutting.
4. Cryo-sections 5–10 μm thick are cut using a suitable cryostat (we use a CM 1900, Leica, *see* also **Note 3**).
5. Tissue sections can be stored at −80 °C or immediately used for labeling.

3.2 Labeling Protocol

Ryanodine receptor, junctophilin, and WGA labeling of cardiac tissue for both super resolution and confocal imaging (*see* **Note 4**).

1. Cryo-sections to be labeled are removed from the freezer and allowed to warm to room temperature before opening the sealed slide box (preventing condensation forming on sections).
2. Coverslips with attached sections are attached to Perspex slide chambers with nail varnish. This provides a convenient open chamber for labeling as well as replacement of switching buffer (*see* **Note 5** on switching buffer activity) but is only suitable for

an inverted microscope. Alternatively, labeling can be done on coverslips and then mounted on slides at the end of the protocol.

3. Cryo-sections are hydrated in PBS (10 min) and permeabilized with diluted Triton X-100 in PBS for 15 min at room temperature and then washed in PBS for a further 15 min. Triton X-100 is generally used at low concentration of 0.05% (v/v) but for well-anchored structural proteins can be increased (we have used up to 1% (v/v)) but this may require separate experiments (by, for example, electron microscopy) to test if ultrastructure remains unaffected.
4. Sections are then blocked in Image-iT FX signal Enhancer at room temperature for 1 h.
5. Sections are then labeled for ryanodine receptor (RyR2) and junctophilin-2 (JPH2) by incubating overnight (at 4 °C) with mouse monoclonal IgG1 anti-RyR2 (1:100) and a rabbit polyclonal IgG anti-JPH2 (1:100). Both primary and secondary antibodies are diluted in incubation buffer: 0.05% NaN3 (w/v), 1% bovine serum albumin (w/v), 0.05% Triton X100 (v/v) in PBS.
6. Sections are then washed three times for 30 min in PBS at room temperature.
7. Sections are then incubated for 2 h (room temperature) with goat anti rabbit Alexa Fluor 750 (1:100), goat anti mouse Alexa Fluor 680 and Alexa Fluor 488 (1:100 and 1:400, respectively, *see* **Note 6**), and wheat germ agglutinin (WGA) Alexa Fluor 594 (1:50).
8. Sections are washed three times for 30 min in PBS at room temperature.
9. After draining excess PBS sections are mounted in freshly made switching buffer: 10 mM MEA as described below.
10. A negative control consists of sections labeled as above except the primary antibodies are omitted from the incubation solution.

3.3 Preparation of Localization Microscope Switching Buffer

1. Prepare 500 mM stock β-mercaptoethylamine. Weigh out 38.6 mg of MEA into a 2 mL microcentrifuge tube. Add 1 mL 10× PBS; mix thoroughly. Adjust pH to ~8.5–9.0 with 10% HCl (*see* **Note 7**).
2. Pipette 900 μl of glycerol into a 1.5 ml microcentrifuge tube.
3. Then add a suitable quantity of MEA stock solution to achieve final MEA concentrations typically somewhere between 5 and 100 mM (*see* **Note 8**) and mix thoroughly by inversion (*see* **Note 9**).

4. Adjust to desired pH (7–8) with diluted HCl and adjust final volume to 1 ml with 10× PBS.
5. Make fresh before use, allow ~30 min to fully mix constituents.

3.4 Preparation of Bead Slide

1. Dilute 0.2 μm fluorescent microspheres in either methanol or ethanol. A dilution of either 1000- to 5000-fold for the FluoSpheres® or 10- to 50-fold for TetraSpeck™ (*see* **Note 1**).
2. A pipette is then used to streak ~10 μl of the bead colloid onto a coverslip and allowed to dry.
3. After allowing the coverslip with the beads to dry it is then mounted on a slide using 10 μl of 90% glycerol in PBS and sealed with nail varnish. It should last for several months when stored in the fridge between imaging runs.
4. The microspheres are primarily used to correct for chromatic shifts between channels in dSTORM imaging (Alexa Fluor 680 and Alexa Fluor 750) as previously described [13]. They can also be used to calibrate alignment between confocal imaging channels and acquire point-spread function data for the offline deconvolution of confocal data.

3.5 Correlative dSTORM and Confocal Imaging

Details of the microscopes used for dSTORM and confocal imaging have been previously described [7] and are outlined briefly in **Note 10**. In general, the protocol described below can be carried out on any commercial or custom-built setups capable of dSTORM imaging or related SMLM methods and confocal microscopy, respectively. Below we describe an overview of the approach we typically use to collect correlative data sets. With any correlative imaging approach the ability to identify the same cellular region across imaging modalities is essential. Fluorescent beads have been used as fiduciary markers in correlative light and electron microscopy [14] and in sequential labeling and imaging with SMLM [15]. A distinct advantage of tissue imaging is that the shape of the tissue section provides its own inherent fiduciary land marks that can be used to identify target structures between microscopes.

1. The dSTORM system is used in widefield mode to visually inspect labeling of RyR (Alexa 488) and WGA (Alexa 594). WGA labeling provides excellent visualization of tissue structure and cellular organization. Furthermore, diffraction-limited images of the JPH and RyR labeling (Alexa 680 and Alexa 750), not visible to the unaided eye, can be captured with the EMCCD.
2. Once a suitable region is located a widefield image of the cells is recorded. An example for WGA labeling is shown in Fig. 1a. This image is useful in subsequently identifying the same region in the confocal microscope.

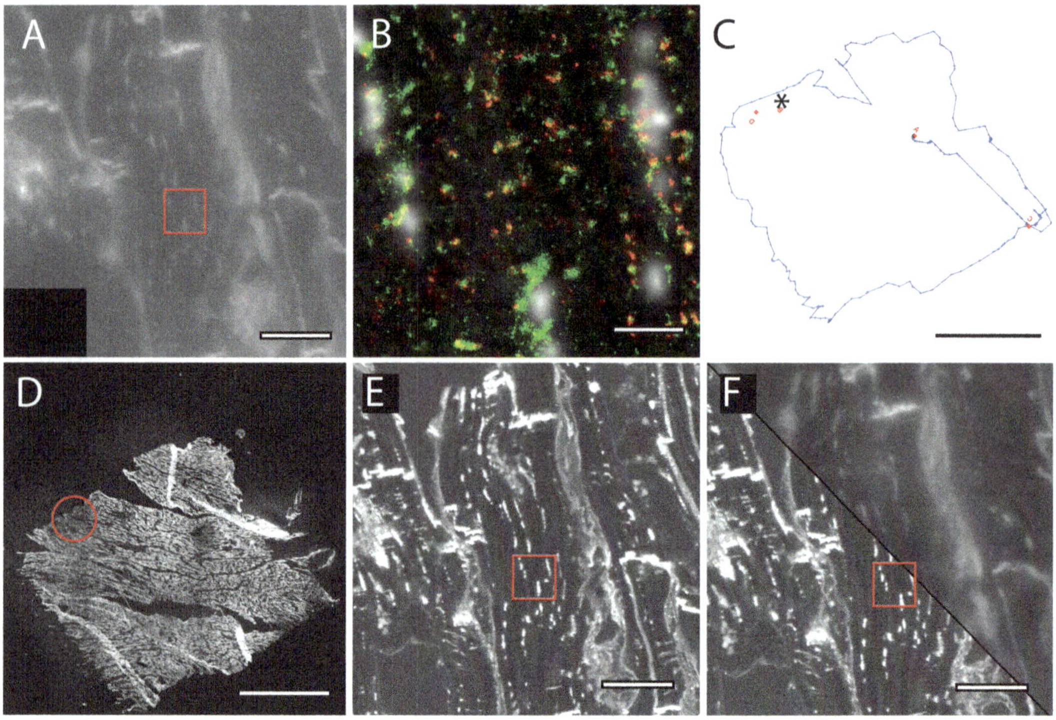

Fig. 1 Process of correlative imaging between dSTORM and confocal microscopy. The procedure is illustrated using identification of target regions across systems within cardiac tissue as an example. Panel (**a**) shows a widefield fluorescence overview image acquired on the dSTORM setup of wheat germ agglutinin (WGA) in the 594 nm excitation channel. Panel (**b**) shows a dSTORM image of the *red square* region found in Panel (**a**) labeled for ryanodine receptors (RyR2, 750 nm) in *red*, junctophilin (JPH2, 680 nm) in *green*, and a diffraction-limited view of the WGA labeling in *white*. Panel (**c**) shows a traced outline of the section being imaged with *red points* and *letters* denoting dSTORM imaged regions. The area marked with an *asterisk* in panel (**c**) represents the target region in panels (**a**) and (**b**). Panel (**d**) shows a stitched low magnification (10×) overview of the sample in the JPH channel acquired with 633 nm on the confocal microscope. Red circled region represents the region of interest shown in previous panels. Panel E shows a confocal image of the region of interest corresponding to panels (**a**) and (**b**). Panel (**f**) shows a comparison between the widefield image in (**a**) with the confocal image in (**b**). Scale bars: (**a**)—15 μm, (**b**)—2 μm, (**c**, **d**)—1 mm, (**e**, **f**)—15 μm

3. A super-resolution dSTORM image of the identified region is then recorded for RyR (Alexa 750) and JPH (Alexa 680, Fig. 1b, *see* **Note 10**).
4. As an additional aid to cell identification, a map of the tissue section, including the location of cells imaged at super resolution, is created (Fig. 1c, *see* **Note 11**).
5. The slide is then transferred to the confocal microscope system. To identify the location of super-resolution imaged cells a low magnification lens (10×) is initially selected to survey the tissue section while referencing the map generated in **step 4**. At this stage a confocal tile scan of JPH labeling (or another suitable

channel) can be used to record tissue and cellular organization of the entire section (Fig. 1d, **Notes 12–14**).

6. Once the approximate location of the cell is identified a high numerical aperture lens is used to confirm the location and to capture confocal image stacks for all desired channels (e.g., WGA-Alexa 594 as shown in Fig. 1e).

The resulting confocal and super-resolution data sets can be correlated and registered using standard registration techniques [16], as long as at least one channel is shared across confocal and widefield/super-resolution data sets. In our application this includes WGA, as well as the RyR and JPH2 channels. Fig. 1f shows the aligned widefield and confocal WGA data which is used to align confocal and super-resolution data.

Figure 2 shows typical results and the tissue scales that are accessible by combining confocal and SMLM information (Fig. 2a–e). The correlative information can be exploited in correlative 3D rendering that overlays the membrane structures from the confocal data sets (WGA labeling) with the super-resolution channels, here showing JPH2 (Fig. 2f) and RyR (Fig. 2g) that demonstrates the tight association of these protein distributions with elements of the internal membrane system in cardiac myocytes, the transverse tubules.

4 Notes

1. The FluoSpheres® are available with single excitation/emission peaks including blue (365/415 nm), yellow-green (505/515 nm), and dark-red (660/680 nm). Conveniently, TetraSpeck beads have four excitation/emission peaks in each bead (blue (360/430 nm), green (505/515 nm), orange (560/580 nm), and dark red (660/680 nm) allowing alignment to be checked between confocal channels. FluoSpheres® are much more resistant to photobleaching than TetraSpeck beads and are therefore more appropriate for intensity calibration. Different microspheres can be combined on the same slide.
2. Human cardiac tissue from transplant recipients was obtained directly from an operating theater in accordance with ethical approval from the Health and Disabilities Ethics Committee of New Zealand. Note that the staining and imaging protocol will also work with animal tissue. In order to maximize event density during dSTORM imaging, and hence the resolution achieved, a high density of antibody labeling is desirable. In general, we have found high density antibody labeling is facilitated by gentle fixation and have confirmed that it provided

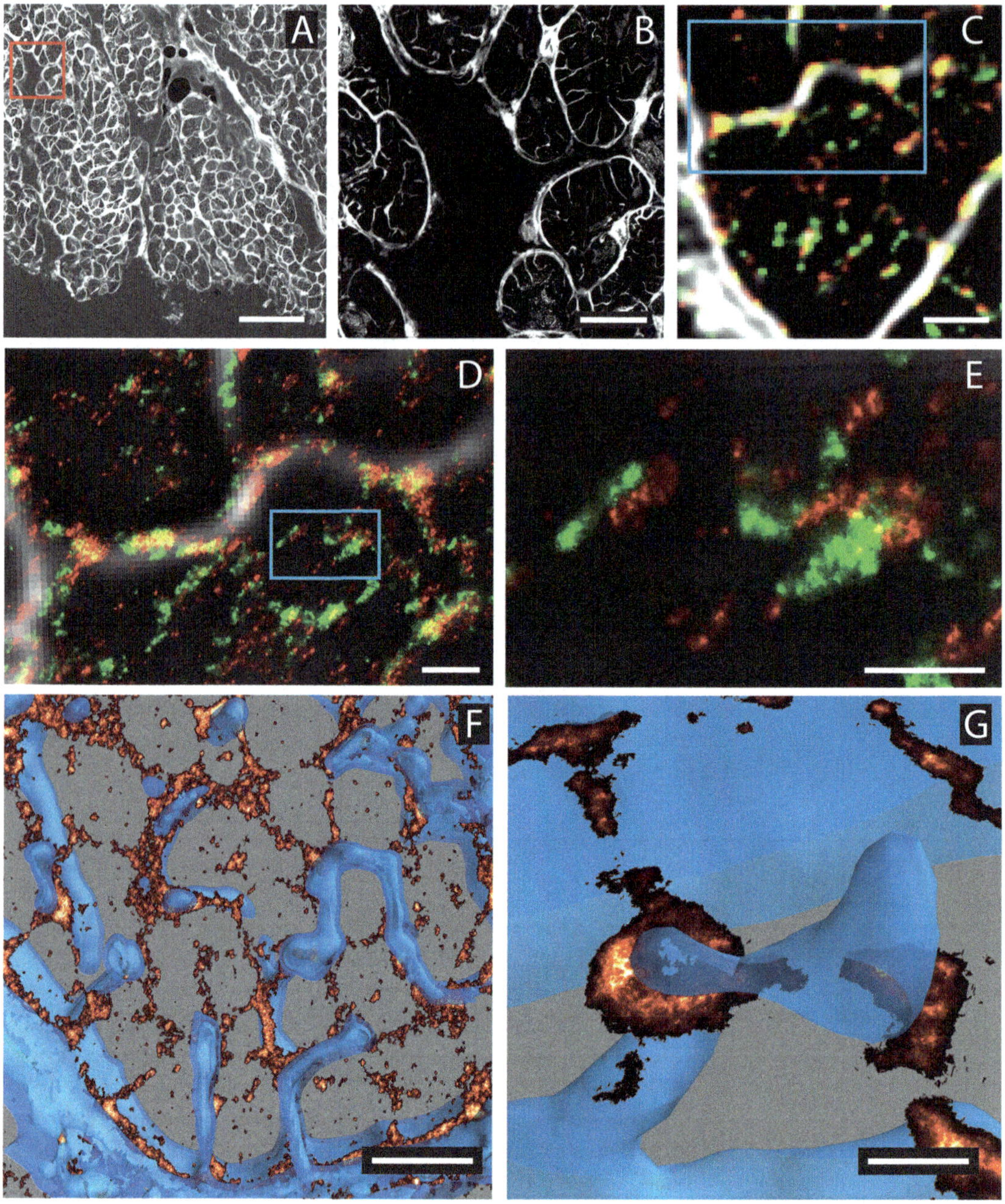

Fig. 2 Demonstration of multi-scale imaging with the correlative approach in cardiac tissue. Panel (**a**) shows a confocal image of a tissue region labeled using WGA and acquired with the confocal microscope. Panel (**b**) shows a magnification (*boxed* region in **a**) of WGA labeling detailing local cellular arrangements of the fluorescent labeling. Panel (**c**) shows a high magnification confocal closeup of a myocyte labeled with WGA in the 594 nm excitation channel (*white*), RyR2 in the 488 nm excitation channel (*red*), and JPH2 in the 680 nm excitation channel (*green*). Panel (**d**) shows a further magnification of the boxed region in (**c**) imaged with a combination of RyR2 (*red*) and JPH2 (*green*) dSTORM super-resolution and WGA (*white*) confocal data. Panel (**e**) demonstrates the resolution in the super-resolution data (corresponding to the *boxed* region in **d**). Panel (**f**) shows a 3D rendering of a confocal data stack of WGA (*blue*) with overlayed 2D dSTORM data of JPH2 labeling (*orange*) at the plane of imaging. Panel (**g**) is an alternative detail view of the data shown in panel (**f**) detailing the arrangement of JPH2 around the membranous WGA labeling. Scale bars: **a**—100 μm, **b**—20 μm, **c**—2 μm, **d**—1 μm, **e**—500 nm, **f**—2 μm, **g**—500 nm

comparable organelle structure preservation to tissue prepared for electron microscopy [17].

3. We conduct cryo-sectioning with a chamber temperature of ~−25 °C (adjust for optimal cutting results according to local conditions; a local pathologist may be able to provide advice). Sections are mounted on acid-washed and poly-lysine-coated coverslips to facilitate tissue adherence and are allowed to dry. Sections are mounted directly to the coverslip (rather than the slide) to ensure the tissue is close to the coverslip surface that reduces spherical aberration and avoids issues with the often limited working distance of high NA objectives.
4. Localization microscopy is critically dependent on labeling density. Generally speaking, the higher the density the better is the resolution (until sufficient spatial sampling is achieved). To achieve high density labeling only high affinity and specific antibodies should be used that give rise to dense labeling of the target of interest. As a rule of thumb only labeling results that are considered excellent in confocal imaging are likely to work well for super-resolution imaging. Users with experience in confocal imaging of immuno-labeled samples will generally know which of their antibody agents produce consistently strong and specific labeling. Note that all of this assumes that the confocal microscope is well aligned as a lot of signal can be lost by a poorly aligned pinhole. Even in well-maintained imaging facilities users should regularly establish this using multi-color fluorescent beads samples such as Tetraspeck beads (see above) that can also help determine subpixel shifts between imaging channels [18]. Two suitable antibodies that provide strong and specific staining in our samples are the mouse monoclonal antibody generated against the ryanodine receptor and a rabbit polyclonal antibody generated against junctophilin [19, 20].
5. The dye switching of samples can degrade during an imaging run. Often the switching performance can be recovered by replenishing the slide with fresh switching buffer. This procedure is greatly simplified in samples mounted in an open slide chamber/dish with coverslip bottom and can be done with the sample still on the microscope when using an inverted configuration. To inexpensively make such slide chambers we get those laser-cut from Perspex/acrylic sheets to our specification and glue coverslips to the bottom using fast-setting two-component silicone. We have "ready to send away to your favorite laser cutting service" artwork that can be used to order chambers. For work requiring sterile slide chambers commercial products with suitable coverslip bottom can be used—note to match the coverslip thickness with the design thickness of your microscope objectives, typically 170 μm. To

check suitability of the buffer or to verify the composition if dye switching is poor, the following steps can be used to check the composition of switching buffer.

(a) Check the pH is 7–8 by blotting a small amount of buffer on universal indicator paper (Fluka, Sigma-Aldrich). Adjust pH as required.

(b) Check oxygen concentration in the solution in the slide chamber. This can be conveniently done with a fiber-based oxygen probe (e.g., FOXY-R probe available from Ocean Optics).

(c) Check the concentration of thiols (e.g., MEA) using the spectrophotometric based Ellman's test for free thiols, *see* also [12]. The Ellman's reagent and protocol can, for example, be obtained from Thermo Scientific Pierce. The test can be conducted fully quantitatively but we find that for brief qualitative checks the degree of yellowing can be well judged by a moderately experienced user and the test can be completed in less than a minute.

6. In this protocol, we have dual labeled RyR with both Alexa Fluor 488 and Alexa Fluor 750. This allows the RyR labeling to be imaged in the visible spectrum by confocal (or inspected by eye) and in the far red by dSTORM. The ratio of confocal to super-resolution secondary antibody concentration is typically chosen so that more fluorochrome is available for super-resolution imaging (e.g., Alexa Fluor 750). The effective density of the confocal fluorophore can be relatively lower as it has little effect on image quality at this resolution (provided that signal-to-noise ratios are not too low).

7. The stock solution is prepared for convenience and stored frozen at −20 °C to stabilize MEA. This facilitates preparation of the switching buffer immediately before super-resolution imaging.

8. The final MEA concentration can be changed to adjust fluorochrome switching properties, the existing literature gives various concentrations for different fluorochromes. We have used concentration between 5 and 100 mM of MEA. In particular, the combination of Alexa 680 and Alexa 750 works well at the low end of concentrations between 5 and 10 mM. Note that the effective MEA concentration is the relevant quantity here, *see* **Note 5** for how to estimate the concentration of thiol that remains active. The mixing (*see* also **Note 9**) should be gentle to avoid aerating the solutions any more the necessary as this will lead to oxidation of the thiol MEA.

9. We use a glycerol-based buffer to improve refractive index matching in tissue samples. The mountant will have a streaky appearance when held up to light if incompletely mixed which

is a good visual check of mixing. This streaky appearance (schlieren) is due to refractive index differences between glycerol and water. The solution is well mixed when schlieren have disappeared.

10. Super-resolution imaging was performed on a customized Nikon TE2000 inverted fluorescence microscope. This included a stage modified to reduce drift, which incorporated a piezo focusing unit (P-725, Physik Instrumente, Germany). Excitation was achieved with a solid-state 671 nm laser (Viasho, China) that was focused onto the sample in a highly inclined light beam [15] with a 60× 1.49NA oil-immersion TIRF objective (Nikon) and achieved a strong illumination (typically using 100–300 mW of 671 nm light coupled into the microscope) within a 20 μm wide area up to a depth of several microns. This system provided dual color imaging (e.g., RyR and JPH) by simultaneous excitation of both Alexa Fluor 680 and 750 dyes. Emission light was passed through a Q680LP dichroic mirror (Chroma Technology) and a XF3104-690ALP emission filter (Omega optical) prior to being split into two spectral channels using a custom-built splitter device [13] . The emission light in the two channels was recorded onto the two halves of the cooled EM-CCD chip of an IXon DV887DCS-BV camera (Andor Technology, Belfast) for 20,000–40,000 frames at 20 frames/s. The fluorophores species were identified with a high level of discrimination by determining the emission intensity ratio in the two detection channels and using an algorithm as described [13]. Super-resolution images were captured by selecting tissue regions containing transversely orientated human ventricular myocytes and a z-line was brought into focus taking care to focus deep enough to avoid cutting artifacts at the bottom edges of the tissue.
11. The widefield tissue map is created with our freely available SMLM software (https://bitbucket.org/david_baddeley/python-microscopy) that allows graphical recording of the current stage position via mouse click. By moving around the section the location of imaged cells and the outline of the tissue section can be traced. The generated map facilitates identification of the general location of imaged cells for confocal imaging. The exact location is provided by diffraction-limited images recorded by the dSTORM system. Note the orientation of the cells may appear rotated or flipped between microscopes depending on the configuration of the cameras or photo detectors. However, this is easily adjusted for by rotating and flipping images digitally to match.
12. Confocal microscopy was performed with a Zeiss LSM 710 confocal laser scanning microscope. This instrument was used

to perform three channel sequential scanning of markers for RyR2 (Alexa Fluor 488), WGA (Alexa Fluor 594), and JPH (Alexa Fluor 680) using 488 nm (Argon), 561 nm (diode solid state), and 633 nm (helium-neon) laser illumination. 3D image stacks were collected for each channel with pinhole diameter set to 1 Airy unit and collected at a pixel sampling of 0.09, 0.09, and 0.3 μm in *x*, *y*, and *z*-directions using a 63× (1.4 NA, Apochromat) oil immersion lens. Tiled overview scans of entire tissue sections were collected with a 10× (0.45 NA) dry lens using the tiling function in the Zeiss Zen confocal control software.

13. We have previously used the reverse imaging order, performing confocal imaging first and dSTORM imaging after that. Both approaches provide largely equivalent data. However, we prefer to capture the dSTORM images first because this provides a more stringent selection of well-labeled sites. In practice some cells that appear well-labeled in confocal fluorescence do not contain the desired labeling density. Further, considerable time can be saved by only collecting confocal data sets from regions already successfully imaged at super resolution. A potential issue with this approach is that bleaching of Alexa 680 can occur affecting the confocal imaging. However in our hands sufficient unbleached fluorophore remains after dSTORM imaging so that it provides adequate intensity for confocal imaging (note labeling with Alexa 488 (RyR) and Alexa 594 (WGA) are unaffected). This point is illustrated by confocal image quality that is preserved even with comparatively large loss (~90%) of labeling [12]. In addition, confocal imaging itself can result in substantial bleaching of Alexa 680 that we have found to adversely impact the resolution and the nanostructure in super-resolution images. In order to prevent this loss of fluorophore intensity a substantial drop in 633 laser intensity is required essentially negating the benefit of imaging with the confocal first.

14. As additional check confocal imaging typically reveals regional bleaching in the Alexa 680 channel (JPH) from the dSTORM imaging process.

Acknowledgments

This work was supported by Health Research Council of New Zealand (HRC.govt.nz) grant 12/240, a Human Frontier Science Programme Award and EPSRC grant EP/N008235/1 to CS.

References

1. Hell SW (2007) Far-field optical nanoscopy. Science 316:1153–1158. doi:10.1126/science.1137395
2. Huang B, Bates M, Zhuang X (2009) Super-resolution fluorescence microscopy. Annu Rev Biochem 78:993–1016. doi:10.1146/annurev.biochem.77.061906.092014
3. Schermelleh L, Heintzmann R, Leonhardt H (2010) A guide to super-resolution fluorescence microscopy. J Cell Biol 190:165–175. doi:10.1083/jcb.201002018
4. Leung BO, Chou KC (2011) Review of super-resolution fluorescence microscopy for biology. Appl Spectrosc 65:967–980. doi:10.1366/11-06398
5. Barna L, Dudok B, Miczán V et al (2016) Correlated confocal and super-resolution imaging by VividSTORM. Nat Protoc 11:163–183. doi:10.1038/nprot.2016.002
6. Hou Y, Jayasinghe I, Crossman DJ et al (2015) Nanoscale analysis of ryanodine receptor clusters in dyadic couplings of rat cardiac myocytes. J Mol Cell Cardiol 80:45–55. doi:10.1016/j.yjmcc.2014.12.013
7. Jayasinghe I, Crossman D, Soeller C, Cannell M (2012) Comparison of the organization of T-tubules, sarcoplasmic reticulum and ryanodine receptors in rat and human ventricular myocardium. Clin Exp Pharmacol Physiol 39:469–476. doi:10.1111/j.1440-1681.2011.05578.x
8. Soeller C, Baddeley D (2013) Super-resolution imaging of EC coupling protein distribution in the heart. J Mol Cell Cardiol 58:32–40. doi:10.1016/j.yjmcc.2012.11.004
9. Takeshima H, Komazaki S, Nishi M et al (2000) Junctophilins: a novel family of junctional membrane complex proteins. Mol Cell 6:11–22
10. Soeller C, Crossman D, Gilbert R, Cannell MB (2007) Analysis of ryanodine receptor clusters in rat and human cardiac myocytes. Proc Natl Acad Sci 104:14958–14963
11. Crossman DJ, Ruygrok PR, Soeller C, Cannell MB (2011) Changes in the organization of excitation-contraction coupling structures in failing human heart. PLoS One 6:e17901
12. Crossman DJ, Hou Y, Jayasinghe I et al (2015) Combining confocal and single molecule localisation microscopy: a correlative approach to multi-scale tissue imaging. Methods. doi:10.1016/j.ymeth.2015.03.011
13. Baddeley D, Crossman D, Rossberger S et al (2011) 4D super-resolution microscopy with conventional fluorophores and single wavelength excitation in optically thick cells and tissues. PLoS One 6:e20645
14. Kukulski W, Schorb M, Welsch S et al (2012) Precise, correlated fluorescence microscopy and electron tomography of lowicryl sections using fluorescent fiducial markers. Methods Cell Biol. doi:10.1016/B978-0-12-416026-2.00013-3
15. Tam J, Cordier GA, Borbely JS et al (2014) Cross-talk-free multi-color STORM imaging using a single fluorophore. PLoS One 9:e101772. doi:10.1371/journal.pone.0101772
16. Zitová B, Flusser J (2003) Image registration methods: a survey. Image Vis Comput 21:977–1000. doi:10.1016/S0262-8856(03)00137-9
17. Wong J, Baddeley D, Bushong E et al (2013) Nanoscale distribution of ryanodine receptors and caveolin-3 in mouse ventricular myocytes: dilation of t-tubules near junctions. Biophys J 104:L22–L24. doi:10.1016/j.bpj.2013.02.059
18. Cole RW, Jinadasa T, Brown CM (2011) Measuring and interpreting point spread functions to determine confocal microscope resolution and ensure quality control. Nat Protoc 6:1929–1941. doi:10.1038/nprot.2011.407
19. Baddeley D, Jayasinghe ID, Lam L et al (2009) Optical single-channel resolution imaging of the ryanodine receptor distribution in rat cardiac myocytes. Proc Natl Acad Sci U S A 106:22275–22280. doi:10.1073/pnas.0908971106
20. Jayasinghe ID, Baddeley D, Kong CHT et al (2012) Nanoscale organization of junctophilin-2 and ryanodine receptors within peripheral couplings of rat ventricular cardiomyocytes. Biophys J 102:L19–L21. doi:10.1016/j.bpj.2012.01.034

Chapter 18

Correlative Fluorescence Super-Resolution Localization Microscopy and Platinum Replica EM on Unroofed Cells

Kem A. Sochacki and Justin W. Taraska

Abstract

Platinum replicas of unroofed mammalian cells can be imaged with a transmission electron microscope (TEM) to produce high contrast, high resolution images of the structure of the cytoplasmic side of a plasma membrane. A complementary approach, super-resolution fluorescence localization microscopy, can be used to localize labeled molecules with better than 20 nm precision in cells. Here, we describe a correlative method that couples these two techniques and produces images where localization microscopy data can be used to highlight specific proteins of interest within the structural context of the platinum replica TEM image. This combined method is uniquely suited to investigate the nanometer-scale structural organization of the plasma membrane and its associated organelles and proteins.

Key words Electron microscopy, CLEM, Super-resolution, Localization microscopy, PREM, Platinum replica, Unroofed cells, Fluorescence, Correlative microscopy, Nanoscopy

1 Introduction

In platinum replica electron microscopy (PREM), a biological sample is shadowed with platinum and coated with carbon to create a cast of the surface. The replica is then removed from the sample, and imaged in a transmission electron microscope. This metal replica provides high contrast images of the structure of macromolecular complexes at the surface of the sample. The resolution of PREM, however, can be limited by the platinum granule size and structures must be accessible to the platinum during vacuum-vapor deposition of the metal. PREM has been used to image freeze-fractured cells [1], whole cells [2], single proteins or protein complexes [3], isolated cytoskeletons [4], or unroofed cells [5]. Unroofed cells, like those used in the method described here, are cells that are attached to a substrate and the top of the cell is removed by various mechanical means leaving behind only the adhered membrane and its associated proteins.

Holger Erfle (ed.), *Super-Resolution Microscopy: Methods and Protocols*, Methods in Molecular Biology, vol. 1663,
DOI 10.1007/978-1-4939-7265-4_18,

Identifying proteins in a PREM image has been challenging because immunogold labeling, the classical EM protein localization technique, relies on the same contrast mechanism as the rest of the metal replica. An alternative protein localization method combines PREM with fluorescence microscopy [4]. With this combination, one can highlight regions on the electron micrograph that are enriched in a specific protein. However, correlative fluorescence and PREM was previously limited to diffraction-limited fluorescence microscopy that cannot exceed a resolution of 200 nm. Thus, signals from the fluorescence image cannot be assigned with certainty to specific EM structures at the nanometer scale. Super-resolution localization fluorescence microscopy offers a method to exceed the diffraction limit of fluorescence and localize fluorophores with better than 20 nm precision. Correlation of super-resolution localization microscopy data with PREM uniquely allows one to highlight proteins with high precision without obstructing the high contrast structural background of the electron micrograph [6].

Here, we present methods to extend correlative diffraction-limited fluorescence PREM with localization microscopy and reproducibly obtain correlation with better than 20 nm accuracy across a 20 × 20 μm image. One key step in obtaining highly accurate and reproducible correlation is the use of stable spatial fiducial markers that do not move with respect to the sample through all of the imaging and processing. The method presented here uses glass coverslips embedded with gold nanorods that are visible during fluorescence imaging [7]. These gold nanorods adhere to the metal replica when it is removed from the sample so that they are also visible in the EM image. A spatial image transformation can then be used to map the fluorescence image to the EM image using the gold nanorods as an independent reference.

In this protocol we describe two-dimensional image correlation. The method can, however, be extended for three-dimensional correlation with three-dimensional fluorescence localization and EM tomography [6]. When doing 3D imaging, we use a fluorescent membrane marker (myristoylated-psCFP2) to align the fluorescent membrane with the membrane visible in the EM tomogram in the axial dimension. Our method uses critical point drying to dry the sample in the preparation for PREM. This works well for unroofed cells because the sample is thin and tightly adhered to the coated-glass substrate. However, delicate objects or objects that project into the cell move slightly during the drying process and have increased correlation error. Therefore, this method is likely not suitable for correlating tall or structurally unstable objects. One could also use freeze-etching to dry the samples but this may result in increased processing error and a higher failure frequency.

We have used this technique to study specific proteins associated with clathrin-mediated endocytosis. After unroofing, the distinct honeycomb structure of clathrin on the plasma membrane is exposed. These 100–500 nm cellular structures vary greatly in shape, size, and depth which are visible in the EM image as well as irregularities in the clathrin meshwork, neighboring objects, or membrane deformations. Most proteins associated with clathrin are underneath the meshwork and not obviously visible in the micrograph but their position can be highlighted using localization microscopy.

2 Materials

2.1 Cell Culture and Fluorescence Preparation

1. Gold nanorod embedded coverslips (hestzig.com, part #600-200AuF) (*see* **Note 1**).
2. Cleaning Solution: 200 mL water, 40 mL 30% Hydrogen Peroxide, 40 mL 30% ammonia.
3. Ethanol: 200 proof.
4. Growth medium without phenol red (*see* **Note 2**).
5. Stabilization Buffer: 70 mM KCl, 30 mM HEPES brought to pH 7.4 with 5 M KOH, 5 mM $MgCl_2$, 3 mM EGTA. Sterile Filtered. This can be stored at room temperature.
6. Sonication Media: 0.5% PFA (fresh from 16% stock amber ampule, Electron Microscopy Sciences) in unroofing stabilization buffer (*see* **Note 3**).
7. Primary Fixative: 2% PFA in unroofing stabilization buffer.
8. Sonicator: Branson Sonifier 450 with a 1/8″ tapered microtip.
9. Alexa Fluor 647 labeled antibody or label of choice (*see* **Note 4**).

2.2 Fluorescence Localization Imaging

1. Blinking Buffer: 10% w/v glucose in PBS, 0.8 mg/mL Glucose oxidase (from 80 mg/mL frozen aliquot in PBS), 0.04 mg/mL Catalase (from 10 mg/mL frozen aliquot in PBS), 100 mM 2-mercaptoethanol. This must be made fresh within minutes of using. After applying to the sample, it needs to be sealed from air (*see* **Note 5**).
2. 25 mm round coverslips.
3. Attofluor cell chamber (Life Technologies) (*see* **Note 6**).
4. Inverted fluorescence microscope preferably with localization microscopy software installed and an automated stage. We use a 100× TIRF objective with 1.49NA on a Nikon Eclipse Ti microscope running the Nikon Elements NSTORM module.
5. Diamond objective marker (Leica). Our diamond objective marker is too short for our microscope so we have machined a 14 mm spacer tapped with m25 to receive the marker and m25 threading to fit into our microscope turret.
6. 2% glutaraldehyde (fresh from 8% stock in an amber ampule, Electron Microscopy Sciences) in PBS.

2.3 EM Preparation and Imaging

1. 0.1% w/v tannic acid in water. This needs to be made fresh.
2. 0.1% w/v uranyl acetate (Electron Microscopy Sciences). Dilute in water from a 1% stock solution stored in an amber vial in the refrigerator.
3. 15%, 30%, 50%, 70%, 80%, 90%, 100% ethanol solutions (Warner-Graham) (*see* **Note 7**).
4. Critical point drying sample holder. We machined a coverslip holder like Tousimis #8767 but with an extended diameter to accommodate the 25 mm coverslip. Wavy washers (Tousimis 8767-01) can be used as coverslip spacers.
5. Dehydration beaker assembly: two 100 mL beakers with a stir bar on the bottom and metal mesh bent to create a shelf in the beakers.
6. Critical point dryer (Tousimis Samdri 795) equipped with liquid carbon dioxide.
7. Freeze fracturer (JEOL JFD-V or JEOL/RMC 9010 FF machine).
8. Light microscope with 10× phase contrast objective.
9. Glow discharge (Pelco EasiGlow).
10. Electron microscope. We use a JEOL 1400 equipped with a high tilt specimen retainer and Serial EM Freeware [8].
11. 5% Hydrofluoric Acid (Fisher Scientific) in water.
12. Platinum 4–5 mm loop (Electron Microscopy Sciences).
13. Diamond scriber (Electron Microscopy Sciences).
14. Formvar/carbon-coated 75 mesh copper TEM grids (Ted Pella) (*see* **Note 8**).
15. Filter paper.
16. 12-well plate (Corning 3512).
17. Plastic bulb transfer pipettes (Fisher 13-711-7M) (*see* **Note 9**).
18. IMOD freeware [9].

3 Methods

3.1 Cell Culture and Fluorescence Preparation

1. Boil the gold-embedded coverslips in the cleaning solution for 10 min to clean, rinse with clean water, and store in 100% ethanol (*see* **Note 10**).
2. Coat the gold-embedded coverslips with a surface to which your cells adhere and grow. We use poly-L-lysine for PC12 and HeLa cells.
3. Let cells grow on this surface overnight. You may also want to transfect your cells with a plasmid for your gene of interest and a fluorescent protein. When we are doing 3D imaging, we transfect with myristoylated PS-CFP2 for membrane alignment (*see* **Note 11**).

3.2 Cell Unroofing and Immunolabeling

1. Prepare sonicator with tip that is in good condition and tightly secured to the horn of the sonicator. We set ours to give a 400 ms pulse of the lowest output power setting. This will need to be optimized for your specific sonicator.
2. Rinse samples with stabilization buffer.
3. Sonicate in sonication media with a tip sonicator. We sonicate within 10 s of immersion in the media and position the tip 5 mm above the coverslip. This will need to be optimized for your specific sonicator (*see* **Note 12**).
4. Move coverslip to a fresh well containing primary fixative for 20 min.
5. Antibody label. Any fluorescence labeling scheme that works for localization microscopy will generally work for this technique. This includes antibody labeling, snap-tag labeling, or photo-activatable protein labeling. We use a nanobody bound with Alexa Fluor 647. Post-fix with PFA after labeling (*see* **Notes 13** and **14**).
6. Rinse cells with PBS. Store overnight in PBS at 4 °C or image immediately. Membranes will degrade with longer storage.

3.3 Fluorescence Localization Imaging

1. Place your coverslip in an AttoFluor imaging chamber with freshly made blinking buffer and seal immediately with another 25 mm coverslip on the top lip of the chamber (*see* **Note 15**).
2. Place the diamond marking objective in the turret of your microscope prior to imaging (Fig. 1a). It should be set to make a 4 mm diameter circle around your area of interest.
3. Create a map (~1 mm^2) of the area to be imaged. This can be done with a 10× DIC objective (*see* **Note 16**).
4. Create a localization image of 5–9 different regions within your 1 mm^2 mapped area. Mark on your map each region that has been imaged (*see* **Note 17**).
5. For Alexa Fluor 647, our localization images are generally made with 30,000 frames, 10 ms each, with 10 kW/cm^2, 647 nm laser intensity. You may require shorter or longer imaging depending on the density of your labeling (*see* **Note 18**).
6. After localization imaging, move your stage to the center of your 1 mm^2 area and swing the turret to the diamond objective marker. Hold the sample with one hand as you slowly bring the diamond objective marker up to mark the coverslip. When you start to see the spring-loaded tip recess, rotate the objective collar to make a 4 mm diameter etched circle around the area of interest (Fig. 1b).
7. Clean the oil off of the sample well with ethanol. Rinse away imaging buffer in PBS. Store your coverslip in 2% glutaraldehyde (in PBS) at 4 °C overnight (minimum 20 min) prior to EM preparation.

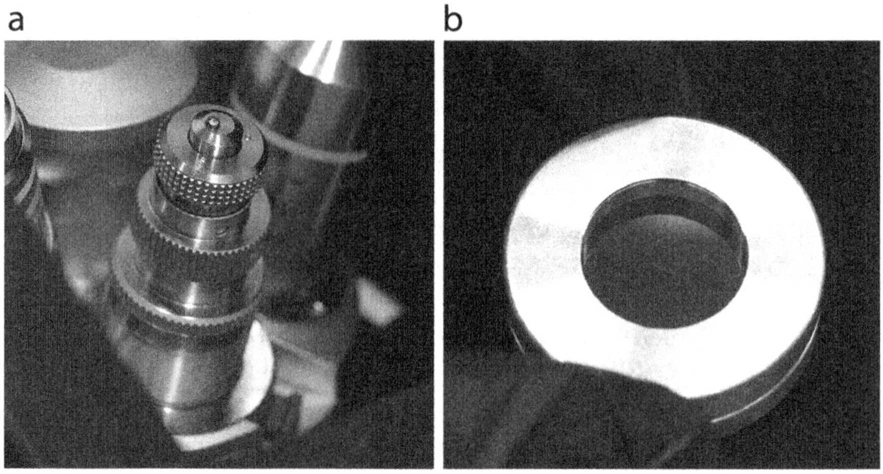

Fig. 1 The imaged region is marked with a diamond objective marker. (**a**) The diamond objective marker is shown in the microscope turret with our custome-made spacer attached at the *bottom*. (**b**) A 4 mm *etched circle* is shown on the bottom of the coverslip

3.4 Electron Microscopy Preparation and Imaging

1. From the 2% glutaraldehyde, transfer your sample into 0.1% w/v freshly made tannic acid in water for 20 min.
2. Rinse your sample in water four times.
3. Stain with 0.1% w/v uranyl acetate (diluted from 1% stock stored in amber vial at 4 °C) for 20 min.
4. Rinse with water twice.
5. Prepare your sample in a critical point drying (CPD) chamber with wavy washers above and below each sample. We generally process 2–9 samples at one time.
6. Dehydrate your samples by placing the CPD chamber through increasing concentrations of ethanol: 15%, 30%, 50%, 70%, 80%, 90%, 100%. Use the dehydration beaker assembly to agitate each ethanol bath for 4 min. Use the second dehydration beaker assembly to move the CPD chamber quickly to the new concentration without allowing the samples to dry. Bathe the samples in 100% ethanol three times to ensure complete removal of water.
7. Move the chamber to the critical point dryer that is prefilled with 100% ethanol. Perform the critical point drying using manufacturer's instructions. We use a 20 min purge to ensure all ethanol is replaced with liquid carbon dioxide prior to heating to critical point.
8. Remove sample from the critical point dryer and place in a petri dish on filter paper.

9. Use a fine-tipped marker to mark dots on the bottom of the coverslip around the etched circle. This will just help you see the etching easily during the next few steps.
10. Cut out the area of interest with a diamond pen. We use a razor blade to guide our pen marks and cut a ~8–10 mm square.
11. Use two small (~1 mm) pieces of double-sided tape to place the sample on the freeze fracturer stage.
12. Place the sample in the freeze-fracturer and coat with Platinum/Carbon at 17° with rotation. There is no thickness monitor on our freeze fracturer but our images generally have less than 5 nm of Pt/C coating objects of well-known size. Stabilize the Pt/C layer with carbon at 90 °C.
13. Remove the sample from the freeze fracturer and carefully remove from the double-sided tape.
14. Take an image of your area of interest with a 10× phase contrast objective. It should be easy to find in the center of your etched circle. This image will help you find all of your regions after you have lifted your replica onto the grid.
15. Cut your sample again before lifting (Fig. 2a–c). During lifting the sample should be a 4–5 mm diameter circle (*see* **Note 19**).
16. Prepare the formvar/carbon-coated grids by glow discharging (*see* **Note 20**).
17. Lift your replica from your coverslip by gently setting the coverslip (replica side up) on the top surface of 5% hydrofluoric acid (HF) in a 12-well plate (*see* **Note 21**) (Fig. 2d). Warning: HF is highly corrosive.
18. When you see the coverslip fall to the bottom of the well, the thin replica will remain floating at the top surface (Fig. 2e).
19. Replace the HF with water by sucking out all but ~1 mL from the well with a plastic pipette and replacing the volume with water (Fig. 2f). Do this 20 times (*see* **Note 22**).
20. Place your EM grid (glow discharged within the last 12 h) on filter paper.
21. Use a loop to lift your replica from the water and set down on your EM grid (Fig. 2g–i). If the coverslip was cut to the size of your loop, your area of interest should land near the center of your grid.
22. Image the grid with a 10× phase contrast objective and compare it to your pre-lifted image. You should easily be able to find your regions of interest.
23. Image each region of interest in the electron microscope. We use SerialEM on a JEOL 1400 transmission electron microscope at 15,000× magnification to make 8 × 8 montages. This creates a 22 × 22 μm square area (*see* **Note 23**).

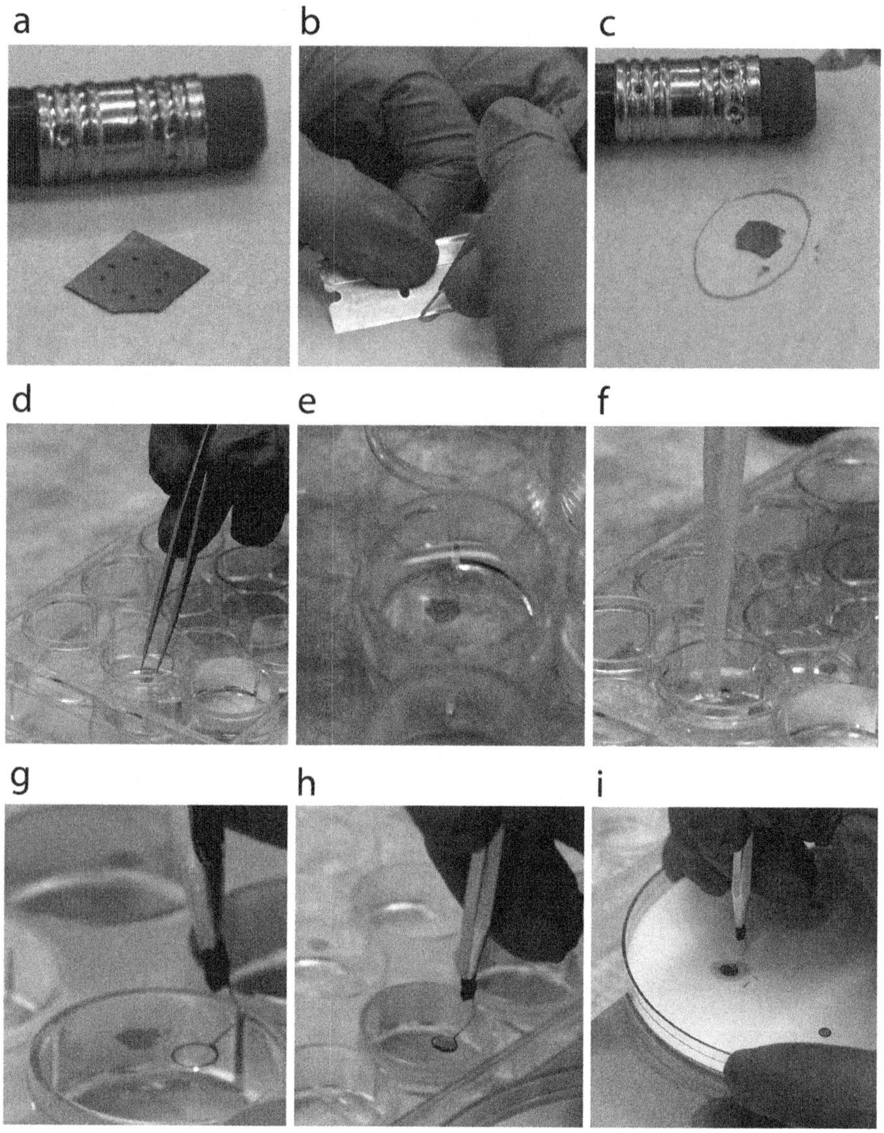

Fig. 2 Lifting the replica onto a grid. (**a**) The coverslip fragment is shown after platinum shadowing and carbon coating. (**b–c**) The coverslip is cut to approximately the size of the loop that will be used to lift the replica. (**d**) The coverslip is placed on the top surface of 5% hydrofluoric acid. (**e**) The glass coverslip falls away leaving the platinum/carbon replica floating on the top of the surface. (**f**) The hydrofluoric acid is replaced with water by serial dilutions with a plastic pipette. (**g, h)** The replica is lifted from below and (**i**) placed over the top of the TEM grid

3.5 Image Processing

1. SerialEM montages should be stitched together using IMOD blending.
2. A super-resolution image can be created with any localization software package. We use Nikon Elements to create an image with 5 nm pixel resolution.
3. Find 3–10 gold nanorods that are visible in both your EM image and your fluorescence image. The nanorods should be

well spaced across your EM image and not in a line. You may have to overlay the two images in Adobe Photoshop to find corresponding spots.

4. Find the centroid position of each of the three spots in your fluorescence image.
5. Find the center of the gold nanoparticle for each that you found in your EM image.
6. Use these coordinates to do an affine image transformation on your fluorescence image to map it onto the EM image (*see* **Note 24**).
7. Overlay the two images and check how well the other gold nanoparticles (not the ones you used for your transformation) aligned. This will allow you to check the quality of your correlation. If the distance between the fluorescence and EM of the other gold nanoparticles is greater than 20 nm, try the transformation again using different gold fiducials. If correlation is still unsuitable, check the EM montaging (*see* **Notes 25** and **26**).

4 Notes

1. These coverslips can also be made by coating a clean coverslip with poly-L-Lysine, adding gold nanorod solution and allowing gold to stick to surface, drying, and sputter coating with SiO_2 [6, 10].
2. Purchase phenol red free DMEM with no glutamine and supplement with glutamax for best results.
3. Make all fixative fresh from an ampule.
4. Any other fluorescence label that works for fluorescence localization microscopy can be used for this method. Alexa fluor 647 on a nanobody gives excellent precision localization.
5. This blinking buffer is specific for dSTORM imaging of Alexa Fluor 647. In the case that you are using a photoactivatable protein, blinking buffer would not be necessary. In the case that you are using a different dye, the blinking buffer may need to be different for that dye [11]. If your fluorophore is sensitive to pH, you should use a tris buffer as the base instead of PBS.
6. You may use any other coverslip chamber or slide but we find that this chamber has a useful upper lip where an additional coverslip can sit and seal in the STORM blinking buffer.
7. This needs to be dry ethanol. The 100% ethanol solution should be opened fresh.
8. If you coat your own grids, 50 mesh with thin bars will be even better than 75-mesh.

9. We have found that some plastic pipettes cause problems during the replica lifting process. If your replicas start sinking to the bottom of the well instead of floating, try using other plastic. Poly-propylene works better than poly-styrene and non-graduated pipettes are better than graduated pipettes.
10. The coverslips will float during boiling if they are not anchored in a coverslip holder. Always keep track of which side contains gold.
11. Before unroofing, you want your cells to be very dense. This will allow you to collect as much data as possible from a single coverslip.
12. If having trouble optimizing your sonication with fixative present, try using no fixative during sonication to start. Also, a brief poly-L-lysine treatment and brief hypo-osmotic shock will cause the membranes to stick to the coverslip better during sonication. [5]
13. We use Alexa Fluor 647 (AF647) on a nanobody because it gives very good localization precision (<20 nm).
14. Do not use detergent (like Triton-X) when labeling.
15. We seal the chamber by simply sucking the remaining solution from above the coverslip and allowing the two coverslips to make a vacuum seal.
16. A map could alternatively be made using a 100× fluorescence objective to make a 15 × 15 fluorescence montage.
17. Not all of your regions will make it all of the way through the correlative process. Some regions will land on the grid bar of the TEM grid. Always image at least twice as many cells as you expect to have in your final data.
18. You should see the gold nanorods during imaging (they do not blink like the Alexa Fluor 647).
19. We cut the sample twice because we have found that the edges of the coverslip get coated with platinum and carbon and may cause breaking during the lifting process. This also allows us to have a larger sample while coating so our tweezers avoid scraping the area of interest while handling.
20. Glow discharging the grids will keep the replica from wrinkling as it settles on the grids.
21. If your replica sinks with the glass, you probably have a contaminant in your HF solution. This can happen if it has come into contact with certain plastics as discussed in **Note 9**.
22. Doing these successive dilutions with your replica at the surface is a fragile process and you need to avoid touching and breaking the replica. The replica also needs to stay away from the wall of the well.

23. We do not make correlative images larger than this because we have found that there is increased registration error with larger images. This could be from montaging errors our large scale bending in the grid film.
24. We use Matlab function imtransform.m to do this transformation but many other software packages will do image transformations as well.
25. Alternatively, if all of the gold in the image was used to do the transformation, the residuals from the fit are informative about the quality of correlation.
26. The procedure needs to be tested using a protein of known position. Clathrin is a perfect control because it is easily visible in PREM (Fig. 3). There are some places where there will be reproducibly lower correlation like high up vesicles that are

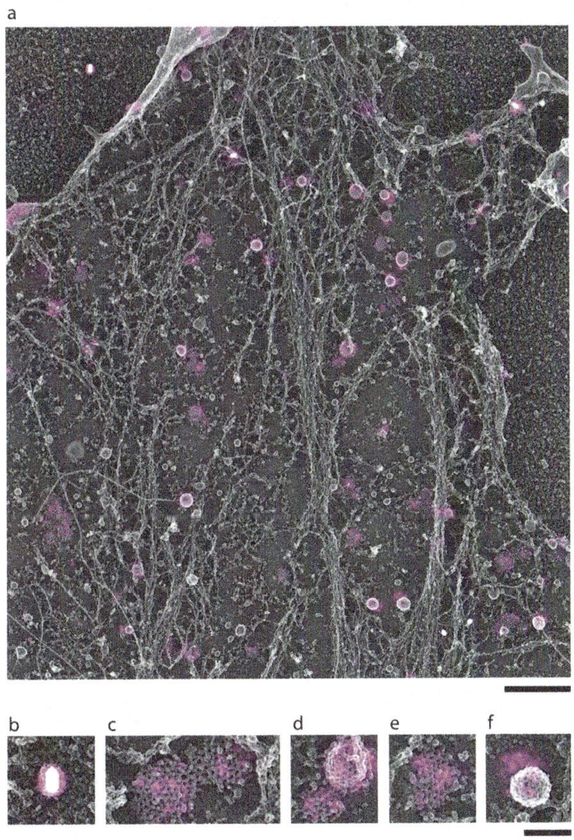

Fig. 3 Correlative data from GFP-Clathrin light chain expressing cells labeled with GFP-nanobody conjugated to Alexa Fluor 647. (**a**) Large zoomed out view of unroofed HeLa cell after correlation. Fluorescence signal is shown in magenta. Scale bar is 1 μm. (**b**) Gold nanoparticle from (**a**). (**c**–**e**) Examples of clathrin correlation from (**a**). (**f**) Example of vesicle that moved during critical point drying due to a weak attachment to the membrane. Scale bar in **b**–**f** is 200 nm

only loosely tethered to the membrane or regions at the very edge of the membrane where there may have been some rolling of the membrane during the drying process. Using a control, one can identify what data are trustable for analysis.

References

1. Moor H, Mühlethaler K (1963) Fine structure in frozen-etched yeast cells. J Cell Biol 17 (3):609–628
2. Lung B (1968) Whole-mount electron microscopy of chromatin and membranes in bull and human sperm heads. J Ultrastruct Res 22 (5):485–493
3. Heuser J, Keen J (1988) Deep-etch visualization of proteins involved in clathrin assembly. J Cell Biol 107(3):877–886
4. Svitkina T (2015) Imaging cytoskeleton components by electron microscopy. Meth Mol Biol (Clifton, NJ) 1365:99–118
5. Heuser J (2000) The production of 'cell cortices' for light and electron microscopy. Traffic 1 (7):545–552
6. Sochacki KA, Shtengel G, van Engelenburg SB, Hess HF, Taraska JW (2014) Correlative super-resolution fluorescence and metal-replica transmission electron microscopy. Nat Methods 11(3):305–308
7. Shtengel G, Galbraith JA, Galbraith CG, Lippincott-Schwartz J, Gillette JM, Manley S, Sougrat R, Waterman CM, Kanchanawong P, Davidson MW (2009) Interferometric fluorescent super-resolution microscopy resolves 3D cellular ultrastructure. Proc Natl Acad Sci 106 (9):3125–3130
8. Mastronarde DN (2005) Automated electron microscope tomography using robust prediction of specimen movements. J Struct Biol 152 (1):36–51
9. Kremer JR, Mastronarde DN, McIntosh JR (1996) Computer visualization of three-dimensional image data using IMOD. J Struct Biol 116(1):71–76
10. Shtengel G, Wang Y, Zhang Z, Goh WI, Hess HF, Kanchanawong P (2013) Imaging cellular ultrastructure by PALM, iPALM, and correlative iPALM-EM. Methods Cell Biol 123:273–294
11. Dempsey GT, Vaughan JC, Chen KH, Bates M, Zhuang X (2011) Evaluation of fluorophores for optimal performance in localization-based super-resolution imaging. Nat Methods 8(12):1027–1036

Chapter 19

In Situ Super-Resolution Imaging of Genomic DNA with OligoSTORM and OligoDNA-PAINT

Brian J. Beliveau, Alistair N. Boettiger, Guy Nir, Bogdan Bintu, Peng Yin, Xiaowei Zhuang, and C.-ting Wu

Abstract

OligoSTORM and OligoDNA-PAINT meld the Oligopaint technology for fluorescent *in situ* hybridization (FISH) with, respectively, Stochastic Optical Reconstruction Microscopy (STORM) and DNA-based Point Accumulation for Imaging in Nanoscale Topography (DNA-PAINT) to enable in situ single-molecule super-resolution imaging of nucleic acids. Both strategies enable ≤20 nm resolution and are appropriate for imaging nanoscale features of the genomes of a wide range of species, including human, mouse, and fruit fly (Drosophila).

Key words Single-molecule, Super-resolution, Genome, Chromosomes, Chromatin, FISH, Oligopaint, STORM, DNA-PAINT, OligoSTORM, OligoDNA-PAINT

1 Introduction

Single-molecule-based super-resolution imaging methods can offer resolutions that represent a ten-fold or greater improvement as compared to the diffraction limit, reaching a few tens of nanometers or even below 10 nm [1–4]. These techniques rely on the stochastic switching of individual fluorophores between fluorescent and dark states or the binding and unbinding of individual labeled molecules in order to separate the spatially overlapping images of single molecules in time. This strategy allows the position of numerous individual fluorescent molecules to be determined with high precision from the centroid positions of their images. A super-resolution image is then rendered from a collection of single-molecule localizations. Here, we describe how two forms of single-molecule-based super-resolution imaging can be used for the in situ visualization of chromosomal DNA [5, 6]. These

Brian J. Beliveau, Alistair N. Boettiger, Guy Nir and Bogdan Bintu contributed equally to this work.

Holger Erfle (ed.), *Super-Resolution Microscopy: Methods and Protocols*, Methods in Molecular Biology, vol. 1663,
DOI 10.1007/978-1-4939-7265-4_19, © Springer Science+Business Media LLC 2017

technologies, called OligoSTORM and OligoDNA-PAINT, use Oligopaint technology [7, 8] to generate single-stranded oligonucleotide probes for fluorescent in situ hybridization (FISH) and then use the probes in combination with, respectively, Stochastic Optical Reconstruction Microscopy (STORM) [9–11] and DNA-based Point Accumulation for Imaging in Nanoscale Topography (DNA-PAINT) [12–14] to generate high-resolution images of genomic DNA. Below, we provide brief descriptions of Oligopaints, OligoSTORM, and OligoDNA-PAINT, followed by protocols for generating Oligopaint probes from a library of oligonucleotides (oligos) and preparing samples for imaging by OligoSTORM or OligoDNA-PAINT. As for the actual steps of image acquisition, we refer the reader to published literature for STORM and OligoSTORM [5, 6, 9, 10, 11, 15] and DNA-PAINT and OligoDNA-PAINT [5, 13, 14, 16, 17].

1.1 Oligopaints

Oligopaints are short single-stranded oligos which, when labeled directly or indirectly with a fluorophore, can be used to visualize genomic regions as small as a few kilobases (kbs) to as large as tens of megabses (Mbs) or more [7, 8]. The oligos feature a short region of genomic homology (typically, 30–42 bases) flanked by nongenomic sequences, called Mainstreet and Backstreet, and are generated through amplification of oligo libraries that have been computationally designed and then synthesized by traditional column-based or massively parallel array-based methods—in either case, libraries can be readily purchased from commercial vendors (Fig. 1A). The region of genomic homology targets the oligos to genomic loci, wherein the typical number of binding sites in single-copy genomic regions ranges from ~1 to 20 per kb, depending on the parameters used to select the target sites ([7]; unpublished data). Mainstreet and Backstreet are used for a variety for applications, including amplifying, and thus renewing, libraries (e.g., via the incorporation of primer sequences), multiplexing libraries (e.g., through the use of multiple primer pairs), and colocalizing complementary secondary oligos through hybridization [5–7, 18, 19]. Thus, Oligopaint oligos can be made into imaging probes by conjugating them with fluorophores during amplification (Fig. 1B) or targeting them with labeled secondary oligos that are complementary to sequences on Mainstreet or Backstreet (Fig. 1C). Below, we include a protocol for synthesizing Oligopaint probes.

1.2 OligoSTORM

STORM enables super-resolution imaging through the switching and localization of single fluorescent molecules [9–11]. It achieves this by permitting only a stochastic subset of fluorophores to be switched on at any given time, such that their images do not substantially overlap. This strategy allows the positions of fluorophores to be determined with very high precision. A variety of photoswitchable fluorophores can be used to achieve stochastic

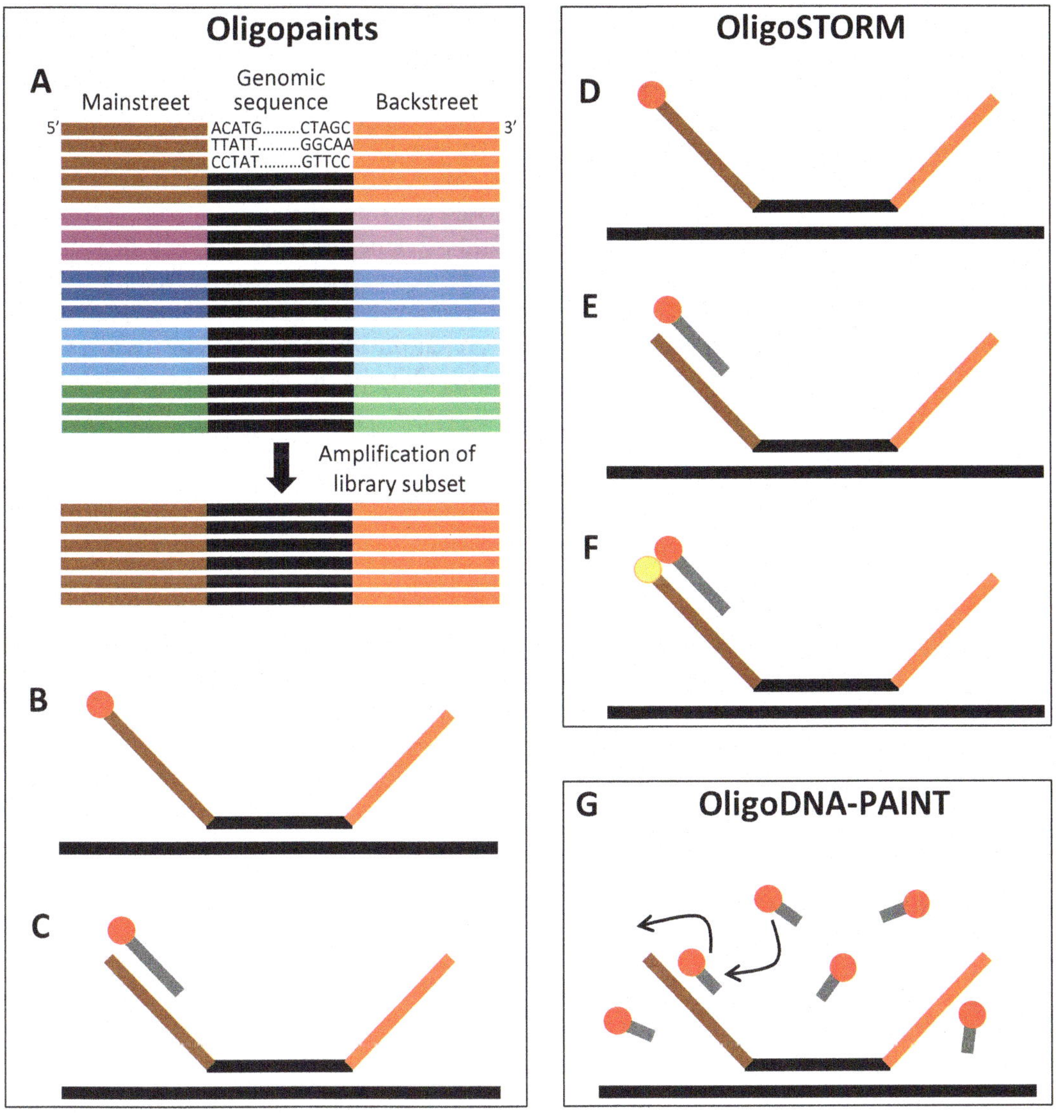

Fig. 1 Implementation of OligoSTORM and OligoDNA-PAINT using Oligopaints. (**A**) The Mainstreets and Backstreets of Oligopaint oligos consist of nongenomic sequences that can be used for a variety of purposes. For example, inclusion of primer sequences enables oligo amplification, including selective amplification of subsets of the oligo library. (**B**) Oligopaint oligos can be directly labeled with a fluorophore (*red circle*) via amplification with primers conjugated with a fluorophore. (**C**) Oligopaint oligos can be indirectly labeled via hybridization of labeled secondary oligos to Mainstreet and/or Backstreet. (**D–F**) OligoSTORM can be achieved via a variety of strategies, including labeling Oligopaint oligos (**D**) or secondary oligos (**E**) with a photoswitchable reporter (*red*) or pairing an activator (*yellow*) conjugated to Mainstreet or Backstreet with a reporter (*red*) that is carried by a secondary oligo hybridized to Mainstreet or Backstreet (**F**). (**G**) OligoDNA-PAINT can be achieved via the repeated, transient binding to Mainstreet or Backstreet of short imager strands conjugated with a fluorophore

switching, examples that have been used in OligoSTORM including AlexaFluor 647 and AlexaFluor 750 and their respective analogs, Cy5 and Cy7, all of which can exist in two states: a dark state, which does not fluoresce, and a bright or "on" state, which emits fluorescence upon excitation [11, 20, 21]. While these photoswitchable fluorophores, which we call reporters, can be activated from the dark to the fluorescent state through direct UV or near-UV excitation, the rate and efficiency of activation can be substantially increased when they are placed in close proximity to an activator molecule, such as AlexaFluor 405. In OligoSTORM, a variety of labeling and excitation strategies can be used [5, 6], including (1) conjugating Oligopaint oligos with a reporter (Fig. 1D), (2) hybridizing secondary oligos labeled with a reporter to Mainstreet or Backstreet (Fig. 1E), and (3) using a dual-fluorophore system in which, for example, an activator is conjugated to the 5′ end of Mainstreet and a secondary oligo bearing a reporter is hybridized to Mainstreet such that the reporter and the activator are in close proximity (Fig. 1F). In the first two strategies, a 405 nm laser is used to directly switch the reporter to the fluorescent state, while in the third strategy, the 405 nm laser excites the activator, which facilitates the activation of the reporter. We have found that the third strategy gives a greater number of localizations and have used it to image chromosomal regions ranging in size from 5 kb to more than a megabase ([5, 6]; unpublished data). Below, we provide a protocol for preparing samples for OligoSTORM.

1.3 OligoDNA-PAINT

DNA-PAINT achieves super-resolution imaging through the transient binding of fluorescently labeled oligos, called imager strands, that are freely diffusing in the imaging buffer to complementary strands, called docking strands, present on the target to be imaged [12–14]. Here, one round of binding and dissociation of an imager strand is detected as a single event of fluorescence, wherein the transient spikes in intensity of fluorescence at sites of hybridization permit users to discriminate between the fluorescence of imager strands bound to their target from the fluorescence of unbound imager strands diffusing in the imaging buffer. Imager strands, which are continuously replenished from the buffer, are designed to be short (e.g., 9–10 bases) in order to promote the transient nature of their binding. Additionally, as the concentration of imager strands and thermodynamic stability of the hybridized duplex control the rates of, respectively, binding and dissociation, users can tune the conditions of the imaging buffer to attain a desired blinking rate [13, 14, 16, 17]. Moreover, users can achieve numerous pseudocolors using only one spectral channel via the assignment of orthogonal docking strand sequences to distinct targets and, using a version of DNA-PAINT, called Exchange-PAINT, enable multiple targets to be distinguished via serial imaging [14]. In OligoDNA-PAINT, docking sites for the imager strands can be included in Mainstreet or Backstreet, such that binding of Oligopaint oligos to a genomic target renders that target amenable to super-resolution imaging when imager strands are introduced (Fig. 1G). Using this approach, we have generated super-resolution images of genomic regions ranging in size from 5 kb to more than a megabase [[5]; unpublished data]. Below, we present a protocol for preparing samples for OligoDNA-PAINT.

1.4 Organization of the Protocols

In keeping with the format of the journal, the protocols for synthesizing Oligopaint probes and preparing samples for imaging by OligoSTORM or OligoDNA-PAINT are integrated, such that the Materials for the three protocols are presented together in one section (Subheading 2), the Methods for the three protocols are presented together in one section (Subheading 3), and the Notes for the three protocols are presented together in one section (Subheading 4). Thus, the organization of the protocols appears as follows:

Materials

1. Materials for synthesizing Oligopaint probes.
2. Materials for preparing samples for OligoSTORM.
3. Materials for preparing samples for OligoDNA-PAINT.

Methods

1. Methods for synthesizing Oligopaint probes.
2. Methods for preparing samples for OligoSTORM.
3. Methods for preparing samples for OligoDNA-PAINT.

Notes

1. Notes for synthesizing Oligopaint probes.
2. Notes for preparing samples for OligoSTORM.
3. Notes for preparing samples for OligoDNA-PAINT.

2 Materials

2.1 Materials for Synthesizing Oligopaint Probes

Oligopaint probes typically consist of ~100–100,000 or more species of oligos, each carrying either a fluorophore or a binding site for a fluorescently labeled secondary oligo (Figs. 1A–C). In general, they are generated from libraries of oligos that are computationally designed by the user and then synthesized either in-house or by commercial vendors; for libraries with fewer than ~100–200 species of oligos, users might wish to order the oligos individually (that is, have them synthesized through column-based methods), as this strategy allows the user to control the concentration of each oligo species in the probe and, furthermore, can preclude the need for library amplification. Once a library is obtained, Oligopaint probes can be generated from the library in a variety of ways, all of which produce products that are single-stranded; single-stranded products are preferable to those that can become double-stranded, as the latter can renature after the denaturation step and thus become unavailable to their genomic targets during the hybridization step. There are several strategies for producing Oligopaint probes, including PCR amplification followed by asymmetric nicking and gel purification [5, 7, 8], PCR amplification followed by exonuclease digestion [5], T7 amplification followed by reverse transcription (RT; [5, 6, 18, 19, 22, 23]), and rolling-circle amplification followed by cleavage [5, 24]. Depending on the strategy used, the final oligos may carry both Mainstreet and Backstreet or just one. Below, we outline a protocol using T7 amplification, after which the products are reverse transcribed using labeled primers to produce oligos with both Mainstreet and Backstreet and a fluorophore conjugated at the 5′ end [modified from refs. 5–8, 18, 19, 22, 23].

All solutions should be prepared using distilled, deionized water (ddH_2O). Take care to avoid nuclease contamination, particularly from DNases. The recipes provided here for stock solutions will support many samples, while the rest are intended for a single sample.

2.1.1 Oligopaints: PCR Amplification of Oligopaints Library

1. Phusion Hot-start Master Mix (New England BioLabs).
2. Forward primer.
3. Reverse primer with T7 promoter sequence (e.g., TAATACGACTCACTATAGGG) appended to its 5′ end.
4. Oligopaints library (synthesized either in-house or by a commercial vendor).
5. A fluorescent intercalating dye for real-time PCR, such as EvaGreen Dye (20× in ddH_2O; Biotium) or equivalent.

2.1.2 Oligopaints: PCR Cleanup and Gel Electrophoresis

1. DNA Clean & Concentrator-5 (DCC-5) kit (Zymo Research).
2. 6× DNA loading dye.
3. LE Agarose.
4. Borax solution: 1 g anhydrous sodium tetraborate in 1 l of ddH_2O.

2.1.3 Oligopaints: T7 Reaction

1. HiScribe T7 Quick High Yield RNA Synthesis Kit (New England BioLabs).
2. RNasin Plus RNase Inhibitor (Promega).

2.1.4 Oligopaints: Reverse Transcription (RT)

1. Maxima H Minus RT Transcriptase (ThermoFisher Scientific).
2. dNTP Mix—10 mM each.

3. 15% Mini-PROTEAN TBE-Urea Precast Gels (Bio-Rad).
4. Urea-loading buffer.

2.1.5 Oligopaints: Probe Cleanup

1. DNA Clean & Concentrator-25 (DCC-25) kit (Zymo Research).
2. Oligo binding buffer (Zymo).

2.2 Materials for Preparing Samples for OligoSTORM

As illustrated in Figs. 1D–F, OligoSTORM can be implemented in a variety of ways. The following protocol describes the dual-fluorophore version, in which an activator, conjugated to the 5′ end of an Oligopaint oligo, is paired with a reporter carried by a secondary oligo bound to Mainstreet. It outlines the steps of sample preparation, application of Oligopaint oligos, and then the introduction of labeled secondary oligos [modified from refs. 5, 6].

All solutions should be prepared using distilled, deionized water (ddH_2O). Take care to avoid nuclease contamination, particularly from DNases. The recipes provided here for stock solutions will support many samples, while the rest are intended for a single sample.

2.2.1 OligoSTORM: Sample Fixation

1. Coverslip #1.5 (Electron Microscopy Science).
2. Tissue culture cell suspension.
3. 1× phosphate-buffered solution (PBS) pH 7.4 solution: Dilute a 10× PBS pH 7.4 stock solution (137 mM NaCl, 2.7 mM KCl, 8 mM Na_2HPO_4, and 2 mM KH_2PO_4) tenfold in ddH_2O.
4. Fixation solution: Combine 4 ml 10× PBS pH 7.4, 10 ml 16% (vol/vol) paraformaldehyde, and 26 ml ddH_2O (*see* **Note 4.2.1**). Vortex vigorously to mix. Always prepare fresh.

2.2.2 OligoSTORM: Sample Permeabilization

1. Borohydride solution: 1 mg/ml sodium borohydride in dH_2O, make fresh.
2. 1× PBS.
3. Permeabilization solution: 1× PBS + 0.5% (vol/vol) Triton X-100 (10 ml 1× PBS + 50 μl Triton X-100). Always prepare fresh.
4. Kimwipes.
5. 0.1 N HCl (425 μl of 36% HCl in 50 ml = 0.1 M).
6. RNaseA solution: 100 μg/ml RNaseA in PBS.
7. 2× SSC solution.
8. Formamide (*see* **Note 4.2.1**).
9. 0.1% Tween-20.

2.2.3 OligoSTORM: Sample Denaturation and Hybridization

1. Hybridization buffer: 50% formamide, 2× SSC, 0.1% Tween-20, 10% Dextran Sulfate (*see* **Notes 4.2.1** and **4.2.2**).
2. Oligopaint oligos (assuming a single sample): ~100 pmol of Oligopaint oligos per 100 kb of genomic target (*see* **Note 4.2.3**).
3. Reporter-labeled secondary probe: 100 μM solution (*see* **Note 4.2.4**).
4. Heat source for denaturation (*see* **Note 4.2.5**).
5. Hybridization incubator (*see* **Note 4.2.6**).
6. Humidified chamber (*see* **Note 4.2.7**).
7. Glass slide.

2.2.4 OligoSTORM: Post-hybridization Washes

1. 2× SSCT solution: 2× SSC + 0.1% vol/vol Tween-20.
2. 2× SSC.

2.2.5 OligoSTORM: Mounting Sample

1. Standard glass slide.
2. STORM imaging buffer:

   400 μl of 2× SSC.

   100 μl of 50% glucose.

   10 μl of GLOX buffer: 175 μl 2× SSC, 10 mg glucose oxidase powder, and 25 μl catalase at 16 mg/ml. Can be stored up to 1 week at 4 °C.

   5 μl β-mercaptoethanol.
3. Nail polish.

2.3 Materials for Preparing Samples for OligoDNA-PAINT

OligoDNA-PAINT entails repeated, transient binding of labeled imager strands to complementary docking sites within Mainstreet and/or Backstreet (Fig. 1G). Here, we describe sample preparation using either coverglass chamber slides or coverslips, application of Oligopaint oligos, and then the introduction of imager strands [modified from refs. 5, 12, 13]. Although not described here, sequential application of different species (sequences) of imager strands in the context of Exchange-PAINT [14] will enable pseudocolors.

All solutions should be prepared using distilled, deionized water (ddH_2O). Take care to avoid nuclease contamination, particularly from DNases. The recipes provided here for stock solutions will support many samples, while the rest are intended for a single sample.

2.3.1 OligoDNA-PAINT: Sample Fixation

1. Coverglass chamber slide or #1.5 coverslip (*see* **Note 4.3.1**).
2. Tissue culture cell suspension (*see* **Note 4.3.2**).

3. 1× PBS pH 7.4 solution: Dilute a 10× PBS pH 7.4 stock solution (137 mM NaCl, 2.7 mM KCl, 8 mM Na_2HPO_4, and 2 mM KH_2PO_4) ten-fold in ddH_2O.
4. Fixation solution: Combine 4 ml 10× PBS pH 7.4, 10 ml 16% (vol/vol) paraformaldehyde, and 26 ml ddH_2O (*see* **Note 4.3.3**). Vortex vigorously to mix. Always prepare fresh.

2.3.2 OligoDNA-PAINT: Sample Permeabilization

1. Permeabilization solution: 1× PBS + 0.5% (vol/vol) Triton X-100 (10 ml 1× PBS + 50 μl Triton X-100). Always prepare fresh.
2. PBT solution: Combine 100 ml 10× PBS pH 7.4, 899 ml ddH_2O, and 1 ml Tween-20.
3. 2× SSCT solution: Combine 100 ml 20× SSC stock solution (3 M NaCl, 300 mM sodium citrate), 899 ml ddH_2O, and 1 ml Tween-20.
4. 0.1 N HCl (425 μl of 36% HCl in 50 ml = 0.1 N).
5. 4× SSCT solution: Combine 200 ml 20× SSC stock solution (3 M NaCl, 300 mM sodium citrate), 798 ml ddH_2O, and 2 ml Tween-20.
6. Deionized formamide (*see* **Note 4.3.3**). Store at 4 °C.
7. 2× SSCT +50% (vol/vol) formamide solution: combine 4× SSCT and deonized formamide in equal proportions. Always prepare fresh (*see* **Note 4.3.3**).

2.3.3 OligoDNA-PAINT: Sample Denaturation and Hybridization

1. 4× hybridization buffer: 40% (wt/vol) dextran sulfate, 8× SSC, 0.8% (vol/vol), Tween-20 (*see* **Note 4.3.4**).
2. Oligopaint oligos (assuming a single sample): enough to reach a concentration of ~1.6 μM in the hybridization mix (*see* **Note 4.3.5**).
3. 10 μg/μl RNase A.
4. Heat source for denaturation (*see* **Note 4.3.6**).
5. Hybridization incubator (*see* **Note 4.3.7**).
6. Humidified chamber (*see* **Note 4.3.8**).
7. Rubber cement if using a coverslip in place of a coverglass chamber slide.
8. A glass microscope slide if using a coverslip in place of a coverglass chamber slide.

2.3.4 OligoDNA-PAINT: Post-hybridization Washes and Preparation for Imaging

1. 2× SSCT, as described above.
2. DAPI staining solution: 998 μl 2× SSCT +2 μl 10 μg/μl 4′,6-diamidino-2-phenylindole.
3. DNA-PAINT Buffer C: Combine 3 ml 10× PBS, 3 ml 5 M NaCl, and 24 ml ddH_2O.

4. Au particle solution: Combine 20 μl 40 nm Au nanoparticle stock solution (Sigma) and 480 μl DNA-PAINT Buffer C.
5. Centrifuge with swinging-bucket rotor.
6. Imager strands (*see* **Note 4.3.9**): dilute from concentrated stocks directly into Buffer C to a final concentration of ~0.1–10 nM.
7. A glass microscope slide if using a coverslip in place of a cover-glass chamber slide.

3 Methods

3.1 Methods for Synthesizing Oligopaint Probes

All steps are carried out at room temperature, unless indicated otherwise.

3.1.1 Oligopaints: Limited Cycle PCR (50 μl Reaction; See Notes 4.1.1 and 4.1.2)

1. Prepare on ice:

   15 μl ddH_2O.

   25 μl Phusion Hot-start Master Mix.

   2.5 μl 10 μM Forward primer.

   2.5 μl 10 μM Reverse primer with T7 promoter sequence.

   2.5 μl 1:50 dilution of Oligopaints library (*see* **Note 4.1.3**).

   2.5 μl 20× EvaGreen.
2. Run reactions on a real-time PCR machine, stopping the reaction for each tube while it is still in the exponential growth phase (i.e., before it reaches saturation). A calibration run may be necessary to identify when this will happen (*see* **Notes 4.1.4** and **4.1.5**).
   (a) Incubate at 98 °C for 3 minutes (min).
   (b) Incubate at 98 °C for 5 seconds (s).
   (c) Incubate at 72 °C for 20 s.
   (d) Cycle through **steps b** and **c** approximately 15–25 times (*see* **Note 4.1.4** regarding number of cycles).
3. Recommended: validate PCR reaction by running 2 μl out on a 2% agarose borax gel.
   (a) Dissolve 2 g agarose in 100 ml borax solution by boiling for 30 s.
   (b) Cast the hot solution and let it sit until gelled.
   (c) A satisfying result will feature a single band, which will be more likely to happen if the PCR reaction does not saturate.

3.1.2 Oligopaints: PCR Cleanup (50 μl Reaction)

1. In a 1.5 ml tube, add 350 μl Zymo DNA binding buffer to ~50 μl of PCR product.
2. Transfer to a Zymo DCC-5 column and spin at >10,000 × *g* for 30 s.
3. Add 200 μl DNA Wash Buffer to column and spin at >10,000 × *g* for 30 s.
4. Repeat the wash step.
5. Discard flow-through by aspirating the collection tube dry.
6. Spin at >10,000 × *g* for 1 min to remove residual buffer.
7. Transfer column to clean 1.5 ml tube.
8. Incubate at room temperature for 1 min.
9. Elute DNA in 11 μl ddH_2O.

3.1.3 Oligopaints: T7 Reaction (30 μl Reaction)

1. Prepare on ice in RNase-free environment:

   10 μl PCR product.

   10 μl NTP buffer mix (from HiScribe kit).

   2 μl T7 Pol Mix (from HiScribe kit).

   1 μl RNase inhibitor.

   7 μl ddH_2O.
2. Incubate at 37 °C for 4–16 h (*see* **Note 4.1.6**).

3.1.4 Oligopaints: RT Reaction (70 μl Reaction) (See Note 4.1.7)

1. Prepare on ice in RNase-free environment:

   7 μl dNTP (10 mM each).

   14 μl Maxima H Minus RT buffer.

   10 μl 100 μM unlabeled or labeled primer (*see* **Note 4.1.8**).

   5 μl or less Maxima H Minus RT enzyme.

   4 μl RNase inhibitor.

   30 μl RNA from T7 reaction.
2. Incubate at 50 °C for 1–2 hours (h) (no pre-denaturing step required).
3. Digest remaining RNA:
   (a) Mix 1:1 0.5 M EDTA and 1 M NaOH.
   (b) Add 50 μl of the NaOH-EDTA solution.
   (c) Heat at 95 °C for 10 min.
4. Recommended: Test reaction product by running 2 μl out on a 15% urea TBE gel. Use urea-loading buffer.
   (a) Run gel in 60 °C water bath for ~40 min to ensure probe is denatured. This allows for clear separation between the RT primer and the desired ssDNA probe.
   (b) 80% or more incorporation of the RT primer would be a satisfying outcome.

3.1.5 Oligopaints: Probe Cleanup

1. Add 240 μl Oligo binding buffer and mix. (The volume of Oligo binding buffer added should be 2× the volume of the sample after **step 3.1.4.3**; 2 × 120 μl = 240 μl).
2. Split volume evenly into two 1.5 ml tubes (180 μl per tube). To each tube add 480 μl of 100% ethanol and mix. (The total volume of ethanol added should be 8× the volume of the sample after **step 3.1.4.3**; 8 × 120 μl = 960 μl = 2 × 480 μl).
3. Transfer the solution of one tube to a Zymo DCC-25 column and spin at ≥10,000 × *g* for 30 s. (Note, the maximum capacity of the column is 750 μl.) Dispose the flow-through, add contents of the second tube to the column, and spin again at ≥10,000 × *g* for 30 s (*see* **Note 4.1.9**).
4. Add 750 μl DNA Wash Buffer.
5. Spin at >10,000 × *g* for 30 s and discard flow-through.
6. Spin at ≥10,000 × *g* for 30 s to remove excess buffer.
7. Transfer column to a clean 1.5 ml tube.
8. Elute in 30 μl ddH_2O.
9. Measure final probe concentration.

3.2 Methods for Preparing Samples for OligoSTORM

All steps are carried out at room temperature, unless indicated otherwise. Conduct all RNaseA work such that RNase A does not contaminate work areas and equipment.

3.2.1 OligoSTORM: Sample Fixation

1. Prepare a sample of tissue culture cells adhered at desired confluency to a coverslip.
2. Gently rinse cells with 1× PBS by briefly immersing coverslip in a small petri dish containing the buffer and then placing into a fresh, dry petri dish.
3. Immediately add enough freshly prepared fixation solution to cover the cells and let it stand for 10 min (*see* **Notes 4.2.1** and **4.2.8**).
4. Remove fixation solution and rinse the sample once or twice with 1× PBS (*see* **Note 4.2.1**).

3.2.2 OligoSTORM: Sample Permeabilization (See Note 4.2.9)

1. Incubate fixed cells for 10 min in permeabilization solution.
2. Incubate for 2 min in 1× PBS.
3. Incubate for 5 min in 0.1 M HCl.
4. Wash 3× for 1 min in PBS.
5. For DNA FISH only: incubate for 1 h at 37 °C in RNaseA solution.
6. Wash 3× for 1 min in 2× SSC.
7. Incubate for 35 min in 2× SSC + 50% vol/vol formamide and 0.1% Tween-20 (*see* **Note 4.2.1**).

8. Use sample directly in **step 3.2.3** or store at 4 °C. Coverslips may be stored at 4 °C for several days.

3.2.3 OligoSTORM: Sample Denaturation and Hybridization (Assuming One Sample)

1. Prepare hybridization mix on ice (*see* **Note 4.2.1**):

   ~100 pmol of Oligopaint oligos (*see* **Note 4.2.3**).

   1 μl of 100 μM Reporter-labeled secondary oligo.

   Enough hybridization buffer to achieve a final volume of ~20–70 μl (*see* **Notes 4.2.3** and **4.2.10**).
2. Add ~20–70 μl hybridization mix to a clean glass slide. It is important that enough solution is added such that, after the following step (3.2.3.3), all cells are exposed to the solution; too little solution results in air bubbles, where cells will not be exposed to hybridization mix.
3. Gently tap coverslip of cells dry on a Kimwipe. Invert coverslip carrying the cells onto the solution on the glass slide.
4. Seal edges of coverslip with rubber cement and allow the rubber cement to dry for 5 min (*see* **Note 4.2.11**).
5. Denature the slides at 78 °C for 3 min (*see* **Notes 4.2.5** and **4.2.12**).
6. Hybridize overnight (*see* **Note 4.2.13**) at the hybridization temperature appropriate for the Oligopaint probe(s) being used (*see* **Notes 4.2.6** and **4.2.14**). If an air incubator is being used, place the sample in a humidified chamber (*see* **Note 4.2.7**).

3.2.4 OligoSTORM: Post-hybridization Washes

1. Pre-warm 2× SSCT solution to 60 °C.
2. Remove the coverslip and wash 2× for 10 min each in 2× SSCT at 60 °C.
3. Incubate in 2× SSCT for 10 min.
4. Wash in 2× SSC.
5. Optional: Store in 2× SSC at 4 °C for up to 2 weeks.

3.2.5 OligoSTORM: Mounting Sample (See Notes 4.2.15 and 4.2.16)

1. Make chambered slide:
   (a) Adhere two strips of double stick tape along both long edges of a standard slide.
   (b) Gently tap off excess fluid from coverslip and mount it on the doublestick tape (cells facing the slide).
2. Add STORM imaging buffer to side of coverslip and allow channel to fill by capillary action (*see* **Note 4.2.17**).
3. Seal coverslip to slide with nail polish.

3.3 Methods for Preparing Samples for OligoDNA-PAINT

Carry out all procedures at room temperature unless indicated otherwise. Conduct all RNaseA work such that RNase A does not contaminate work areas and equipment. If using a coverglass chamber slide, ~75% maximum well volume can be used for all steps unless otherwise indicated; if using a coverslip, incubations and washes can be performed in a mini-staining jar or in a well of a 6-well tissue culture plate using a volume sufficient to fully immerse the coverslip unless otherwise indicated.

3.3.1 OligoDNA-PAINT: Sample Fixation

1. Prepare a sample of tissue culture cells adhered at desired confluency to a coverglass chamber slide or a coverslip (*see* **Notes 4.3.1** and **4.3.2**).
2. Add enough 1× PBS to cover the cells and let it stand for 1 min.
3. Aspirate the 1× PBS and then add enough freshly prepared fixation solution to cover the cells and let it stand for 10 min (*see* **Notes 4.3.3** and **4.3.10**).
4. Remove the fixation solution and rinse the sample once or twice with 1× PBS (*see* **Note 4.3.3**).

3.3.2 OligoDNA-PAINT: Sample Permeabilization (See Note 4.3.11)

1. Add freshly prepared permeabilization solution and incubate for 10 min.
2. Remove the Triton solution and incubate the sample with PBT for 2 min.
3. Remove the PBT and incubate the sample in 0.1 N HCl for 5 min.
4. Remove the HCl and rinse the sample twice with 2× SSCT.
5. Add freshly prepared 2× SSCT + 50% (vol/vol) formamide solution and incubate for 2 min (*see* **Note 4.3.3**).
6. Remove the solution from **step 3.3.2.5**, then add pre-warmed 2× SSCT +50% (vol/vol) formamide and incubate at 60 °C for 20 min (*see* **Notes 4.3.3** and **4.3.12**).

3.3.3 OligoDNA-PAINT: Sample Denaturation and Hybridization

1. During 3.3.2.6, prepare a 100 μl hybridization mix (assuming one sample; if using a coverslip in place of a coverglass chamber slide, the reaction can be scaled down four-fold to a final volume of 25 μl.

   50 μl deionized formamide (*see* **Note 4.3.3**).

   25 μl 4× hybridization buffer.

   4 μl 10 μg/μl RNase A.

   Enough Oligopaint oligos to reach a concentration of ~1.6 μM (*see* **Note 4.3.5**)

   ddH_2O needed to raise the total volume to 100 μl.

2. When 3.3.2.6 is complete, remove the 2× SSCT +50% (vol/vol) formamide solution and add the hybridization mix (*see* **Note 4.3.3**). If using a coverslip in place of a coverglass chamber slide, pipet the hybridization mix directly onto the coverslip, place a glass microscope slide on the top of the coverslip, carefully invert the slide, and seal the coverslip to the slide with rubber cement.
3. Denature the sample at 78 °C for 3 min (*see* **Notes 4.3.6** and **4.3.13**).
4. Hybridize overnight (*see* **Note 4.3.14**) at the hybridization temperature appropriate for the Oligopaint probe(s) being used (*see* **Notes 4.3.7** and **4.3.15**). If an air incubator is being used, place the sample in a humidified chamber (*see* **Note 4.3.8**).

3.3.4 OligoDNA-PAINT: Post-Hybridization Washes and Preparation for Imaging

1. Pre-warm 2× SSCT solution to 60 °C.
2. Remove coverslip and then remove the hybridization mix (*see* **Note 4.3.16**).
3. Add pre-warmed 2× SSCT, incubate at 60 °C for 5 min, and then remove 2× SSCT.
4. Repeat **step 3.3.4.3** three additional times (for a total of four 5 min washes).
5. Add 2× SSCT, incubate for 2 min at room temperature, and then remove 2× SSCT.
6. Repeat **step 3.3.4.5** one additional time.
7. Add DAPI solution and incubate for 5 min.
8. Remove the DAPI solution and incubate the sample in DNA-PAINT Buffer C for 2 min.
9. Remove the DNA-PAINT Buffer C and add Au particle solution (*see* **Note 4.3.17**).
10. Spin the sample at 500 RCF in a swinging bucket centrifuge for 3 min (*see* **Note 4.3.18**).
11. Remove the Au particle solution and incubate the sample in DNA-PAINT Buffer C for 2 min.
12. Remove the DNA-PAINT Buffer C and add DNA-PAINT Buffer C containing DNA-PAINT imager strand. Optionally, the solution can include an oxygen-scavenging system (*see* **Notes 4.3.9** and **4.3.19**).
13. Place the sample on a microscope stage for imaging or store at 4 °C for up to 2 weeks. Coverslips will need to be mounted onto a glass microscope slide prior to imaging (*see* **Note 4.3.20**). Take care to avoid adding antifade mounting media at this stage (*see* **Note 4.3.21**).

4 Notes

4.1 Notes for Synthesizing Oligopaint Probes

1. This step also serves as an early check on the quality of the Oligopaint library.
2. We recommend designing oligos with the minimum length that can satisfy the labeling requirements. This may improve the overall success of probe generation, including the step of amplification.
3. Oligo libraries vary in concentration of full product, which is not perfectly correlated with the reported concentration. However, these minor variations in concentration, which are due to library-synthesis and uncertainties in dilution, are generally not critical at this step.
4. For best results, users may wish to optimize the PCR conditions (e.g., template and primer concentrations), including determining at which rounds of amplification the exponential and saturation phases begin. In our experience, halting the reaction after 15–25 rounds has generally given satisfactory results.
5. You can increase the quantity of your stock of the Oligopaints library via PCR amplification of your raw library (e.g., the initial library received from a commercial source), using as few cycles as necessary (with $\leq$25 cycles as a guideline).
6. The T7 reaction saturates after approximately 4–6 h. The RNA product remains stable for 16 h. Thus, the suggested 4–16 h reaction time allows for convenient time management of the protocol.
7. We did not include an RNA purification step between **steps 3.1.3** and **3.1.4** as we have not found such a step to improve the outcome of reverse transcription. However, if RNA degradation is observed, we recommend purifying the RNA followed by polyacrylamide gel electrophoresis. Otherwise, users may wish to proceed immediately to the RT reaction after the in vitro transcription step has been completed, as RNA is more prone than ssDNA to degradation.
8. Use unlabeled primer when generating Oligopaint oligos for use in OligoDNA-PAINT (Fig. 1G) and labeled primer when generating Oligopaint oligos that carry a 5′ fluorophore, such as would be case for the dual-fluorophore variation of Oligo-STORM (Fig. 1F). Fluorophores that we have found to be compatible with the RT step include AlexaFluor 405, AlexaFluor 488, AlexaFluor 647, Cy3, and Cy5.
9. For large-scale probe cleanup, use the Zymo DCC-100 columns (1–10 nmol of primer). Scale up washes as directed by the manufacturer and elute probe in 100 μl of ddH_2O. Concentrate sample using a SpeedVac concentrator or vacuum lyophilizer or by ethanol precipitation.

4.2 Notes for Preparing Samples for OligoSTORM

1. Paraformaldehyde and formamide are toxic and should be disposed of using appropriate hazardous waste streams.
2. Dextran sulfate can be difficult to get into solution, and thus users may wish to prepare the hybridization buffer with commercially available aqueous solutions of 50% dextran sulfate. (*See* **Note 4.3.4** for another approach, which we have found useful when the desired final concentration of dextran sulfate is relatively high.)
3. The concentration of Oligopaint probe in the hybridization mix may need to be adjusted in order to optimize the ratio of signal to noise (SNR). Thus, while we have achieved good SNR using ~100 pmol of probe per sample per ~100 kb of genomic target (=5 μM Oligopaint oligos when dissolved in 20 μl hybridization mix), users may wish to adjust probe concentrations, keeping in mind that higher concentrations for any particular oligo species (targeting a single 30–42 base genomic target) may increase target occupancy, but overall higher concentration of all oligos, together, could increase background. (*See* **Note 4.3.5** for our experience with respect to OligoDNA-PAINT.)
4. With regard to choice of reporter and activator fluorophores, we recommend Alexa647 as the reporter and Alexa405 as the activator for single color experiments and, for multicolor imaging, we recommend Alexa647 and Alexa750 as reporters and Alexa405 as the activator for both.
5. The heat source for denaturation can simply be an anodized aluminum heat block placed in a benchtop heating unit or immersed in a water bath. For more precise control, a flat-block thermocycler can be used.
6. A flat-block thermocycler can be used for the hybridization step and is the instrument of choice, as its heated lid prevents the evaporation of the hybridization mix.
7. A humidified chamber can help reduce sample drying. A simple humidified chamber can be prepared by placing 3–5 damp lab tissues in the bottom of a 10 cm diameter tissue culture dish. The coverglass chamber slide or microscope slide can get wet, provided the sample side does not come into contact with the lab tissues.
8. While the protocol includes a standard fixation time, users may wish to empirically determine optimal fixation times.
9. We recommend being gentle with the washes, as cells tend to lose adherence to the coverslip under flow even when the coverslips have been pretreated with poly-D-lysine. Note that some cells adhere well on their own and, in such cases, it is not necessary to pretreat the coverslips.

10. The final volume will depend on the size of the coverslip. It is important that enough solution is added such that, after **step 3.2.3.3**, all cells are exposed to the solution; too little solution results in air bubbles, where cells will not be exposed to hybridization mix.
11. Sealing the edges of the coverslip with rubber cement is not an essential step and can be skipped, if desired.
12. While users may consider varying the temperature of denaturation, care should be taken to avoid temperatures that might excessively perturb chromatin structure.
13. Although not carefully documented, there is a general sense that long hybridization times, even up to 24 or 48 h, may be beneficial when targeting large (on the order of Mbs) single copy regions.
14. The optimal hybridization temperature will vary from probe to probe and typically will range from 42 °C for shorter probes (e.g., with ≤32 bases of genomic homology) to 52 °C (e.g., with ≥40 bases of genomic homology).
15. Particularly when imaging large genomic targets, we recommend adding fiducial beads (e.g., 200 nm 540/560 FluorSpheres, ThermoFisher) to the samples prior to adding STORM imaging buffer in order to correct for stage drift during image acquisition. This can be achieved by rinsing the coverslip with a 2× SSC solution containing 100 nm beads labeled with a fluorophore that is spectrally distinct from the reporter.
16. For brighter and thus more accurate localization, we recommend adding 3 μl of 2 M cyclooctatetraene in DMSO to the STORM imaging buffer [25] prior to adding the buffer to the sample. Note, this buffer ages very quickly.
17. Use the buffer that is appropriate for your fluorophore. For example, when using fluorophores that require thiol for switching, use an oxygen-scavenging buffer with a thiol [21].

4.3 Notes for Preparing Samples for OligoDNA-PAINT

1. With respect to sample preparation, we recommend using 8-well coverglass chambers, such as Lab-Tek II Chambered Coverglass (ThermoFisher) or iBiDi μ-slide (iBiDi), as these permit easy fluid exchange. Alternatively, a #1.5 coverslip (17 mm optimal thickness) can be used—in this case, a fine-tipped diamond scribe can be used for marking which side of the coverslip contains cells and the identity of the sample. In either case, the critical issue is simply that the FISH must be performed on a glass support that is very thin (17 mm optimal thickness). For those using coverslips, note that some microscope objectives include correction collars to account for different thicknesses of the glass support, in which case

experienced users can opt to use a different thickness. Cells should not be placed on the surface of a slide and then mounted with a coverslip, as this configuration will place the cells too far from the microscope objective to observe single-molecule fluorescence events.

2. Tissue cultures of adherent cell lines can be distributed into 8-well coverglass chamber slides and further cultured. Alternatively, ~100–200 μl of a 1×10^5 cells/ml suspension of cells can be deposited into the well of a coverglass chamber slide or onto the surface of a coverslip, after which cells should be allowed to adhere for 1–3 h as described [5, 7, 8]. For loosely adherent and suspension cell lines, pretreatment of the glass surface with poly-D-lysine [5, 7] may improve cell adherence.
3. Paraformaldehyde and formamide are toxic and should be disposed of using appropriate hazardous waste streams.
4. Dextran sulfate adds a considerable amount of powder volume upon solvation, so care needs to be taken when preparing this solution. First, add the appropriate volume of 20× SSC stock to reach a final dilution of 8× SSC (e.g., 8 ml of 20× SSC for 20 ml total volume of 4× hybridization buffer). Next, add ~70% of the ddH_2O that would be needed (e.g., 8.5 ml for 20 ml total volume). Then, add the dextran sulfate powder to this solution and mix overnight by inversion on a rotary shaker (e.g., 8 g dextran sulfate powder for 20 ml total volume). Once the dextran powder is fully dissolved, add the appropriate amount of Tween-20 to reach 0.8% (vol/vol) in the total volume (e.g., 160 μl for 20 ml) and enough ddH_2O until the final volume (e.g., 20 ml) of hybridization buffer is reached. Mix again overnight by inversion on a rotary shaker.
5. The optimal hybridization mix will likely vary, as the optimal probe concentration may vary from probe to probe and target to target. If 1.6 μM does not lead to optimal results, concentrations as high as 32 μM can be tried. For example, dense samples and tissues often need ~10× more probe than typical tissue culture samples. (*See* **Note 4.2.3** for our experience with respect to OligoSTORM.)
6. The heat source for denaturation can simply be an anodized aluminum heat block placed in a benchtop heating unit or immersed in a water bath. For more precise control, a flat-block thermocycler can be used.
7. A flat-block thermocycler can be used for the hybridization step and is the instrument of choice, as its heated lid prevents the evaporation of the hybridization mix. If using an air incubator, larger hybridization volumes may be required to offset the evaporation that occurs during extended high-temperature incubation when using coverglass chamber slides. A 50% increase in hybridization volume is generally sufficient in our hands.

8. A humidified chamber can help reduce sample drying. A simple humidified chamber can be prepared by placing 3–5 damp lab tissues in the bottom of a 10 cm diameter tissue culture dish. The coverglass chamber slide or microscope slide can get wet, provided the sample side does not come into contact with the lab tissues.
9. Imager strands, added in DNA-PAINT Buffer C, are typically 9–10 base ssDNA oligos containing a 3′ Cy3B or ATTO655 fluorophore and are used at concentrations ranging from 0.1 to 10 nM. A detailed discussion of imager strands is provided elsewhere [12–14].
10. While the protocol includes a standard fixation time, users may wish to empirically determine optimal fixation times.
11. We recommend being gentle with the washes, as cells tend to lose adherence to the coverslip under flow even when the coverslips have been pretreated with poly-D-lysine.
12. If using a coverslip in combination with a glass staining jar, place the staining jar containing 2× SSCT + 50% (vol/vol) formamide in the water bath at room temperature and allow the jar and water bath to heat up together. Likewise, after the step is complete, allow the jar and water bath to cool down together. Glass staining jars typically are not made of borosilicate glass and thus are heat labile and sometimes break upon large shifts in temperature.
13. While users may consider varying the temperature of denaturation, care should be taken to avoid temperatures that might excessively perturb chromatin structure.
14. Although not carefully documented, there is a general sense that long hybridization times, even up to 24 or 48 h, may be beneficial when targeting large (on the order of Mbs) single copy regions.
15. The optimal hybridization temperature will vary from probe to probe and typically will range from 42 °C for shorter probes (e.g., with $\leq$32 bases of genomic homology) to 52 °C (e.g., with $\geq$40 bases of genomic homology).
16. If using coverglass chamber slides, adding ~100–200 μl (space-allowing) of pre-warmed 2× SSCT makes it easier to remove the viscous hybridization mix.
17. The inclusion of fiducial markers, such as Au nanoparticles, enables the user to correct for drift during image acquisition.
18. Coverglass chamber slides can be taped to bucket attachments for multi-well plates. Coverslips can be placed in the well of a 6-well tissue culture plate and spun in the same bucket attachment.

19. An oxygen-scavenging system such as glucose oxidase/catalase or protocatechuic acid/protocatechuate-3,4-dioxygenase [26] may increase imaging performance.
20. A mounting chamber can be prepared by placing two strips of double-sided tape roughly 75% of the width of the coverslip apart on a microscope slide, such that the coverslip can sit between them while still leaving the majority of the surface untouched by the tape (cells should face downward toward the microscope slide). Buffer can be pipetted into this channel, which can eventually be sealed on the open edges (i.e., the ones not contacting the tape) with nail polish or epoxy.
21. Note that OligoDNA-PAINT is typically performed in aqueous buffers such as 1× PBS, and that binding of the imager strand to the docking site may be impeded in the high viscosity antifade mounting media commonly added to diffraction-limited FISH samples being imaged with diffraction-limited light microscopy. Note, also, that some microscope autofocus systems (e.g., those found on the Nikon N-STORM) are incompatible with antifade mounting media due to its high refractive index.

Acknowledgments

The authors would like to thank members of the Wu, Zhuang, and Yin laboratories for extensive conversations and experience. This work was supported by grants from the NIH to the laboratories of C.-t W. (GM085169, DP1GM106412, RM1HG008525), X.Z. (R01GM105637), and P.Y. (1R01EB018659, 1-U01-MH106011), X.Z. is a Howard Hughes Medical Institute investigator. In addition, B.J.B. and A.N.B. were supported by Damon Runyon Postdoctoral Fellowships.

References

1. Huang B, Babcock H, Zhuang X (2010) Breaking the diffraction barrier: Super-resolution imaging of cells. Cell 143:1047–1058
2. Sengupta P, Van Engelenburg S, Lippincott-Schwartz J (2012) Visualizing cell structure and function with point-localization superresolution imaging. Dev Cell 23:1092–1102
3. Thompson MA, Lew MD, Moerner WE (2012) Extending microscopic resolution with single-molecule imaging and active control. Ann Rev Biophys 41:321–342
4. Godin AG, Lounis B, Cognet L (2014) Super-resolution microscopy approaches for live cell imaging. Biophys J 107:1777–1784
5. Beliveau BJ, Boettiger AN, Avendaño MS, Jungmann R, McCole RB, Joyce EF, Kim-Kiselak C, Bantignies F, Fonseka CY, Erceg J, Hannan MA, Hoang HG, Colognori D, Lee JT, Shih WM, Yin P, Zhuang X, CT W (2015) Single-molecule super-resolution imaging of chromosomes and in situ haplotype visualization using Oligopaint FISH probes. Nat Commun 6:7147. doi:10.1038/ncomms8147
6. Boettiger AN, Bintu B, Moffitt JR, Wang S, Beliveau BJ, Fudenberg G, Imakaev M, Mirny LA, C-t W, Zhuang X (2016) Super-resolution imaging reveals distinct chromatin folding for different epigenetic states. Nature 529:418–422

7. Beliveau BJ, Joyce EF, Apostolopoulos N, Yilmaz F, Fonseka CY, McCole RB, Chang Y, Li JB, Senaratne TN, Williams BR, Rouillard JM, CT W (2012) Versatile design and synthesis platform for visualizing genomes with Oligopaint FISH probes. Proc Natl Sci U S A 109:21301–21306
8. Beliveau BJ, Apostolopoulos NA, Wu CT (2014) Visualizing genomes with Oligopaint FISH probes. Curr Protoc Mol Biol 105:Unit 14.23
9. Rust M, Bates M, Zhuang X (2006) Sub-diffraction-limit imaging by stochastic optical reconstruction microscopy (STORM). Nat Methods 3:793–796
10. Huang B, Wang W, Bates M, Zhuang X (2008) Three-dimensional super-resolution imaging by stochastic optical reconstruction microscopy. Science 319:810–813
11. Bates M, Huang B, Dempsey G, Zhuang X (2007) Multicolor super-resolution imaging with photo-switchable fluorescent probes. Science 317:1749–1753
12. Sharanov A, Hochstrasser RM (2006) Wide-field subdiffraction imaging by accumulated binding of diffusing probes. Proc Natl Acad Sci U S A 103:18911–18916
13. Jungmann R, Steinhauer C, Scheible M, Kuzyk A, Tinnefeld P, Simmel FC (2010) Single-molecule kinetics and super-resolution microscopy by fluorescence imaging of transient binding on DNA origami. Nano Lett 10:4756–4761
14. Jungmann R, Avendaño MS, Woehrstein JB, Dai M, Shih WM, Yin P (2014) Multiplexed 3D cellular super-resolution imaging with DNA-PAINT and Exchange-PAINT. Nat Methods 11:313–318
15. Bates M, Jones SA, Zhuang X (2010) Stochastic optical reconstruction microscopy (STORM) – a method for superresolution fluorescence imaging. In: Yuste R (ed) Imaging: a laboratory manual. Cold Spring Harbor Laboratory Press, Cold Spring Harbor, NY
16. Dai M, Jungmann R, Yin P (2016) Optical imaging of individual biomolecules in densely packed clusters. Nat Nanotechnol 11:798–807
17. Jungmann R, Avendaño MS, Dai M, Woehrstein JB, Agasti SS, Feiger Z, Rodal A, Yin P (2016) 2016 Quantitative super-resolution imaging with qPAINT. Nat Methods 13:439–442
18. Chen KH, Boettiger AN, Moffitt JR, Wang S, Zhuang X (2015) Spatially resolved, highly multiplexed RNA profiling in single cells. Science 348:aaa6090
19. Wang S, J-H S, Beliveau BJ, Bintu B, Moffitt JR, CT W, Zhuang X (2016) Spatial organization of chromatin domains and compartments in single chromosomes. Science 353:598–602
20. Bates M, Blosser TR, Zhuang X (2005) Short-range spectroscopic ruler based on a single-molecule optical switch. Phys Rev Lett 94:108101
21. Dempsey GT, Vaughan JC, Chen KH, Bates M, Zhuang X (2011) Evaluation of fluorophores for optimal performance in localization-based super-resolution imaging. Nat Methods 8:1027–1036
22. Murgha Y, Beliveau Bm Semrau K, Schwartz D, Wu CT, Gulari E, Rouillard JM (2015) Combined in vitro transcription and reverse transcription to amplify and label complex synthetic oligonucleotide probe libraries. BioTechniques 58:301–307
23. Moffit JR, Zhuang X (2016) RNA imaging with multiplexed error-robust fluorescence in situ hybridization. Methods Enzymol 572:1–49
24. Schmidt TL, Beliveau BJ, Uca YO, Theilmann M, Da Cruz F, CT W, Shih WM (2015) Scalable amplification of strand subsets from chip-synthesized oligonucleotide libraries. Nat Commun 6:8634. doi:10.1038/ncomms9634
25. Olivier N, Keller D, Gönczy P, Manley S (2013) Resolution doubling in 3D-STORM imaging through improved buffers. PLoS One 8:e69004
26. Aitken EC, Marshall RA, Puglisi JD (2008) An oxygen scavenging system for improvement of dye stability in single-molecule fluorescence experiments. Biophys J 94:826–183

Chapter 20

Super-Resolution High Content Screening and Analysis

K. Soliman

Abstract

Revealing the subcellular phenotypes at the molecular scale is critical to understand the mechanisms by which the cells function and respond to chemical treatments. Super-resolution microscopy and robust analysis tools enabled biologists to reveal and quantify phenotypes at unprecedented resolution. Developing automated imaging analysis solutions for super-resolution imaging will make high-content-screening (HCS) applicable for super-resolution microscopy, which will give access to new complex information. Here, I provide an instant automated analysis procedure for analyzing super-resolution images via *CellProfiler* (www.cellprofiler.org) platform.

Key words Super-resolution microscopy, Imaging analysis, *CellProfiler*, *Gaussian*

1 Introduction

Super-resolution microscopy allowed scientists to visualize biomolecules that are apart with a distance below the diffraction limit of light (~200 nm). There are several super-resolution microscopy modalities available. They can be divided into three major common techniques: (a) Geometrically engineered point spread function (PSF) microscopy, e.g., Stimulated Emission Depletion Transmission (STED) and its relatives [1]. (b) Single Molecule Localization Microscopy (SMLM), e.g., Photoactivatable Light Microscopy (PALM) and Stochastic Optical Reconstruction Microscopy (STORM) [2]. (c) Pattern grid microscopy that works by creating subdiffraction interference light patterns using nanometer grids to light subdiffraction spaced patterns on a specimen at different angles, then a super-resolution image is obtained by mathematical reconstruction of the interfering patterns, e.g., Structured-Illumination Microscopy (SIM) [3].

High Content Screening and Analysis of data obtained by super-resolution imaging techniques is a highly demanding task, which requires advanced computational capabilities. Correlative microscopy and multi-scale imaging with integrated super-

Holger Erfle (ed.), *Super-Resolution Microscopy: Methods and Protocols*, Methods in Molecular Biology, vol. 1663,
DOI 10.1007/978-1-4939-7265-4_20, © Springer Science+Business Media LLC 2017

resolution microscopy intended for high content screening were recently developed [4, 5]. Currently, automated analysis solutions for super-resolution microscopy are neither commercially nor publically available and often require the user (i.e., biologist) to have good basis of computer programming. Conventional parametric imaging analysis algorithms in readily available user-friendly automated platforms were not rediscovered with super-resolution imaging, until recently [6]. I describe here an automated pipeline implemented in *CellProfiler* (www.cellprofiler.org) software, based on Gaussian blurring of images and Voronoi segmentation [6–8]. A two-color STED image of the nuclear pore complex (NPC) is used to show analysis flow and outcomes. The STED image described reveals the previously known subdiffraction structures of the gp210 and pan-FG scaffold proteins in ovarian amphibian cells (gift from Prof. Hell's Lab) (Fig. 1a) [9].

2 Materials

1. Images obtained by super-resolution optical microscope to be processed and analyzed. Data size does not decide for the operation time. In my lab, we often analyze several hundreds of images per experiment. This can be scaled up or down depending on each experiment. *CellProfiler* accepts different file formats. In case of SMLM microscopy, it is possible to generate images from the localization data using available plug-in in ImageJ. This can be done automatically using Batch processing and macro programming in ImageJ. Afterward, the produced images can be used for the analysis (*see* **Note 1**).
2. Computer with CellProfiler v2.1.1 or latest version installed, which can be downloaded for free (www.cellprofiler.org).
3. Data analysis and statistics software for data analysis and visualization.

3 Methods

3.1 Image Loading and Preprocessing

1. In the new version of CellProfiler (v2.1.1) it is possible to drag and drop image folders directly to CellProfiler Input modules window (Images).
2. In the input module there are Names and Types: If images are composed of several channels, set the type of image to color. In case of single channel images set it to grayscale. Images with multi-channels (color) must be converted into grayscale images by "ColorToGray" module. This module splits the images into respective channels depending on the number of recorded

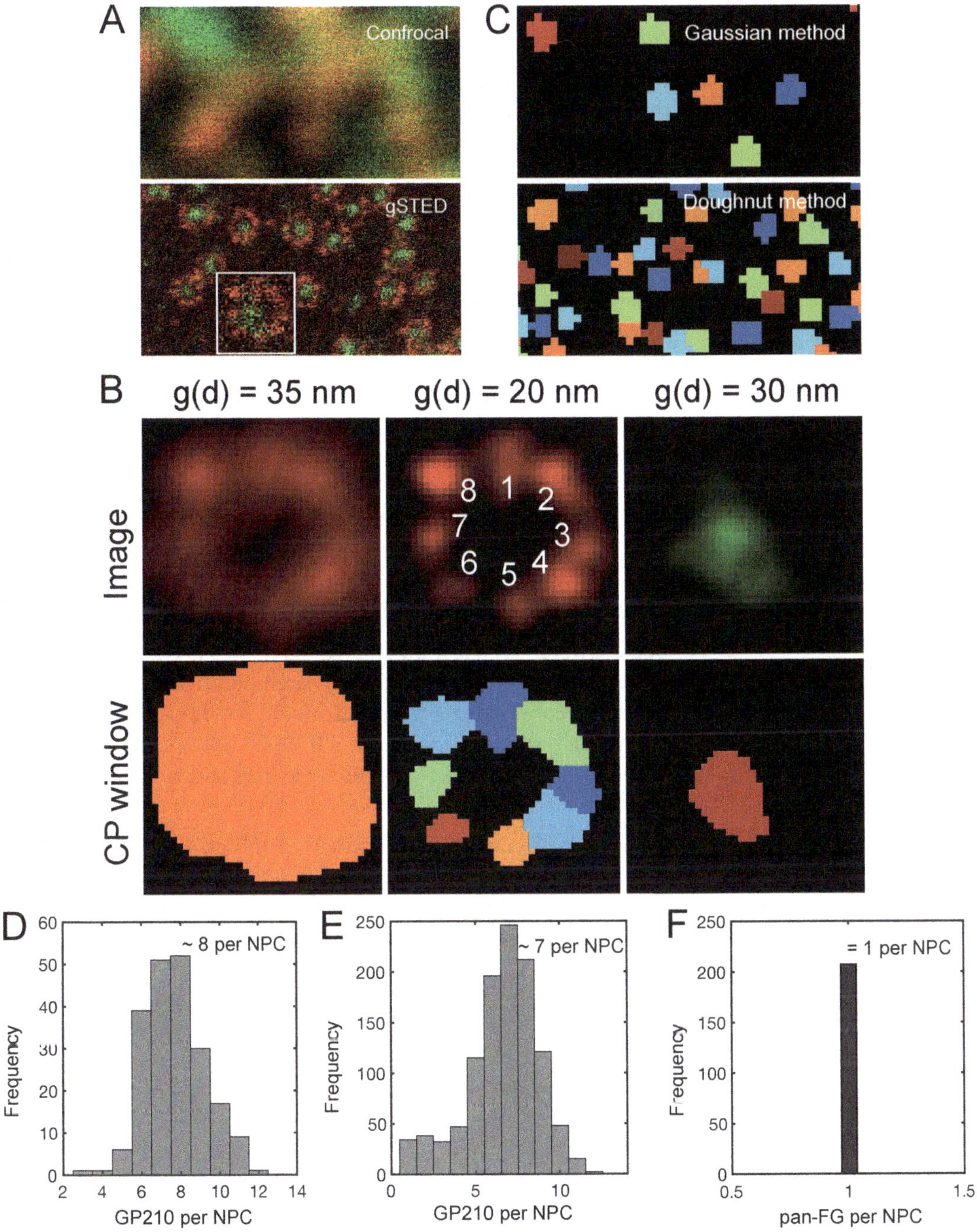

Fig. 1 *CellProfiler* (CP) automated counting of NPC's gp210 and pan-FG subunits. (**a**) Confocal view (*top*) and two-color STED view (*bottom*) of gp210 (*red*) and pan-FG (*green*) primary stained by indirect immunofluorescence. (**b**) Close view of Gaussian filter applied to the STED image shows ring structure of gp210 dimers and CP segmentation result (*left*), Smaller Gaussian radius show enhanced visualization of gp210 dimers and their segmentation results (*middle*), pan-FG Gaussian image and segmentation result (*right*). (**c**) *Top view* shows the Gaussian method segmentation results of the gp210 rings and *bottom view* shows the results of the doughnut approach. **(d)** Histogram plot of gp210 quantification using Gaussain blurr method. Average gp210 dimers per NPC indicated. **(e)** Histogram plot of the gp210 quantification using doughnut method. Average gp210 dimers per NPC indicated. (**f**) pan-FG objects histogram showing filtered counts identified per NPC by the Gaussian method

channels. Rename the channels with the appropriate names. In the shown example, we selected color image, renamed red channel to "GP210" and green channel to "pan-FG."

3. Inspect image for background and signal to noise (S/N). Most super-resolution microscopy techniques do not show a background illumination problem compared to conventional microscopy; however, preprocessing of images with smooth function can enhance image contrast and eliminate background noise artifacts (*see* **Note 2**).

3.2 Image Segmentation and Object Identification

3.2.1 Gaussian Method

1. To facilitate single NPC identification, I used a Gaussian function "$g(d)$" to average gp210 clusters into rings that can be further segmented and counted (Fig. 1b, *left column*). To implement a Gaussian smooth function in CellProfiler window, I added the "smooth" module and Gaussian diameter (d) was manually adjusted on the example image to 35 nm (*see* **Note 3**).
2. To segment individual rings, I used primary object module, i.e., IdentifyPrimary. The module was modified as follows: (a) Size diameter between 25 and 40 pixels was used to discard big or small objects that failed correct segmentation. (b) Threshold settings were changed from automatic to Otsu with allowable threshold between 0.02 and 1.0 (min and max). (c) Shape information was used to segment objects and to draw dividing lines between objects.
3. Single gp210 clusters (subdiffraction dimers) are identified by smoothing gp210 image using a Gaussian $g(d)$ smooth module with a diameter of 20 nm and a primary object module to identify single gp210 clusters (Fig. 1b, *middle column*) (*see* **Note 4**). IdentifyPrimary module was adjusted as follows: (a) We set the threshold to Maximum Correlation Threshold (MCT) algorithm, but you can test best thresholding strategy to your images. In case of MCT, an additional smooth filter of 1 pixel can be applied prior to segmentation to enhance thresholding and segmentation. Therefore, in total the example image was smoothed over a ~25 nm diameter. (b) Intensity (local-maxima) information was used to segment objects and local-maxima were suppressed if the distance between local-maxima is less than 1 pixel.
4. To segment pan-FG primary marker another smooth module was used with a $g(d)$ function with a diameter of 5 pixels, i.e., 30 nm. In this pipeline another IdentifyPrimary module is implemented to segment single pan-FG objects (Fig. 1b, *right column*). The following modifications were implemented into this module: (a) Size diameter set between 10 and 30 pixels. (b) Threshold strategy was set to automatic. (c) Intensity (local-maxima) was used to segment objects.

3.2.2 Voronoi "Doughnut" Method

1. This method, like the Gaussian method, requires images to be in grayscale and pan-FG structures were segmented as previously mentioned (*see* Subheading 3.2.1.4). I implemented two methods in the same pipeline.
2. In this approach, pan-FG marker is used as a seed to search the gp210 clusters of individual NPCs via a secondary object identification module (i.e., "IdentifySecondary"). This method offers better segmentation of dense regions compared to the Gaussian-based method (Fig. 1c); however, it can only work with color image that includes at least two distinct overlapping features (*see* **Note 5**).
3. This method does not require Gaussian blurring of gp210 structures into rings, because it does not use shape information to segment gp210 rings.

3.2.3 Relating Objects and Data Analysis

1. Single NPCs are identified and drawing a parent-child relationship between the Gaussian gp210 rings and gp210 clusters/ dimers can reveal the number gp210 clusters per NPC (Fig.1d). This can be achieved by adding RelateObject module to the pipeline. Adjust the gp210 rings as parent objects and the gp210 clusters as children. It is possible to calculate the distance between parent and children (*see* **Note 6**).
2. In case of the doughnut method, I added another RelateObject module. Secondary_Rings objects were set as parent objects, whereas gp210_clusters objects were set as children. It is obvious that more counts are found using this method compared to the Gaussian-based method.
3. Finally, results are analyzed using a graphing software, e.g., Matlab (www.mathwork.com). Only gp210 rings with one pan-FG object were included in the analysis (Fig. 1f).

4 Future Directions

Developing automated analysis procedures for super-resolution imaging data will unlock new opportunities for high-content-screening in combination with super-resolution microscopy. This will allow biologists to address more complex and subtle cellular phenotypes that may have not been addressed earlier. For drug research and screening, it may ultimately advance our understanding of favorable drug actions and underlying cellular mechanisms.

5 Notes

1. Currently, *CellProfiler* does not offer tools to process single-molecule point localization data. Therefore, it must be

converted into *.tiff* or any other compatible image *format*. Please note that image segmentation counts cannot resemble the number of localizations, especially in dense regions.

2. Super-resolution images are noisier compared to conventional fluorescence microscopy, especially when scanning a large field of view. Noise can lead to poor image segmentation efficiency. There are other ways than Gaussian smooth filters that can be used to overcome noise artifacts, e.g., bandpass filters that can enhance contrast of the edge pixels by cutting-off low frequencies. Unfortunately, bandpass filters are not applicable to implement in *CellProfiler*, except via the RunImageJ module. The user can design a macro in ImageJ and run it on *CellProfiler*, in order to automate the whole procedure when multiple preprocessing steps are necessary. Please refer to ImageJ macros (https://imagej.nih.gov/ij/developer/index.html) and *CellProfiler* RunImageJ module for more details.
3. The Gaussian diameter is chosen based on the resolution of the super-resolution microscope, which can be estimated by measuring the Full-Width-Half-Maximum (FWHM) of Gaussian fit line scans through subdiffraction structures centroids. In our example image, it was found that the smallest gp210 structure/cluster shows a FWHM ~25 nm. Therefore, local-maxima with distance below 25 nm will not be real subdiffraction structures, but probably due to pixel to pixel intensities variations.
4. Gaussian diameter should be carefully adjusted based on the distances between structures that are ought to be together. Too large diameter can lead to poor segmentation (more clumped objects ought to be separated) and a small diameter can be insufficient and lead to a hyper-segmented image (too many objects ought to be together).
5. I have had success experience using the doughnut method by relying only on the distance (i.e., Distance-N) method. This method implies the expansion of the edges of primary (pan-FG) objects to a specified distance to create secondary objects. Other methods exist in this module that can be used. The best method should be tested for each experiment depending on the local distribution, signal, and pattern of the imaged subdiffraction structures.
6. In RelateObjects module, the distance between parent and objects can be measured in CellProfiler and knowing the pixel size; the net result can be expressed in nanometer scale.

Acknowledgment

This work was supported financially by the German Research Foundation (DFG) and the Open Access Publication Funds of Göttingen University. I wish to thank CellProfiler team for providing clear explanation tutorials of their open source software image analysis platform. Special thanks to Dr. Fabian Göttfert from Prof. Hell's Lab for providing the STED image.

References

1. Hell SW (2007) Far-field optical nanoscopy. Science 316:365–398. doi:10.1126/science.1137395
2. Betzig E, Patterson GH, Sougrat R et al (2006) Imaging intracellular fluorescent proteins at nanometer resolution. Science 313:1642–1645. doi:10.1126/science.1127344
3. Song L, Lu-Walther H-W, Förster R et al (2016) Fast structured illumination microscopy using rolling shutter cameras. Meas Sci Technol 27:55401. doi:10.1088/0957-0233/27/5/055401
4. Flottmann B, Gunkel M, Lisauskas T et al (2013) Correlative light microscopy for high-content screening. BioTechniques 55:243–252. doi:10.2144/000114099
5. Gunkel M, Flottmann B, Heilemann M et al (2014) Integrated and correlative high-throughput and super-resolution microscopy. Histochem Cell Biol 141:597–603. doi:10.1007/s00418-014-1209-y
6. Soliman K (2015) CellProfiler: novel automated image segmentation procedure for super-resolution microscopy. Biol Proced Online 17:11. doi:10.1186/s12575-015-0023-9
7. Fairfield J (1983) Segmenting dot patterns by voronoi diagram concavity. IEEE Trans Pattern Anal Mach Intell 5:104–110
8. Jones TR, Carpenter A, Golland P (2005) Voronoi-based segmentation of cells on image manifolds. Springer, Berlin, pp 535–543
9. Löschberger A, van de Linde S, Dabauvalle M-C et al (2012) Super-resolution imaging visualizes the eightfold symmetry of gp210 proteins around the nuclear pore complex and resolves the central channel with nanometer resolution. J Cell Sci 125:570–575. doi:10.1242/jcs.098822

INDEX

A

Actin ... 33–35, 38, 39, 66, 86, 89–92, 98, 101, 112, 190
Acute brain slices ... 47, 51, 54, 59, 60
Antibodies ... 83–86, 90, 92, 98, 99, 141, 147, 154, 155, 158, 160, 206–208, 211, 213, 214, 221, 223
Atomic force microscopy (AFM) ... 109, 112, 190

B

Bioorthogonal chemistry ... 66
Biosensor ... 27
Brain slices ... 46, 47, 51, 54, 58–60, 153–161

C

CellProfiler ... 140, 254–259
Chromatic aberrations ... 33, 93
Chromatin ... 3–7, 86, 248, 250
Chromosomes ... 1
Cluster analysis ... 8, 120, 124, 150
Co-localization ... 232
Confocal imaging ... 24, 28, 143, 147–149, 151, 206, 207, 209–213, 215, 216
Confocal laser scanning microscope ... 29, 31–33, 37, 40, 215
Correlative light and electron microscopy (CLEM) ... 96, 163–165, 209
Correlative microscopy ... 95–102, 109, 110, 112, 253
Cranial window ... 46, 54, 58, 59
Cytoskeletons ... 4, 89, 219

D

Dendritic spines ... 32, 45–47, 58–61
Diffusion ... 130, 135
Dimerization ... 128–130, 135
DNA-PAINT ... 232, 233, 239, 240, 245, 250

E

Electron microscopy (EM) ... 1, 5, 7, 46, 96, 161, 163, 171, 176, 182, 206, 208, 209, 213, 219, 221, 222, 224, 225, 237

F

Fluorescence ... 2, 15, 29, 46, 67, 80, 95, 105, 115, 128, 139, 154, 163, 179, 190, 205, 220, 231, 255
Fluorescence microscopy ... 2, 15, 29, 46, 105, 106, 108, 115, 139, 163, 179, 220, 258
Fluorescent in situ hybridization (FISH) ... 3, 232, 242, 248, 251
Fluorescent proteins ... 3, 6, 16, 21, 46, 54, 84, 85, 92, 115–118, 124, 128, 165, 171, 173, 179, 191, 222
Fluorophores ... 2, 4, 6, 30, 39, 42, 48, 53, 65, 66, 80, 85, 86, 93, 102, 106–108, 110, 115, 116, 118, 120, 123, 147, 164, 170, 179, 190–192, 195, 197, 201, 205, 215, 220, 231, 232, 234, 247, 248
Focal adhesions (FAs) ... 80, 89, 90, 92

G

Gaussian ... 25, 37, 47, 55, 60, 61, 100, 101, 129, 145, 150, 254–258
Genome ... 7, 95
Green fluorescent protein (GFP) ... 6, 7, 53, 54, 85, 89, 90, 95, 165, 171, 175, 179, 229

H

HaloTag ... 128, 130–134, 136, 179–188
Heart ... 156, 206
High content screening (HCS) ... 257, 258
High pressure freezing (HPF) ... 164–166, 168, 169, 171, 173, 174
Human cardiac tissue ... 205, 206, 210
Hydrogen peroxide ... 27, 221
HyPer2 ... 27

I

Imaging analysis ... 254
Inverse electron demand Diels-Alder reaction ... 66
In-resin fluorescence ... 176

K

Konstanz Information Miner (KNIME) ... 140–142, 145, 146, 148–151

Holger Erfle (ed.), *Super-Resolution Microscopy: Methods and Protocols*, Methods in Molecular Biology, vol. 1663,
DOI 10.1007/978-1-4939-7265-4, © Springer Science+Business Media LLC 2017

L

Labeling protocol ... 128, 129, 180
Live cell imaging ... 1, 3, 16, 28, 134, 140, 182, 186
Localizations ... 2, 6, 23, 30, 32, 79–81, 90, 95, 100, 106, 109, 110, 115, 120, 122–124, 144, 148, 151, 153–155, 160, 161, 164, 170, 173, 190, 197, 199, 200, 210, 212, 216, 219–232, 234, 248, 253, 254, 257
Localization microscopy ... 106, 117, 120, 125, 144, 148, 151, 153, 154, 161, 164, 190, 210, 212, 216, 226, 229, 253

M

Membranes ... 2, 7, 46, 66, 116–122, 125, 127–129, 160, 165, 168, 171, 173, 174, 191, 206, 211, 219–222, 228–230
Membrane protein ... 117, 119, 120, 125
Microglial processes ... 47, 59
Microglia-synapse interactions ... 46
Molecular counting ... 116, 118, 120, 121, 124
Mouse cortex ... 46
Multi-scale imaging ... 212, 253

N

Nanoscale neural morphology ... 60
Nanoscopy 5, 31–37, 40, 42, 45–63, 83, 105–107, 112, 115, 116, 120, 140, 189–201, 205, 220, 221, 227, 229, 232, 252, 258

O

OligoDNA-PAINT ... 232, 233, 235, 238–240, 244–251
Oligopaints ... 232–247
OligoSTORM ... 232–235, 237, 238, 242–244, 246, 247, 249, 251
Online feedback image analysis ... 140
Organic chemistry ... 182, 184, 185

P

Platinum replica ... 226, 229
Platinum replica electron microscopy (PREM) ... 219, 220, 229
Protein-protein interactions ... 127, 135
Protein conformation ... 191

S

Signaling ... 46, 135
Single-molecule imaging ... 180, 182, 186
Single-molecule localization microscopy (SMLM) ... 2, 4, 6–9, 106–112, 115–125, 139–151, 154, 164, 165, 167, 168, 170–172, 175, 194, 205–216, 253, 254
Single-particle tracking ... 129, 135
Site-specific labeling ... 191
SNAP-tag ... 128, 130–134, 136, 181, 185, 188, 223
Spatially modulated illumination (SMI) microscopy ... 2–4
STED microscopy ... 2, 4–6, 21, 22, 27, 30, 39, 40, 46–48, 51–54, 56, 59–61, 67, 70, 73
Stimulated emission depletion (STED) ... 2, 16, 21, 30, 46, 65, 80, 96, 253
Stochastic optical reconstruction microscopy (STORM) ... 6, 16–19, 65, 80, 96, 97, 102, 123, 153, 182, 190, 191, 193, 205, 227, 232, 238, 243, 248, 251, 253
Structured-illumination microscopy (SIM) ... 3, 4, 16–19, 79–93, 96, 97, 102, 253
Super-resolution ... 9, 15, 16, 18, 19, 21, 25, 46, 48, 59, 65, 81, 84, 89, 92, 93, 96, 109, 112, 153, 163, 206, 210–214, 216, 220, 231–251, 253–258
Super-resolution imaging *in vivo* ... 46, 47, 59
Super-resolution microscopy ... 1–9, 15–18, 21, 46, 65, 80, 106, 110, 116, 153, 253, 254, 256, 257

T

Tannic acid (TA) ... 164, 166, 171, 172, 222, 224
Time gating ... 37, 40, 42
Tissue imaging ... 209
Transmission electron microscopy (TEM) ... 164, 165, 167, 170–172, 175, 176, 219, 222, 225, 226, 228
Two-photon microscopy ... 61

U

Unroofed cells ... 226, 229

CPSIA information can be obtained
at www.ICGtesting.com
Printed in the USA
LVOW06*1404210917
549539LV00011B/8/P

9 781493 972647